Communications in Computer and Information Science 1007

Commenced Publication in 2007
Founding and Former Series Editors:
Phoebe Chen, Alfredo Cuzzocrea, Xiaoyong Du, Orhun Kara, Ting Liu, Krishna M. Sivalingam, Dominik Ślęzak, and Xiaokang Yang

More information about this series at http://www.springer.com/series/7899

Vadim Ermolayev · Mari Carmen Suárez-Figueroa
Vitaliy Yakovyna · Heinrich C. Mayr
Mykola Nikitchenko · Aleksander Spivakovsky (Eds.)

Information and Communication Technologies in Education, Research, and Industrial Applications

14th International Conference, ICTERI 2018
Kyiv, Ukraine, May 14–17, 2018
Revised Selected Papers

Springer

Editors
Vadim Ermolayev
Zaporizhzhya National University
Zaporizhzhia, Ukraine

Mari Carmen Suárez-Figueroa
Universidad Politecnica de Madrid
Boadilla del Monte, Madrid, Spain

Vitaliy Yakovyna
Lviv Polytechnic National University
Lviv, Ukraine

Faculty of Mathematics and Computer Science
University of Warmia and Mazury in Olsztyn
Olsztyn, Poland

Heinrich C. Mayr
Institute of Applied Informatics
Alpen-Adria-Universität Klagenfurt
Klagenfurt, Austria

Mykola Nikitchenko
Department of Theory and Technology of Programming
Taras Shevchenko National University of Kyiv
Kyiv, Ukraine

Aleksander Spivakovsky
Kherson State University
Kherson, Ukraine

ISSN 1865-0929 ISSN 1865-0937 (electronic)
Communications in Computer and Information Science
ISBN 978-3-030-13928-5 ISBN 978-3-030-13929-2 (eBook)
https://doi.org/10.1007/978-3-030-13929-2

Library of Congress Control Number: 2019931876

This Springer imprint is published by the registered company Springer Nature Switzerland AG
The registered company address is: Gewerbestrasse 11, 6330 Cham, Switzerland

Preface

This volume contains a number of selected refined and extended contributions to ICTERI 2018, the 14th International Conference on Information and Communication Technologies (ICT) in Education, Research, and Industrial Applications. The conference was held in Kiev, Ukraine, during May 14–17, 2018, with a focus on research advances in ICT, business or academic applications of ICT, and design and deployment of ICT infrastructures.

ICTERI 2018 continued the tradition of hosting co-located events this year by offering five workshops and a PhD Symposium. The workshops addressed:

(1) 3L-Person 2018: new and emerging technologies in education, learning environments and methods that aimed at satisfying the life-long learning needs of a person based on the use of a person-oriented approach
(2) DSEDU 2018: the state of the play in data science education, challenges and expectations in the field, emerging opportunities for universities and ICT industry, and future trends
(3) ITER 2018: research advances, business and academic applications of ICT related to solving practical economic problems, and pushing forward economic research
(4) RMSE 2018: rigorous methods used in different fields of software engineering. In particular, the following aspects were in focus: specification, verification, and optimization of software; software analysis, modeling, business rule extraction; software testing; re-engineering problems.
(5) TheRMIT 2018: building an efficient and effective bridge between the mathematical reliability and engineering practices in safety-critical industries and domains like energy grids, aerospace, railway and automotive systems, health systems

The PhD Symposium provided the opportunity for PhD candidates to present, listen to, and discuss the research on the topics relevant for ICTERI.

The proceedings of ICTERI 2018 have been published at CEUR-WS as three separate volumes: 2105 for the main ICTERI conference (http://ceur-ws.org/Vol-2105/); 2104 for the workshops (http://ceur-ws.org/Vol-2104/); and 2122 for the PhD Symposium (http://ceur-ws.org/Vol-2122/). These three volumes contained overall 118 papers selected from 251 submissions. The 35 best of these papers were nominated by the program and workshop chairs to be invited for the submission, in substantially extended and revised versions, for the proceedings volume. Out of these, 30 extended and refined submissions were received and reviewed by at least three experts. Finally, the Proceedings Review Panel selected the 14 most mature and interesting papers for publication. The acceptance rate thus was 5.6% regarding the overall number of ICTERI 2018 submissions.

The papers in this volume were grouped in three topical parts.

Part I focused on the advances in ICT research. The contributions discussed and elaborated several novel important aspects in the fields of big data analytics, knowledge extraction from textual data, knowledge representation, logical inference, and parallel software optimization. In their invited paper, Rajendra Akerkar and Minsung Hong discussed the key questions relating methodologies, tools, and frameworks to improve ubiquitous data team effectiveness as well as the potential goals for a data processing methodology in the context of ubiquitous data analytics.

Hennadii Dobrovolskyi and Nataliya Keberle presented a novel text mining approach, based on snowball sampling and topic modeling techniques, which led to the convergence to topically saturated collections of scientific papers. As a result, these saturated collections satisfactorily covered a chosen domain of interest.

Victoria Kosa et al. reported on the refinement of their algorithm for measuring terminological differences between text datasets in automated terms extraction. The refinement was done using the appropriately selected string similarity measure for grouping the terms looking similar as text strings and presumably having similar meanings. After term grouping, terminological saturation was achieved quicker and with fewer source documents.

Ievgen Ivanov and Mykola Nikitchenko proposed a new, and simpler, rule-based technique that allows the classic Floyd–Hoare logic to be extended for making it sound in the case of partial pre- and post-conditions. In their approach, the rules were formulated in a program algebra extended with the composition of predicate complement. The obtained logic was called the Complemented Partial Floyd–Hoare logic.

Grygoriy Zholtkevych, Lyudmyla Polyakova, and Hassan El Zein presented an extension and refinement of their prior work regarding the development of category-theoretic methods for specifying and analyzing models of logical time in distributed systems, including cyber-physical systems.

Anatoliy Doroshenko et al. reported on their mixed method, combining formal and auto-tuning approaches, for minimizing the execution time of parallel programs. Their improvement against the related work was in the use of automatic training of a neural network model on the results of "traditional" tuning cycles. They also elaborated the technique for the subsequent replacement of some auto-tuner calls with an evaluation from the statistical model.

Part II collected the contributions that further elaborate effective and efficient use of ICT in teaching, learning, and education management.

Hennadiy Kravtsov and Vitaliy Kobets presented their approach to the revision system of computer science curriculum based on the evolution of the requirements of employers regarding the competencies of university graduates. They used the expert method to assess the quality of the proposed curriculum revision model.

Olena Kuzminska et al. identified the core factors that reflected the degree of readiness of teachers and students for digital education based on their self-evaluation. They also estimated the level of digital competencies based on the analysis of case studies. Finally, they proposed the methodology and model for evaluating the levels of competencies based on surveying, expert case rating, and statistical analysis.

Marina Zharikova and Volodymyr Sherstjuk presented the latest results of the GameHub project, which was designed to foster cooperation between universities and computer games companies in Ukraine. This cooperation covered monitoring of the competence profiles required by the industrial partners, building the necessary infrastructure, and developing the relevant curricula, study programs, and education resources.

Part III dealt with the application and use of different ICT-based techniques and approaches in various industrial domains.

Ihor Skyrda proposed a novel automated control method for unmanned aerial vehicles (UAV) swarm flight performance. He demonstrated that a multi-UAV system, controlled by this method, was able to autonomously perform shaping and to maintain the expected formation with desired flight parameters.

Anastasiia Strielkina, Vyacheslav Kharchenko, and Dmytro Uzun presented an approach to develop a set of Markov models, for an IoT infrastructure in health care, which take safety and security issues into account. They focused on considering the failures of components. The simulation results presented revealed the most frequent possible failures and attacks on a health-care IoT system. The authors proposed a game theoretical approach to select a countermeasure tool to remedy these most probable failures.

Tetiana Paientko discussed the benefits of the use of geographic information systems in the development and implementation of a reform in the sphere of public finance. She argued that geographic information systems could provide a wide range of analysis, better support for ideas of reforms, and higher transparency for citizens and governments. The discussion was grounded on the results of the case studies in optimizing funding for school education and health care in Ukraine.

Jan-Hendrik Meier et al. presented their study of non-linear correlative and auto-correlative time series properties of the electricity spot prices. They proposed to use non-fully connectionist networks, in relation to fully connectionist networks, to decompose non-linear correlative time series properties. Additionally, they recommended the use of a long short-term-memory network to discover and to deal with autocorrelation effects.

Andriy Belinskiy and Vladimir Soloviev, based on the allusion of Planck theory in physics, proposed an approach to predict the crashes and critical events for crypto-currency markets. The approach was illustrated by its application to the recorded time series of Bitcoin.

Oleksandr Snihovyi, Vitaliy Kobets, and Oleksii Ivanov reported on the use of machine learning in the financial sector for building intelligent robo-advisors, trained on the analysis of several popular financial services. They compared the functionality of these services, formulated the list of critical features, and proposed a high-level architecture design of a general robo-advisor tool for private investors.

This volume would not have been possible without the support of many people. First, we are very grateful to all the authors for their continuous commitment and intensive work. Second, we would like to thank the Program Committee members and additional reviewers for providing timely and thorough assessments. Furthermore, we would like to thank all the people who contributed to the organization of ICTERI 2018. Without their efforts, there would have been no substance for this volume.

December 2018

Vadim Ermolayev
Mari Carmen Suárez-Figueroa
Vitaliy Yakovyna
Heinrich C. Mayr
Mykola Nikitchenko
Aleksander Spivakovsky

Organization

General Chair

Aleksander Spivakovsky	Verkhovna Rada of Ukraine, Kherson State University, Ukraine

Steering Committee

Vadim Ermolayev	Zaporizhzhia National University, Ukraine
Heinrich C. Mayr	Alpen-Adria-Universität Klagenfurt, Austria
Mykola Nikitchenko	Taras Shevchenko National University of Kyiv, Ukraine
Aleksander Spivakovsky	Verkhovna Rada of Ukraine, Kherson State University, Ukraine
Mikhail Zavileysky	DataArt, Russian Federation
Grygoriy Zholtkevych	V. N. Karazin Kharkiv National University, Ukraine

Program Chairs

Vadim Ermolayev	Zaporizhzhia National University, Ukraine
Mari Carmen Suárez-Figueroa	Universidad Politécnica de Madrid, Spain

Proceedings Chair

Vitaliy Yakovyna	Lviv Polytechnic National University, Ukraine, University of Warmia and Mazury in Olsztyn, Poland

Presentations Chair

Heinrich C. Mayr	Alpen-Adria-Universität Klagenfurt, Austria

Workshops Chairs

Vadim Ermolayev	Zaporizhzhia National University, Ukraine
Mari Carmen Suárez-Figueroa	Universidad Politécnica de Madrid, Spain

PhD Symposium Chairs

Grigoris Antoniou	University of Huddersfield, UK
Grygoriy Zholtkevych	V. N. Karazin Kharkiv National University, Ukraine

Poster and Demo Chairs

Agnieszka Ławrynowicz	Poznan University of Technology, Poland
Raul Palma	Poznan Supercomputing and Networking Center, Poland

IT Talks Chairs

Aleksander Spivakovsky	Verkhovna Rada of Ukraine, Kherson State University, Ukraine
Mikhail Zavileysky	DataArt, Russian Federation

Local Organization Chair

Mykola Nikitchenko	Taras Shevchenko National University of Kyiv, Ukraine

Publicity Chair

Nataliya Kushnir	Kherson State University, Ukraine

Web Chair

Hennadii Dobrovolskyi	Zaporizhzhia National University, Ukraine

Proceedings Review Panel

Yevhen Alforov	German Climate Computing Center, Germany
Carlos Badenes-Olmedo	Universidad Politécnica de Madrid, Spain
George Baryannis	University of Huddersfield, UK
Sotiris Batsakis	University of Huddersfield, UK
Dominik Bork	University of Vienna, Austria
Jon Hael Brenas	Health Science Center, University of Tennessee, USA
David Chaves-Fraga	Universidad Politécnica de Madrid, Spain
Lukas Chrpa	Czech Technical University in Prague, Czech Republic
David Esteban	TECHFORCE, Spain
Wolfgang Faber	Alpen-Adria-Universität Klagenfurt, Austria

Additional Reviewer

ICTERI 2018 Sponsors

Oleksandr Spivakovsky's Educational Foundation	http://spivakovsky.fund/
DataArt	http://www.dataart.com/
Taras Shevchenko National University of Kiev	http://www.knu.ua/en
BWT Group	http://www.groupbwt.com/
Springer	http://www.springer.com/
Logicify	http://logicify.com/

Contents

ICT Solutions for Industrial Applications

Advances in ICT Research

Unlocking Value from Ubiquitous Data

Rajendra Akerkar(✉) and Minsung Hong

Western Norway Research Institute, P.O. Box 163, 6851 Sogndal, Norway
{rak,msh}@vestforsk.no

Abstract. Data is growing at an alarming rate. This growth is spurred by varied array of sources, such as embedded sensors, social media sites, video cameras, the quantified-self and the internet-of-things. This is changing our reliance on data for making decisions, or data analytics, from being mostly carried out by an individual and in limited settings to taking place while on-the-move and in the field of action. Unlocking value from data directs that it must be assessed from multiple dimensions. Data's value can be primarily classified as "information," "knowledge" or "wisdom". Data analytics addresses such matters as what and why, as well as what will and what should be done. In recent days, data analytics is moving from being reserved for domain experts to becoming necessary for the end-user. However, data availability is both a pertinent issue and a great opportunity for global businesses. This paper presents recent examples from work in our research team on ubiquitous data analytics and open up to a discussion on key questions relating methodologies, tools and frameworks to improve ubiquitous data team effectiveness as well as the potential goals for a ubiquitous data process methodology. Finally, we give an outlook on the future of data analytics, suggesting a few research topics, applications, opportunities and challenges. This paper is based on a keynote speech to the 14th International Conference on ICT in Education, Research and Industrial Applications. Integration, Harmonization and Knowledge Transfer, Kyiv, Ukraine on 16 May 2018.

Keywords: Ubiquitous data · Big data · Data analytics · Transport sector · Smart city · Emergency management

1 Introduction

Data are limitless and inexhaustible can be re-used multiple times for different uses. Supply of data is growing exponentially and delivers far more riches and value (knowledge) than any other assets. Data growth at an alarming rate is spurred by a variety of sources, such as embedded sensors, social media sites, video cameras, the quantified-self and the internet-of-things [1]. This is changing our reliance on data for decision making or data analytics, from being mostly carried out by an individual and in limited settings, to taking place while on-the-move and in the field of action. In terms of data evolution, data converges into wisdom (or intelligence) by way of information and knowledge through activities

V. Ermolayev et al. (Eds.): ICTERI 2018, CCIS 1007, pp. 3–17, 2019.
https://doi.org/10.1007/978-3-030-13929-2_1

like researching, absorbing, acting, interacting and reflecting [2]. While conducting these activities, persons as well as businesses generally gains understanding, experience and insight, and may come up with innovative ideas.

On the one hand, organisations historically collected copious amounts of data. However, by complexity of data analytics, the organisations were unable to use that data to generate meaningful information in a timely manner, and their business insights were fragmented [3]. According to Fig. 1 that shows a value chain as the foundation of Big Data, these situations mean that there are difficulties predicting and responding to changing business needs and rising chances, as a result, business opportunities and related growth were tied to a much slower roadmap. Some Big Data experts have argued that businesses do not use even a small portion of Big Data that they have, and big does not always mean better [4,5]. In other words, as one of data availability aspects, having tremendous amounts of data might not guarantee businesses a competitive advantage over competitors. Therefore, how effectively and quickly you analyse the data and extract actionable information from it will be critical, since fast and actionable data will become an important part in the usage of Big Data [6,7].

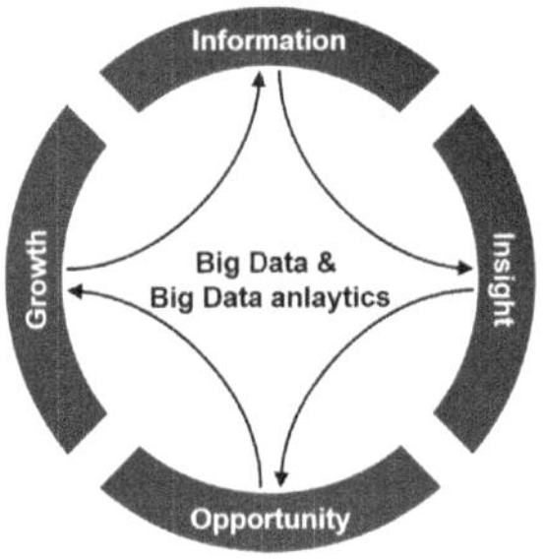

Fig. 1. Value chain in Big Data

In this regard, new data-and-analytics-related businesses and the application of data insights are changing the nature of competition. Data and analytics are also changing the nature of industry competition. Recently, McKinsey reports[1] that seventy percent of all executives think that Big Data and it's analytics have caused changes in their industries' competitive landscapes in recent years. In The fourth industrial revolution, Big Data analytics can support a productivity leap because it generates so many distinct opportunities[2]. One of the simplest forecasts is faster change pace, changes happen more quickly and organizations can do more things in less time. Additionally, higher efficiency means that these

[1] McKinsey&Business: Retrieved from https://www.mckinsey.com/business-functions/operations/our-insights/ops-4-0-fueling-the-next-20-percent-productivity-rise-with-digital-analytics.

[2] McKinsey&Business: Retrieved from https://www.mckinsey.com/business-functions/mckinsey-analytics/our-insights/analytics-comes-of-age.

require fewer resources, while enhanced effectiveness offers the changes greater effect. For instance, Increased predictability, through more accurate forecasting based on unstructured data, lets organizations plan their moves more consistently and respond with greater agility. There are tremendous opportunities for organizations to leverage data in new ways yet many challenges still exist as follows: Many organizations fail to identify the right data or have no idea how to best use them. In addition, insufficient storage capacity and analytical capabilities to handle the massive volume of data. Furthermore, many enterprises lack the right kind of infrastructure or connectivity that can provide seamless access to data. Since data may come in a variety of formats, organizations often find it difficult and expensive to analyse and dig up insights. Last data is generated at the exponential rate of velocity, which makes it difficult to perform real-time analytics that uncovers the intelligence they contain.

In terms of "variety" in characteristics of Big Data, it's analytics can be applied to many different types of data. These data types range from simple atomic types to more complex constructs such as time series, relational data, graphs, images and many others. Over the years, many methods were studied to transform into information and incorporate these different data types to get knowledge. One aspect that has been studied only briefly so far is ubiquity. In this regard, there are following characteristics [8].

- Data is producing asynchronously in a highly decentralized way.
- Data usually is involved to many different data types that make up a whole.
- Data in almost all cases emerges from a very high number of partially overlapping and loosely connected sources.
- Data is produced by many different users distributed all over the world and is often noisy and contradicting, partially overlapping etc.
- Data is multi-dimensional and multi-modal.

We refer to such data as "*Ubiquitous Data*"and define by:

> *the data emerges from pervasive domain specific information provided by static and mobile professional sensors, from social sensors -people- who voluntarily provide information/data, from large-scale data banks of business, municipalities and government, from social media, and open data.*

This paper presents recent examples from work in our research team on LeMO[3] and BDEM[4] projects related to ubiquitous data analytics and open up to a discussion on key questions relating sources, methodologies, tools and frameworks to improve ubiquitous data team effectiveness as well as the potential goals for a ubiquitous data process methodology. Finally, we give an outlook on the future of data analytics, suggesting a few research topics, applications, opportunities and challenges.

---

[3] LeMO: Retrieved from https://lemo-h2020.eu/.

[4] BDEM: Retrieved from https://bdem.squarespace.com/.

2 Related Work

This section shows recent examples such as projects, methodologies, tools, frameworks, related to Ubiquitous Data. MK:smart[5] as a large collaborative initiative has created a 'MK Data Hub' which supports the acquisition and management of vast amounts of data relevant to the city of Milton Keynes, England, systems from a variety of data sources. These have 712 data sets including local/national open data and data streams from both key infrastructure networks (energy, transport, water) and other relevant sensor networks (e.g. weather and pollution data), crowd-source data from social media and mobile applications. It provides a streaming API for timeline and sensor data, as well as an Entity API which aggregates the information available in the hub on general entities from the many data sets contributing to the MK Data Hub by using semantic technology. Especially the Entity API might offer a way of fusing various data sets to generate information from ubiquitous data toward knowledge. UBIMOB[6] develops an adaptive and context sensitive mobility solution in order to make smart decisions taking citizens' personal need into account and to reach equilibrium of mobility services, supply and demand, by smarter resource planning and matchmaking. Aiming to enable a vision for future mobility, this project considers huge amounts of ubiquitous data through sensors, traffic cameras, as well as asynchronous user-generated information, synchronous user-generated data, historic databases and data from mobility companies in real-time. Their solution will be evaluated by people, business and decision makers in 3 cities in Norway. SETA[7] is developing technologies and methodologies based on large, complex and dynamic ubiquitous data from citizens, connected cars, city sensors and distributed databases, in order to change the way mobility is organised, monitored and planned in large metropolitan areas. Their application using smartphone's sensors enable to track users' mobility with considering you an insight into your daily travel. In addition, by integrating with environmental data such as road closures, road works, accidents, delays and levels of pollution, it shows users' daily journeys, the impact they had on the environment, calories burned and any costs associated with the journey. TransformingTransport[8] project is trying to demonstrate measurable, and replicable way the transformations that Big Data will bring to the mobility and logistics market. It addresses seven major domains, such as high-ways, sustainable vehicle, proactive rail infrastructures, intelligent ports, efficient air transport, multi-modal Urban mobility, and dynamic supply chains. Moreover, an open data portal[9] containing 148 data sets related to ubiquitous data has been published in order to provide the community for the reuse of data across the different transport domains. NOESIS[10] aims to offer a methodological

[5] All information can be accessed via http://www.mksmart.org/ and were last accessed on September 3, 2018.
[6] https://www.vestforsk.no/en/project/ubiquitous-data-driven-urban-mobility.
[7] http://setamobility.weebly.com/.
[8] https://transformingtransport.eu/.
[9] https://data.transformingtransport.eu/ was accessed on September 3, 2018.
[10] http://noesis-project.eu/.

framework (i.e. a decision support tool) and data-driven evidence to enable the deployment of a Big Data in transport ecosystem in Europe, by addressing the related technological, institutional/legal, business, and policy challenges. Based on a number of machine learning techniques to associate Ubiquitous Data with a predefined set of use cases, the framework/tool will be used to assess the value generated (i.e. socioeconomic impact) from Big Data investments. In order to enhance the economic sustainability and competitiveness of European transport sector, LeMO[11] studies and analyses Big Data in the European transport domain in particular with respect to five transport dimensions: mode, sector, technology, policy and evaluation. Contrary to the NOESIS project that exploring case studies after implementing a framework, LeMO conducts a series of case studies to provide recommendations and roadmap on the prerequisites of effective Big Data implementation in the transport field, towards data openness, collection, exploitation and data sharing to support European transport stakeholders. In addition, this project considers economic, legal, ethical, political and social issues related to Ubiquitous Data in the field of transport. BDEM[12] aims to develop a international partnership between four countries (i.e. Norway, the USA, Japan and Hong Kong) to share best practices, and strengthen research in Big Data and emergency management. Another major initiative in our research group is Teknoløft project (research-based technology innovation in Sogn og Fjordane region) that aims to nurture and develop a strong knowledge platform on Big Data, which enables local business to be able to use existing and new data for innovation.

So far, some projects related to Ubiquitous Data (especially smart city, smart transport and emergency management) are introduced. In common, the projects argue that Ubiquitous Data should be available, accessible and reusable in order to create more invaluable information and knowledge. In addition, various issues (e.g. economic, legal, social, ethic, environmental issues) in target field also should be considered to appropriately exploit Ubiquitous Data. More detail challenges and open issues will be discussed in Sect. 4.

3 Understanding Ubiquitous Data

The advent of Ubiquitous Data is explained by two parallel movements. One is technological change, for instance cloud computing enables large volumes of data to be stocked at low cost. The improved speed of data update and analysis (real time) permits significant optimisations in all most economic sectors such as transport, energy, construction, health and so on. In addition the advance of many techniques in a variety of fields (e.g. machine learning, data mining and artificial intelligence) are allowing to deal with various data forms and formats likes images, texts, sounds and so on. Other change is in behaviour and use. Nowadays, by the increase in Internet and social media usage, data produced every day on the web represents the equivalent of information contained on 250

[11] https://lemo-h2020.eu/overview/.
[12] https://bdem.squarespace.com/.

million DVDs; On the horizon in 2020, there should be 50 million connected devices globally (smart phones, tablets, smart watches and etc.) These movements for Ubiquitous Data are addressing potential to add value as follows:

1. Data sharing between stakeholders at different levels, but whose interests converge towards certain projects.
2. Crossing of different types and forms of data to keep creating more reliable and easier-to-use services.
3. Interoperability of data, that is the technical capability to connect data of different types and origins.
4. Citizen participation according to their right for the data that they produce and want to put to good use.

These capabilities may be obtained, when following principles [9] of Ubiquitous Data are sufficiently figured out.

- **Data comes from activities.** For instances, every movement of a freight, change of factory temperature, or adjustment of a wing flap is an activity. In most of the machines and buildings of the world, data/information created without these activities is easily lost. Companies are in a race to digitize and 'datafy' them before their rivals do. Digitizing activities means involving sensors or mobile applications in the activities in some way, and datafying activities means extending the observations you capture about them.
- **Data tends to make more data.** As one of examples, algorithms that use accumulated data to predict maintenance schedules for electrical transformers or inventory movements in automated warehouses produce data about their own performance in order to be fed back into the system and improve their future performance. This may becomes a competitive lead that's very hard for others to close. Businesses must develop both infrastructure and new skills in experiment design, data analysis, and interpretation. Managing that process will require new tactics to handle conflicting, data-driven perspectives on the same topic.
- **Platforms being preferred tend to win.** The digitization and datafication of more mechanized activities brings platform competition to industries that have not seen it before. Already, luxury car makers are in a battle to be the preferred platform for connected vehicle services. This same phenomenon will change the terms of competition for wide areas from manufacturers to construction.

Following sections present sources and processing of Ubiquitous Data in order to glance these potentialities.

3.1 Ubiquitous Data Sources

3.1.1 Mobility Data

Among mobility data, mobile data rose in prominence with the advance of smart phones and the interest in civic apps. A call record data containing attributes that are specific to a single instance of a phone call (or data connection) such as the phone numbers and the start/end time and the duration of that call, as well as sensor data which generated embedded sensors to smartphone, even applications installed in users mobile have been utilised in various fields (e.g., analysing individuals' behaviour [10], building a network between different users [11] and inferring other patterns [12]). For instance, this kind of data can be studied to understand large-scale human mobility and trajectory patterns [13]. As another mobility data, data generated by moving vehicle traveling around a city with a GPS sensor is sent to a central/distributed system and matched to a road/rail/path network for deriving and managing traffic status. As becoming this kind of data vary and complex by the emergence of (semi-)automated vehicles and drones, its utilisation is being studied widely for various areas such as transport optimisation [14], emergency management [15], sustainable city [16]. Commuting/public transport data, such as the card usage data for systems of a bus and subway and the ticketing data in parking lots, is generated by people living in cities. This kind of data generated by people passing entrances of a subway station and getting a bus can be used for public transportation systems in a city [17]. Each transaction record contains timestamps and the ID of the station or bus as well as the fare for this move. Payment information created from parking meters in street-side may include the time the parking ticket and the fare. These data indicate the traffic of around a parking place, which can be used to not only manage a city's parking infrastructure [18] but also analyse people's travel patterns [19].

3.1.2 Environmental Data

Meteorological data enable to be crawled from public websites includes various environmental information such as humidity, temperature, barometer pressure, wind speed, and weather conditions [20]. It influences traffic flow and land uses. On the one hand, there are many ways to collect traffic data, such as using loop detectors, magnetometers, infrared sensors, surveillance cameras and so on [21,22]. For example, the loop detectors are typically buried in the major roads (e.g. highways). Such detectors deduces the travel speed of vehicles by using two consecutive (i.e. a pair of) sensors. In addition, the traffic volume on a road can be detected by counting the number of vehicles traversing the loop detectors. However, its cost is usually expensive and area in which the sensors are embedded and limited [23]. Therefore other data can be gathered via infrared or ultrasonic sensors. Additionally, some traffic surveys can be conducted manually [24]. As another instance, surveillance cameras, installed in cities over the last thirty or so years, generate a huge amount of image and video data reflecting traffic patterns. However, it is still challenge to automatically extract traffic information from the collected data using machine learning techniques [25]. Hence, monitoring citywide traffic status is mainly conducted through human resource.

3.1.3 Social Media Data

Due to the advent of the personal mobile devices and web 2.0, Social media data consisting of social network (e.g. relationship, interdependency, or interaction) and user-generated data (e.g., texts, photos and videos) has being utilised in the wide fields such as transport sector [26], smart city [27] and emergency management [28]. It includes enriched information for ordinary or specific context and users' behaviour/interests. In particular, some part of the data have user's location and timestamp. Therefore many researchers and practitioners have been tried to use it for their domains. However, there still are many issues such as reliability, processing unstructured data [29], sparsity of locations [30] and privacy. For instance, it is difficult to obtain reliable data from social media, since it is usually many noisy and contain too broad information from daily life to specific context [31]. In addition, social media, since it is content directly created by end users, might carry personally identifiable information. Therefore, researchers and practitioners should be mindful of any explicit or inferred personal information in the data [32].

3.1.4 Geographical Data

Road network data is one of the most frequently utilised geographical data in various area, for instatnce, transport routing [14], city planning [33] and emergency management [34]. This kind of data is able to be presented by a graph consisting of edges (i.e. road segments) and a nodes (intersections). An edge explaining by two nodes and a geospatial points is related to the length, speed constraint, tyep of road and number of lanes, while a node has geospatial coordinates. On the on hand, a Point of interest (POI), such as a restaurant, a museum and a zoo, usually has a name, address, category and geospatial coordinates. It is difficult to gather POI data due to a characteristic of their changing according to time. For instance, a shop may replace its name, be moved to a new place, or even shut down. The POI data can be generated by thwo methods: one is collected from existing Yellow page data by geocoding algorithm, and the other is manually gathering POI information (e.g., Navinfo and AutoNavi). In case of Foursquare, as a location-based social networking service, end users is able to create a new POI which has not been included in a system. Online map services (e.g., Google maps and Bing) generally combine the aforementioned two approaches to collect a large coverage of POIs. Therefore, issues have been arisen, such as accuracy of POI [35], integrating the POI data from various [36] sources and so on.

3.2 Processing Ubiquitous Data

3.2.1 Architecture for Ubiquitous Data

In this subsection, we propose an architecture of Ubiquitous Data Analytics platform as shown in Fig. 2 and briefly describe each component of the architecture. It consists of four layers such as data collection, analytics, management and providing.

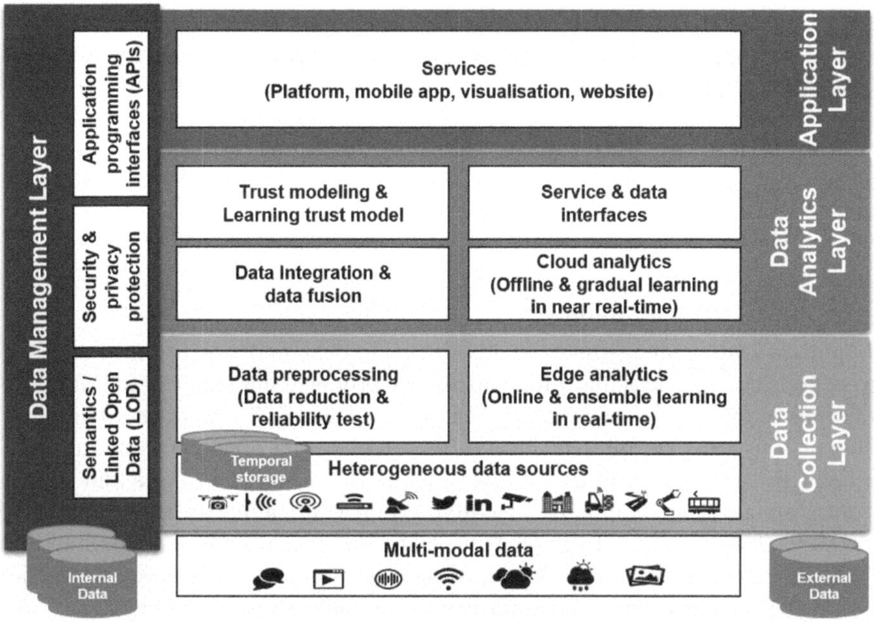

Fig. 2. An architecture for Ubiquitous Data Analytics platform

- **Data collection layer** gathering data is connected closely to physical world and can often be called as edge computing (or network). Heterogeneous data are generated from a variety of sources and transferred to edge clouds (or routers) in order to conduct learning algorithms in real-time. Online or ensemble learning technique has being studied as an appropriate approach for edge analytics. Data preprocessing such as data reduction and reliability test is necessary, since edge nodes have limited resources.
- **Data analytic layer** obtain various types of data from edge nodes to analyse data using powerful resources (i.e., storage, computing power and so on) in near real-time. This layer also performs data integration and fusion through semantic and LOD techniques in order to obtain higher level of information and knowledge. In addition, trust modelling and its learning are conducted to a many amount of data based on the strong resources. The model learnt may be propagated at each edge to test data reliability in real time. Last refined or learnt data/information will be presented through service and data interfaces to the application layer (e.g., platform, mobile app, visualisation, website).
- **Application layer** directly connected with service providers or end users to offer data/information. Here visualisation techniques openning a way of efficient and effective information provision have being studied and utilised.
- **Data management layer** impacts other layers overall. Aforementioned semantic/LOD techniques can also be used in data collection and data provid-

ing layers. For instance, a visualisation generating data network may utilise LOD technique to discover relations (i.e. edges of graph) between data nodes. Another important issue is preserving data security & privacy and must be performed for all process carefully. Application programming interfaces should be also offered to enhance the accessibility and usability of Ubiquitous Data.

Next subsection presents more details for data management and analytics.

3.2.2 Data Management

Ubiquitous Data management involves development and execution of architectures, policies, practices and procedures that properly manage the full data lifecycle needs for applications using Ubiquitous Data [37]. As data comes from different sources with different formats, there is a need for advanced data management features that will lead to recognizing the different formats and sources of data, structuring, managing, classifying and controlling all these types and structures. Ubiquitous data management should also provide scalable handling for massive data to support offline applications as well as low latency processing to serve effectively in real-time applications. In addition, it is a policy-based approach for determining which information should be stored where within an organization's IT environment, as well as when data can safely be deleted. Within a typical enterprise, people with many different job titles may be involved in ubiquitous data management. They are a chief data officer, chief information officer, data managers, database administrators, data architects, data modelers, data scientists, data warehouse managers, data warehouse analysts, business analysts, developers and others.

3.2.3 Data Analytics

Typical analytic algorithms used in regular applications may not be sufficient or efficient enough to handle Ubiquitous Data applications due to their unique requirements and pressing need for high volume high speed processing. These algorithms need to be optimized to handle high data volumes, large variety of data types, time constraints on decision making processes, and distributed components across various geographical locations. Also, these algorithms need to work effectively across heterogeneous environments and be capable of managing and operating in highly dynamic environments. Additionally, they should consider handling hight data volumes, working across large variety of data types, time constraints on decision making processes, geographical distributed components [16]. Unobtrusive and ubiquitous sensing technologies, advanced data management and analytics models, and novel visualization methods, should be connected to create win-win-win solutions that improve urban environment, human life quality, and city operation systems.

4 Challenges and Open Issues

This section explores challenges and open issues on the Ubiquitous Data and its utilisation. As aforementioned, Ubiquitous Data is dynamically changed with respect to time and space and needs to be processed by considering its heterogeneous from various type of sources in real time. Actually, for instance, monitoring traffic flow on a road segment is possible, but continually probing the citywide traffic is challenging as we do not have sensors on every road segment; Additionally, there is limited integration of ubiquitous data analytics and social computing; And there is an absence of effective models to predict behaviours efficiently; Real-time prediction and mobility capabilities have not yet been incorporated into the infrastructure for smart transport management. In this regard, following challenges should be considered in order to unlock real values (range from information to wisdom) of Ubiquitous Data.

Accessibility and Usability of Open Data: The open data movement is based on the three principles of transparency, participation and collaboration [38]. While it is believed that through openness, sharing and coopering, the value of data to society can be truly realized, the rapid opening of government data has not been universally welcomed. However, there are still many open data portals simply providing dump files. These sites do not follow any standard or even give any attention for the usability of the data. One of the reasons is that they have been developed by hackathon enthusiasts or data divers [39]. In these cases, unfortunately, there is often no post-event follow up, maintenance or further development [40]. Openness usually refers a number of aspects that circle around the fact that an element is openly (in the sense of publicly and royalty free) available and reusable, developed in an open process, accessible at minimum costs (in terms of data pure reproduction costs or even no costs).

Processing and Integration of Heterogeneous Data Sources: Overlapping the same data, increasing performance and scalability and enabling real-time data access are one of the challenges related to data processing and integration that should be addressed in the future. For instance, collecting data from different sources, storing and providing data with a unified view, and adjusting structures in semi-structured and unstructured data [41]. Traditional data mining usually deals with data from a single domain. In the ubiquitous data era, we face a diversity of datasets from different sources in different domains. These datasets consist of multiple modalities, each of which has a different representation, distribution, scale, and density. How to unlock the power of knowledge from multiple disparate (but potentially connected) datasets is paramount in ubiquitous data research, essentially distinguishing ubiquitous data from traditional data mining tasks. This calls for advanced techniques that can fuse the knowledge from various datasets organically in a machine learning and data mining task.

Spatio-Temporal Property: Spatio-temporal locality is central feature as well as challenge of Ubiquitous Data. Temporality means whether the data describe phenomena that change with time, while spatiality indicates whether the data are spatially located. All most of Ubiquitous data have characteristics combined of both features. In a typical scenario related to Ubiquitous Data, learning algorithm has access to past and maybe present data from a time-varying distribution, and has to make inferences about the present or future. Additionally, it is assumed that a data stream is constantly generated with a memory constrained environment [42]. Thus, various techniques such as concept drift detectors, data reduction, sliding windows and online/ensemble learners have being studied [43]. In this regard, '*edge analytics*' is relatively new and it is still developing [44]. Once it is sufficiently perfected, it may revolutionize the way we process ubiquitous data. Basically, data is analysed the moment it is collected, so you immediately have a complete analysis. This can be really useful for security cameras so that irrelevant data is no longer stored, for navigation devices and so on.

Privacy and Security: A wide range of smart mobility technologies are being deployed within ubiquitous environment. These technologies are generating huge quantities of data and much of them in real-time. However, generating, processing, analysing, sharing, and storing vast amounts of actionable data also raises several concerns and challenges. For example, data privacy, data protection, and data security issues are caused from the creation of smart mobility [45]. Privacy-related issues arise when a system infers or restores personal information using Big Data analytics tools, although data are anonymized. With the proliferation of Big Data analytics technologies used in Ubiquitous data, the privacy issue has become a core problem in the data mining domain. Another security risk associated with Ubiquitous Data is the heterogeneity of the types of devices used and the nature of generated data (e.g., raw devices, data types and communication protocols). To authenticate these devices, a system should assign and maintain a nonrepudiable identification to each device. These activities results in increased security risks [41].

5 Conclusion

As growing Ubiquitous Data at an alarming rate by varied array of sources, such as embedded sensors, social media sites and the internet-of-things, this is changing our reliance on data for making decisions. Nowadays, data analytics is moving from being conducted by/for domain experts, to becoming necessary for the end-user. However, there are still many challenges such as data availability, processing heterogeneous data and privacy. In this paper, we presented recent examples (e.g. projects, methodologies, tools and frameworks) on Ubiquitous Data analytics and discussed on key question relating Ubiquitous Data to give an outlook on the future of data analytics, a few research topics, challenges and opportunities. In this regard, an architecture of Ubiquitous Data Analytics platform was also proposed by considering recent environments and research trends.

Acknowledgements. This work is partially supported by the European Union's Horizon 2020 research and innovation programme under grant agreement No 770038, UBIMOB project (270785) funded by the Norwegian Research Council in 2017 through the IKTPLUSS programme, and BDEM project funded by the Research Council of Norway (RCN) and the Norwegian Centre for International Cooperation in Education (SiU) through the INTPART programme. Authors contributed equally to this work.

References

1. Akerkar, R.: Processing big data for emergency management. In: Emergency and Disaster Management: Concepts, Methodologies, Tools, and Applications, pp. 980–1000. IGI Global (2019). https://doi.org/10.4018/978-1-5225-2575-2.ch005
2. Akerkar, R., Sajja, P.: Knowledge-Based Systems. Jones & Bartlett Publishers, Burlington (2010)
3. Akerkar, R., Sajja, P.S.: Intelligent Techniques for Data Science. Springer, Cham (2016). https://doi.org/10.1007/978-3-319-29206-9
4. Goes, J.D.: Big data is dead. what's next. Venturebeatcom guest blog post (2013). https://venturebeat.com/2013/02/22/big-data-is-dead-whats-next/?goback=%2Egde_62438_member_217099766
5. Chauhan, R.: Transforming big data into actionable insights (2015). https://www.mastercardadvisors.com/content/dam/advisors/en-us/documents/150513_Transforming_Big_Data.pdf
6. Barnaghi, P.M., Sheth, A.P., Henson, C.A.: From data to actionable knowledge: big data challenges in the web of things. IEEE Intell. Syst. **28**(6), 6–11 (2013). https://doi.org/10.1109/MIS.2013.142
7. Carter, K.B.: Actionable Intelligence: A Guide to Delivering Business Results with Big Data Fast!. Wiley, Hoboken (2014)
8. Hotho, A., Pedersen, R.U., Wurst, M.: Ubiquitous data. In: May, M., Saitta, L. (eds.) Ubiquitous Knowledge Discovery. LNCS (LNAI), vol. 6202, pp. 61–74. Springer, Heidelberg (2010). https://doi.org/10.1007/978-3-642-16392-0_4
9. Insights MTR: The rise of data capital. Oracle (2016). https://www.technologyreview.com/s/601081/the-rise-of-data-capital/
10. Senaratne, H., et al.: Urban mobility analysis with mobile network data: a visual analytics approach. IEEE Trans. Intell. Transp. Syst. **19**(5), 1537–1546 (2018). https://doi.org/10.1109/TITS.2017.2727281
11. Song, Y., Hu, Z., Leng, X., Tian, H., Yang, K., Ke, X.: Friendship influence on mobile behavior of location based social network users. J. Commun. Netw. **17**(2), 126–132 (2015). https://doi.org/10.1109/JCN.2015.000026
12. Xia, D., Lu, X., Li, H., Wang, W, Li, Y., Zhang, Z.: A MapReduce-based parallel frequent pattern growth algorithm for spatiotemporal association analysis of mobile trajectory big data. Complexity **2018**, 2818,251:1–2818,251:16 (2018). https://doi.org/10.1155/2018/2818251
13. Bhattacharya, S., Blunck, H., Kjærgaard, M.B., Nurmi, P.: Robust and energy-efficient trajectory tracking for mobile devices. IEEE Trans. Mob. Comput. **14**(2), 430–443 (2015). https://doi.org/10.1109/TMC.2014.2318712
14. Menouar, H., Güvenç, I., Akkaya, K., Uluagac, A.S., Kadri, A., Tuncer, A.: UAV-enabled intelligent transportation systems for the smart city: applications and challenges. IEEE Commun. Mag. **55**(3), 22–28 (2017). https://doi.org/10.1109/MCOM.2017.1600238CM

15. Chen, L., Englund, C.: Every second counts: integrating edge computing and service oriented architecture for automatic emergency management. J. Adv. Transp. **2018**, 1–13 (2018). https://doi.org/10.1155/2018/7592926
16. Nuaimi, E.A., Neyadi, H.A., Mohamed, N., Al-Jaroodi, J.: Applications of big data to smart cities. J. Internet Serv. Appl. **6**(1), 25:1–25:15 (2015). https://doi.org/10.1186/s13174-015-0041-5
17. Ma, X., Liu, C., Wen, H., Wang, Y., Wu, Y.J.: Understanding commuting patterns using transit smart card data. J. Transp. Geogr. **58**, 135–145 (2017). https://doi.org/10.1016/j.jtrangeo.2016.12.001
18. Bagula, A.B., Castelli, L., Zennaro, M.: On the design of smart parking networks in the smart cities: an optimal sensor placement model. Sensors **15**(7), 15,443–15,467 (2015). https://doi.org/10.3390/s150715443
19. Zhao, Z., Koutsopoulos, H.N., Zhao, J.: Detecting pattern changes in individual travel behavior: a Bayesian approach. Transp. Res. Part B: Methodol. **112**, 73–88 (2018). https://doi.org/10.1016/j.trb.2018.03.017
20. Alam, F., Mehmood, R., Katib, I., Albogami, N.N., Albeshri, A.: Data fusion and iot for smart ubiquitous environments: a survey. IEEE Access **5**, 9533–9554 (2017). https://doi.org/10.1109/ACCESS.2017.2697839
21. Nandury, S.V., Begum, B.A.: Smart WSN-based ubiquitous architecture for smart cities. In: Mauri, J.L., et al. (eds.) Proceedings of the International Conference on Advances in Computing, Communications and Informatics, ICACCI 2015, pp. 2366–2373. IEEE, Kochi (2015). https://doi.org/10.1109/ICACCI.2015.7275972
22. Yi, S., Li, C., Li, Q.: A survey of fog computing: concepts, applications and issues. In: Li, Q., Xuan, D. (eds.) Proceedings of the 2015 Workshop on Mobile Big Data, Mobidata@MobiHoc 2015, pp. 37–42. ACM, Hangzhou (2015). https://doi.org/10.1145/2757384.2757397
23. Thakuriah, P.V., Geers, D.G.: Data sources and management. In: Thakuriah, P., Geers, D.G. (eds.) Transportation and Information. BRIEFSCOMPUTER, pp. 15–34. Springer, New York (2013). https://doi.org/10.1007/978-1-4614-7129-5_2
24. Taylor, N., et al.: The transport data revolution: investigation into the data required to support and drive intelligent mobility (2015). https://ts.catapult.org.uk/wp-content/uploads/2016/04/The-Transport-Data-Revolution.pdf
25. Chen, N., Chen, Y., You, Y., Ling, H., Liang, P., Zimmermann, R.: Dynamic urban surveillance video stream processing using fog computing. In: IEEE Second International Conference on Multimedia Big Data, BigMM 2016, Taipei, Taiwan, 20–22 April 2016, pp. 105–112. IEEE Computer Society (2016). https://doi.org/10.1109/BigMM.2016.53
26. Anantharam, P., Barnaghi, P.M., Thirunarayan, K., Sheth, A.P.: Extracting city traffic events from social streams. ACM Trans. Intell. Syst. Technol. **6**(4), 43:1–43:27 (2015). https://doi.org/10.1145/2717317
27. Costa, D.G., Duran-Faundez, C., Andrade, D.C., Rocha-Junior, J.B., Peixoto, J.P.J.: TwitterSensing: an event-based approach for wireless sensor networks optimization exploiting social media in smart city applications. Sensors **18**(4), 1080 (2018). https://doi.org/10.3390/s18041080
28. Poblet, M., García-Cuesta, E., Casanovas, P.: Crowdsourcing roles, methods and tools for data-intensive disaster management. Inf. Syst. Front. **20**, 1–17 (2017). https://doi.org/10.1007/s10796-017-9734-6
29. Luna, S., Pennock, M.J.: Social media applications and emergency management: a literature review and research agenda. Int. J. Disaster Risk Reduct. **28**, 565–577 (2018). https://doi.org/10.1016/j.ijdrr.2018.01.006

30. Burton, S.H., Tanner, K.W., Giraud-Carrier, C.G., West, J.H., Barnes, M.D.: "Right time, right place" health communication on twitter: value and accuracy of location information. J. Med. Internet Res. **14**(6), e156:1–e156:11 (2012). https://doi.org/10.2196/jmir.2121
31. Kim, J., Hastak, M.: Social network analysis: characteristics of online social networks after a disaster. Int. J. Inf. Manag. **38**(1), 86–96 (2018). https://doi.org/10.1016/j.ijinfomgt.2017.08.003
32. Imran, M., Castillo, C., Diaz, F., Vieweg, S.: Processing social media messages in mass emergency: a survey. ACM Comput. Surv. (CSUR) **47**(4), 67 (2015). https://doi.org/10.1145/2771588
33. Yin, C., Xiong, Z., Chen, H., Wang, J., Cooper, D., David, B.: A literature survey on smart cities. Sci. China Inf. Sci. **58**(10), 1–18 (2015). https://doi.org/10.1007/s11432-015-5397-4
34. Yin, J., Yu, D., Yin, Z., Liu, M., He, Q.: Evaluating the impact and risk of pluvial flash flood on intra-urban road network: a case study in the city center of Shanghai, China. J. Hydrol. **537**, 138–145 (2016). https://doi.org/10.1016/j.jhydrol.2016.03.037
35. Ko, E.B., Lee, J.W.: Accuracy improvement methods for string similarity measurement in poi (point of interest) data retrieval. KIISE Trans. Comput. Pract. **20**(9), 498–506 (2014). https://doi.org/10.5626/KTCP.2014.20.9.498
36. Jiang, S., Alves, A.O., Rodrigues, F., Ferreira Jr., J., Pereira, F.C.: Mining point-of-interest data from social networks for urban land use classification and disaggregation. Comput. Environ. Urban Syst. **53**, 36–46 (2015). https://doi.org/10.1016/j.compenvurbsys.2014.12.001
37. Mosley, M., Brackett, M.H., Earley, S., Henderson, D.: DAMA Guide to the Data Management Body of Knowledge. Technics Publications, Basking Ridge (2010)
38. Lathrop, D., Ruma, L.: Open government: collaboration, transparency, and participation in practice. Govern. Inf. Q. **28**(1), 129–130 (2011). https://doi.org/10.1016/j.giq.2010.08.001
39. Townsend, A.M.: Smart Cities: Big Data, Civic Hackers, and the Quest for a New Utopia. WW Norton & Company, New York (2013)
40. Barkham, R., Bokhari, S., Saiz, A.: Urban big data: city management and real estate markets. GovLab Digest, New York (2018)
41. Marjani, M., et al.: Big IoT data analytics: architecture, opportunities, and open research challenges. IEEE Access **5**, 5247–5261 (2017). https://doi.org/10.1109/ACCESS.2017.2689040
42. May, M., Berendt, B., Cornue, A., et al.: Research challenges in ubiquitous knowledge discovery. In: Next Generation of Data Mining, pp. 154–173. Chapman and Hall/CRC (2008). https://doi.org/10.1201/9781420085877.ch7
43. Ramírez-Gallego, S., Krawczyk, B., García, S., Wozniak, M., Herrera, F.: A survey on data preprocessing for data stream mining: current status and future directions. Neurocomputing **239**, 39–57 (2017). https://doi.org/10.1016/j.neucom.2017.01.078
44. Satyanarayanan, M., et al.: Edge analytics in the Internet of Things. IEEE Perv. Comput. **14**(2), 24–31 (2015). https://doi.org/10.1109/MPRV.2015.32
45. Akerkar, R.: Privacy and security in data-driven urban mobility. In: Utilizing Big Data Paradigms for Business Intelligence, pp. 106–128. IGI Global (2019). https://doi.org/10.4018/978-1-5225-4963-5.ch004

On Convergence of Controlled Snowball Sampling for Scientific Abstracts Collection

Hennadii Dobrovolskyi(✉) and Nataliya Keberle

Department of Computer Science, Zaporizhzhya National University, Zhukovskogo Street 66, Zaporizhzhya 69600, Ukraine
gen.dobr@gmail.com, nkeberle@gmail.com

Abstract. This paper presents evidences concerned to convergence of controlled snowball sampling iterations applied to collecting seminal papers in a selected domain of research. Iterations start from the seed paper selection, plain snowball sampling and probabilistic topic modelling, then greedy controlled snowball sampling and analysis of the collected citation network are performed in rotation until the list of seminal papers becomes stable. The topic model is built on the base of word-word co-occurrence probability with combination of sparse symmetric nonnegative matrix factorization and principal component approximation. Experiments show that the number of topics in the model is determined in natural way and the Kullback-Leibler (KL) divergence provides the upper bound of the cosine similarity calculated from keywords assigned by publication authors. Several citation networks are collected and analysed. The analysis shows that all networks are "small worlds" and therefore the observed saturation of the controlled snowball sampling can provide the complete set of publications in domains of interest. Experiments with KL-divergence, symmetric KL-divergence and Jensen-Shannon divergence show that KL-divergence produces less connected citation network but provides better convergence of snowball iterations. Multiple runs of the sampling confirm the hypothesis that the set of seminal publications is stable with respect to variations of the seed papers. The modified main path analysis allows to distinguish the seminal papers including new publications following main stream of research. The comparison of different ranking criterion is made. It shows that Search Path Count provides better lists of seminal papers than citation index, PageRank and indegree.

Keywords: Text mining · Short text document · Topic modelling · Principal component analysis · Sparse symmetric nonnegative matrix factorization · Citation network · Main path analysis · Convergence · Saturation

V. Ermolayev et al. (Eds.): ICTERI 2018, CCIS 1007, pp. 18–42, 2019.
https://doi.org/10.1007/978-3-030-13929-2_2

1 Introduction

The representative review of a related work is the important part of every scientific work and writing review starts with collecting of the scientific publications concerning the research. Therefore, every author needs to decide if all publications he/she has contain all notable scientific results of the domain of interest and which of them describe the mainstream of the knowledge evolution in the domain of interest.

Currently collecting valuable publications for an arbitrary domain, also called representative bibliography, faces several challenges. First of all, it is the absence of an expert level knowledge in the domain, which is common for entering a new research field, especially if it is a cross-domain research. Second, large and ever increasing number of publications, characterizing actively evolving domains. Third, individual variations of terminology, which is also typical for actively evolving domains with unstable terminology. All aforementioned creates a demand to provide a reliable technique helping researchers to obtain a representative bibliography for an arbitrary domain. The research presented in this paper aims at proving a reliability of the technique [12] proposed by the authors in terms of information quality: authority and reproducibility.

In [12] a complete information technology is proposed. It starts from a small set of papers on some scientific domain and produces a list of references. The core of the technology is the collecting a citation network of scientific publications with controlled snowball sampling and analysis of the collected graph aimed at detecting of seminal papers in a selected domain of research. The distinctive features of the presented method are application of the probabilistic topic model to perform controlled snowball sampling and collect citation network, then the main path analysis is applied to point both the most influential publications and the main path of scientific knowledge evolution. The main path allows detecting the newest publications that follow the mainstream and the outliers that potentially contain the completely new ideas.

The objective of the current work[1] is to present results supporting cumulatively the reliance that the collection of relevant publications can be gathered reproducibly and providing authority of a domain:

- if the proposed snowball control method [13] provides adequate semantic distance between publications;
- if the best age of the seed papers is 2–10 years;
- if the controlled snowball method forms the small-world network that allows reaching every node in small number of steps;
- which measure provides the most relevant publication ranking;
- if the controlled snowball provides saturation of the ranked publication list;
- if the biased seed papers can produce unbiased citation network.

The structure of the paper is as follows. Section 2 overviews publications related to the presented technology, Sect. 3 contains description of the crucial

[1] The presented work is extended version of [12] which is available publicly online at http://ceur-ws.org/Vol-2105/10000179.pdf.

steps of the algorithm, Sect. 4 states the experiment pre-conditions and Sect. 5 discusses the results. Conclusion summarises the main results and discusses future work.

2 Related Work

Below we explore several ways of publications gathering related to the proposed method in different aspects. A list of publications can be obtained from common keyword search engines, recommender systems or domain-specific journals and conferences. However, the most similar to our approach are snowball sampling and peer-to-peer search algorithms that utilize links between documents. Also attention is paid to short text similarity measures and citation network analysis which are crucial parts of the proposed method.

2.1 Common Methods of Publication Gathering

First place to search the publications related to subject of investigation is one of academic search engines. Google Scholar[2], Microsoft Academic[3], Semantic Scholar[4] offer rich set of search features, however they do not evaluate the search results with respect to the coverage of representation of research domain landscape. Moreover, the search services while searching by keywords do not suggest the way to find a possibly relevant topic that is tightly related to search subject but described with different keywords than used in the query. As a result, querying with a set of keywords produces a biased set of publications [40].

Publications on a domain of knowledge can be collected from journals or conference proceedings [17], the study of the co-authorship [34,35], elaboration of keywords and topics [33,38] However, there is not a corresponding conference for every research domain, as well as one author can write papers on different topics. Building maps and ontologies of large scientific domains does not provide the list of references rather the set of interconnected concepts.

2.2 Recommender Systems

Next tool to look for is one of document co-occurrence based, content-based filtering, collaborative filtering, or graph-based recommender system [8,43].

Document co-occurrence based systems utilize the assumption that two adjacently placed items are related. For instance, co-citations [45] or co-views [41] were used. However, arXiv.org library access records show that 2/3 of related publications are never co-cited [41], moreover 90% of publications are never cited [21], so co-occurence statistics cannot be calculated.

The most popular recommender approach, content-based filtering [28], suffers from small diversity similarly to familiar search engines. It suggests only

[2] Google Scholar https://scholar.google.com.ua/.
[3] Microsoft Academic https://academic.microsoft.com/.
[4] Semantic Scholar https://www.semanticscholar.org/.

documents similar to ones already found by user [28]. Also it ignores quality of publications [14].

Collaborative filtering uses hypothesis that a user will be interested in documents which were viewed by other similar users. However the academic recommendations cannot be formed by collaborative filtering, because the number of documents is more than the number of researchers [51], so, in the best scenario, the scientific article is read by few peoples [21], hence we cannot get a reliable estimates of relevance counting document views and co-views.

The more promising approaches treat the inherent links existing between academic papers and build citation networks which are graphs showing how documents cite each other [3,23,27]. Lao et al. [24] suggests a mixture of graph and content-based approach complementing the citation graph with terms from the papers' titles. Other works create connections based on text similarity [22,54], co-citation [22,54,58]. Once a graph is built, it is explored to offer recommendations. Typically, one or several graph vertexes are chosen as starting points of random walks with restarts to find the most popular items in the graph [20,23,24].

2.3 Snowball Sampling

The most appropriate way to collect citation networks is snowball sampling [1,12,13,25,50], when each publication from the current queue is considered then all referenced papers and all papers referencing to the publication are added to the next level queue. The snowball sampling allows collecting publications on the narrow research topic and connect them in the citation network [46].

The high quality of citation-based search algorithm is provided with phenomena of "small world" which is a proved property of citation networks [4,46]. Newman [34] has shown that in the most of the cases it is enough to do three iterations. However, the statistical properties of global citation network including all scientific papers is not known, that is why we need to test if the "small world" assumption is true for the collected subset of citation network and estimate the number of iterations needed to collect most of the papers.

Another point of snowball sampling is dependence on the initial queue called a seed collection. The general advice [25] recommends that the seed papers should be the seminal papers of the knowledge domain pointed by experts or the papers selected by the researcher. Valid seed papers should be 2–10 years old and have to be widely cited. The best seeds are the reviews, foundational or framing articles on the topic of interest. However, the advice also should be checked. Moreover, we need to test if the biased seed papers can produce unbiased citation network.

The snowball sampling cannot be applied directly to publication crawling because the list of references can contain the items that are not directly related to the investigated domain. Therefore the straightforward implementation of snowball publication sampling causes infinite collection inflation and some intelligent techniques should be introduced to decide when it is worth to follow a reference and which candidate publication may be accepted as relevant.

2.4 Search Algorithms for Peer-to-Peer Networks

A lot of methods that can be applied to improve the snowball sampling are suggested by scholars developing the search algorithms for peer-to-peer (P2P) networks. P2P networks are large decentralised computer networks consisting of equal nodes (peers) that can join and disjoin the network at any time. Each peer functions in autonomous manner and doesn't need a central server. The search in P2P systems is performed by querying the adjacent nodes, however routing the query to all neighbours produces too many messages inside network, so a lot of efforts were done to decrease messaging overhead. Studies show that semantic grouping of peers and routing search query to selected adjacent nodes depending on their descriptions makes the search more effective [2,5,10,15,57]. Also it was shown [56], that correct selection of query recipients allow keeping high precision and recall.

The intelligent search algorithms [36] can learn the effective search communication patterns by creating new virtual connections and building semantic communities inside P2P system. The suggested approaches [36] use individual "node-topic" tables stored in each peer and combine querying the adjacent nodes (all or random or related to query topic), querying the semantic communities, updates of "node-topic" table and their posting to the neighbour nodes, messaging own topics when joining P2P network. Mapping peers to publications and "node-topic" table to references we can view that the logical P2P system built as described above is similar to citation network. So we can apply the P2P search methods to snowball sampling. For instance, it is plausible that similarly to P2P search [56] following only most appropriate citations we keep the "small world" property of the collected citation network although we need to check this hypothesis.

2.5 Short Text Similarity

To select the most relevant references to follow while sampling Ahad et al. [1] in their approach use vector document model and cosine similarity, however the vector document model relies on word spelling rather than meaning that causes precision loss when the short texts are considered. Lecy et al. [25] apply PageRank calculated by Google Scholar as a measure of paper significance. However, the used PageRank measure is a property of a global citation network including all topics of knowledge, so it is not suitable in narrow domain. One of the most promising approaches is the probabilistic topic model (PTM).

Probabilistic topic models [52] use a large collection of documents and statistical approach to model words and documents as vectors in a high-dimensional semantic space R^n, where n is much less than number of words and number of documents. The base idea of PTM is to construct few topics which are groups of tightly connected words. Then document words are represented as a result of two-stage random sampling. The most known method of topic modelling is Latent Dirichlet Allocations (LDA) [9] which is successful and simple enough.

A general introduction and survey of the topic modelling can be found in [52] along with a novel approach, called Additive Regularization of Topic Models.

However, in most of the scientific databases, full texts are often protected by copyright. Therefore the only information we can use are paper title, paper abstract, and sometimes the database-specific keywords and topics. So the documents that we analyse are short and common PTMs based on document-word statistics lose their precision. This shortcoming is overcome with approaches utilizing word co-occurrence statistics in Biterm Topic Model (BTM) [55] and Word Network Topic Model (WNTM) [59] instead of counting document-word pairs. Another method of word embedding, called GloVe, is proposed in [39]. It is based on word-word co-occurrence matrix and uses global matrix factorization, so it is close to BTM [55] and WNTM [59] statistical topic modelling.

Also, the vague part of common PTMs is that number of topics cannot be determined with document analysis. To overcome this weakness, handling the word-word co-occurences with principal component analysis (PCA) and sparse symmetric nonnegative matrix factorization (Sparse SNMF) was proposed [13].

2.6 Citation Network Analysis

The collected citation network [46] can be analyzed using citation count and other simple statistics [29], PageRank [26], information retrieval techniques [1], knowledge graph [38], combined supervised machine learning approaches [49] or Main Path analysis [29].

The most appropriate way to highlight the seminal papers of the small scientific domain is main path analysis because the method deals only with the collected citation network and allows to increase the precision of sampled dataset. On the contrary, the citation count and other statistics, supervised machine learning applied by Valenzuela, Ha and Etzioni [49] and PageRank cannot point out the tightly interconnected subset of the citation network. Klink-2 [38] and similar algorithms aim to build the map of knowledge domain but do not seek the most influenced publications.

2.7 Section Summary

Summarising the section, we can highlight that keyword search produces narrow lists that are biased due to peculiarity of personal scholar's vocabulary. Not every topic has the corresponding journals and conferences. The more promising approach suggest the graph-based recommender systems but they do not utilize the advanced graph analysis algorithms. Moreover, all the methods mentioned above do not address the coverage, authority and reproducibility of the obtained list of documents.

From this point of view, snowball sampling suggests the method of collecting the set of documents that meet the coverage requirement and citation network analysis supports the authority aspect. However, the known snowball sampling approaches do not use semantic text similarity measures as do advanced peer-to-peer search techniques.

3 Workflow Proposed for the Scientific Abstracts Collection Gathering

The proposed workflow of the controlled snowball sampling used for scientific abstracts gathering contains the following steps [13]:

1. Collect a set of seed papers and put them in the initial, 0-th, queue.
2. Run several iterations of the uncontrolled snowball sampling to pickup baseline documents. For $n \in 0, 1, 2, 3$
    1. get a portion of papers from the n-th queue;
    2. download the papers referenced by the portion;
    3. download the papers referencing the portion;
    4. add all the downloaded papers to the $(n + 1)$-th queue.
3. Create the PTM using baseline documents:
    1. extract title and abstract from each document of the collection;
    2. split all the titles and abstracts into sentences;
    3. create the dictionary containing all the nouns and adjectives that occur in the sentences;
    4. combine all terms from the reduced dictionary occurring in the same sentence into pairs and build the joint probability matrix;
    5. detect the collection specific stop-words and exclude them from the reduced dictionary;
    6. perform Sparse SNMF to create PTM;
    7. map each of the seed papers to a vector of topic probabilities.
4. Perform the batch controlled snowball sampling: for $n \in 0, 1, 2, 3$
    1. get a portion of papers from the n-th queue;
    2. download the papers referenced by the portion;
    3. download the papers referencing the portion;
    4. extract bag of stemmed words from each of downloaded papers;
    5. map each of the downloaded papers to a vector of topic probabilities;
    6. calculate distance from each downloaded paper to the seed papers;
    7. add to the next level queue only those downloaded papers which are close to the seed papers.
5. Analyse the citation network.

The details of the controlled snowball sampling and probabilistic topic model construction are discussed in [13].

4 Citation Network Analysis

The citation network analysis is a crucual step of the proposed method because the PTM similarity places on the document contents a weak restriction aiming at getting rid of the most inappropriate publications. The analysis is performed in two stages. First, the cycles elimination must be done to clean raw input data and, second, the Simple Citation Path Count algorithm is applied to discover top citation paths and the seminal publications as ones contained in the paths.

4.1 Cycles Elimination

The correctly built citation network must be an acyclic directed graph. However, the publication database errors accidentally can cause cycles. The problem with cycles is that if there is a cycle in a network then there is also an infinite number of paths between some vertices. Since a citation network is usually almost acyclic to transform it into an acyclic network we use the "preprint" transformation described by Batagelj [6]. First, we identify cycles and then each paper from a cycle is duplicated with its "preprint" version and the papers inside cycle cite "preprints".

4.2 Simple Citation Path Count

Our approach is similar to Search Path Count (SPC) algorithm [6,29]. We introduce two pseudo-vertices – source and target. A vertex that does not reference any other publication vertex, gets an edge to the target vertex. A vertex that is not referenced by any publication vertex, gets an edge from the source vertex, so the graph becomes connected. Next step is to calculate all simple paths from the source to the target using Python library NetworkX[5]. The algorithm uses a modified depth-first search to generate the paths [19]. As the result we obtain a set of paths, each of which is a sequence of vertices. Each pair of direct neighbour vertices in such a sequence is an edge in the citation graph. For each edge, its frequency is calculated against all paths – a number of paths through it, simple path count. Next we calculate edge resistance as inverse proportional to edge simple path count – this allows diminishing the difference among the most cited and least cited papers [18,44]. The path resistance is then calculated as the sum of its edge resistances. Finally, we set an order over the paths using the path resistances. Using path resistances is a distinguishing feature of the proposed algorithm.

The difference of the applied algorithm from SPC algorithm is the preservation of the citation graph connectivity. In the known algorithm [7], as soon as the edge SPC scores are calculated the edges having low scores are removed and the citation network can become a disconnected graph.

4.3 Chasing New Ideas in Publications

Path resistance allows detecting new publications in the field, not referenced yet by any other authors but existing in a mainstream of the domain. We can separate all papers into mainstream research and probably new research fields or directions. The smallest (up to a certain threshold) path resistances correspond to the mainstream, whereas the biggest path resistances correspond to the publications that are either brand new, bad written or published in a low impact journal/conference proceedings. We assume those publications are the source of potentially new ideas and topics.

[5] NetworkX, https://networkx.github.io.

5 Analysis of Experimental Results

5.1 Experimental Settings

We took three different corpora: "high energy physics"(HEP),"critical thinking"(CT) and "pronunciation quality assessment"(PQA).

HEP publications [42] are available from the European Laboratory for Nuclear Research. The hep-ex partition of the HEP collection is composed of 2802 abstracts related to experimental high-energy physics that are indexed with 1093 main keywords (the categories), the hep-astroph partition contains 2716 abstracts from astrophysics section and 18114 abstracts on theoretical physics in hep-th metadata. Each publication is manually annotated with keywords.

CT corpus is gathered with our snowball sampling software. The CT domain is characterized with a large noisy publications corpus tightly entangled with publications on psychology, didactics, pedagogy and philosophy. The size of CT corpus is 24040 publication abstracts.

PQA domain is very specific and narrow, with a moderate-size corpus containing 8339 scientific abstracts collected by our snowball sampling software.

The sampling was run with following parameters: percentage of stop words to exclude – 2%; percentage of rare words to exclude – 5%; number of components in PCA which is maximal number of topics – 200; threshold KL-divergence – 0.18; sparsity parameter – 0.05; number of top citation paths – 50; minimal number of citations – 3.

We have chosen the few simplest similarity/difference measures (KL-divergence, symmetric KL-divergence, Jensen-Shannon divergence) to pick out patterns and then draw roadmap of future studies of usage of advanced metrics such as Hellinger distance [37] and S2JSD [16].

5.2 Seminal Publications for PQA Domain

To perform preliminary validation of the approach we did the main path analysis of the PQA and CT citation networks. HEP dataset is the plain list rather than a citation network so it cannot be processed with citation path count algorithm.

A result of the analysis is a reduced citation network, containing the most significant publications in the domain connected with directed links. Number of the kept publications depends on a reader ability to analyse. Assuming that a scolar when writing scientific report develops results from the cited publications and following the citation paths we can get an image of the knowledge evolution.

Figure 1 shows the two reduced citation networks for PQA domain containing one top path and 73 top paths respectively. Result of the CT citation network analysis is similar.

In the part (a) of Fig. 1 we can see that the mainstream of pronunciation assessment contains the publications listed in the Table 1. The evolution of pronunciation assessment starts from application of automatic speech recognition, pays some attention to pedagogical aspects, goes to simple machine learning

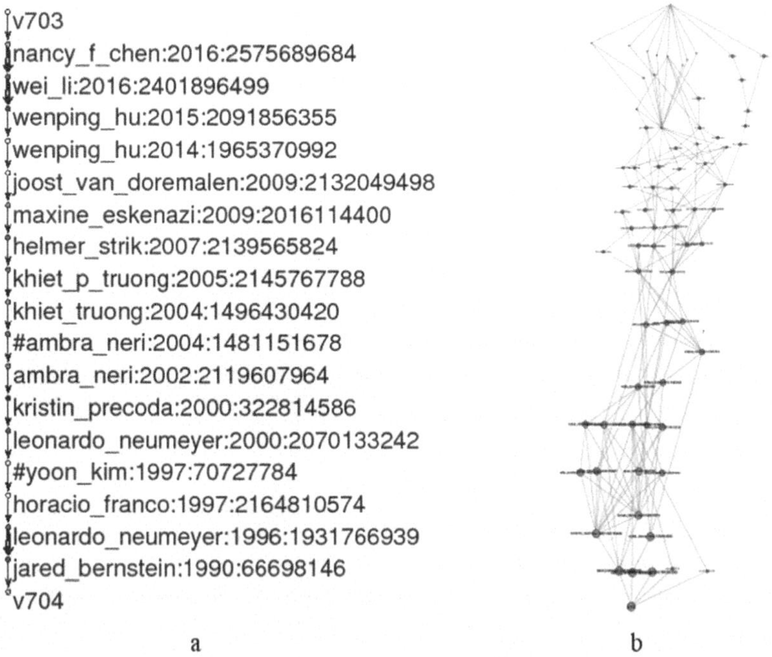

Fig. 1. Top path in PQA citation network (a) and top 73 paths of the citation network (b). Nodes are marked as (first author:year:MS_Academic_Id)

approaches, then to neural networks and to deep learning. Some of the seminal publications are the reviews containing discussions of the feature selection, methods comparison and combination. The part (b) of Fig. 1 shows that the more top paths we keep, the more detailed evolution map we obtain.

5.3 Assumptions Checking

The Proposed Snowball Control Method Provides Adequate Semantic Distance Between Publications

To test the quality of PTM as a controlling tool for the proposed controlled snowball sampling method provides adequate semantic distance between publications we took HEP collection, built PTM for it, and estimate similarity using keywords annotating each publication from HEP collection.

For each pair (a, b) of HEP publications were calculated both symmetric KL-divergence [30] and cosine similarity. The symmetric KL-divergence measures the difference between discrete topic probability distributions $p(t|a)$ and $p(t|b)$ derived from PTM. It is defined as

$$D_{KL}(a,b) = \sum_t \left[p(t|a) \log \frac{p(t|a)}{p(t|b)} + p(t|b) \log \frac{p(t|b)}{p(t|a)} \right]. \tag{1}$$

Table 1. Mainstream of publications concerning pronunciation quality assessment

Year	Reference
1990	Bernstein, Jared, et al. "Automatic evaluation and training in English pronunciation." *First International Conference on Spoken Language Processing.* 1990
1996	Neumeyer, Leonardo, et al. "Automatic text-independent pronunciation scoring of foreign language student speech." *Spoken Language, 1996. ICSLP 96. Proceedings., Fourth International Conference on. Vol. 3. IEEE*, 1996
1997	Franco, Horacio, et al. "Automatic pronunciation scoring for language instruction." *Acoustics, Speech, and Signal Processing, 1997. ICASSP-97., 1997 IEEE International Conference on. Vol. 2. IEEE,* 1997
1997	Kim, Yoon, Horacio Franco, and Leonardo Neumeyer. "Automatic pronunciation scoring of specific phone segments for language instruction." *Fifth European Conference on Speech Communication and Technology.* 1997
2000	Neumeyer, Leonardo, et al. "Automatic scoring of pronunciation quality." *Speech communication* 30.2–3 (2000): 83–93
2000	Precoda, Kristin, Christine A. Halverson, and Horacio Franco. "Effects of speech recognition-based pronunciation feedback on second-language pronunciation ability." *Proceedings of InSTILL 2000* (2000): 102–105
2002	Neri, Ambra, et al. "The pedagogy-technology interface in computer assisted pronunciation training." *Computer assisted language learning* 15.5 (2002): 441–467
2004	Neri, Ambra, Catia Cucchiarini, and Helmer Strik. "Segmental errors in Dutch as a second language: how to establish priorities for CAPT." (2004)
2004	Truong, Khiet. "Automatic pronunciation error detection in Dutch as a second language: an acoustic-phonetic approach." (2004)
2005	Truong, Khiet P., et al. "Automatic detection of frequent pronunciation errors made by L2-learners." *Ninth European Conference on Speech Communication and Technology.* 2005
2007	Strik, Helmer, et al. "Comparing classifiers for pronunciation error detection." *Eighth Annual Conference of the International Speech Communication Association.* 2007
2009	Eskenazi, Maxine. "An overview of spoken language technology for education." *Speech Communication* 51.10 (2009): 832–844.
2009	Van Doremalen, Joost, Catia Cucchiarini, and Helmer Strik. "Automatic detection of vowel pronunciation errors using multiple information sources." *Automatic Speech Recognition & Understanding, 2009. ASRU 2009. IEEE Workshop on. IEEE*, 2009
2014	Hu, Wenping, Yao Qian, and Frank K. Soong. "A new neural network based logistic regression classifier for improving mispronunciation detection of L2 language learners." *Chinese Spoken Language Processing (ISCSLP), 2014 9th International Symposium on. IEEE,* 2014
2015	Hu, Wenping, et al. "Improved mispronunciation detection with deep neural network trained acoustic models and transfer learning based logistic regression classifiers." *Speech Communication* 67 (2015): 154–166
2016	Li, Wei, et al. "Improving non-native mispronunciation detection and enriching diagnostic feedback with DNN-based speech attribute modeling." *Acoustics, Speech and Signal Processing (ICASSP), 2016 IEEE International Conference on. IEEE,* 2016
2016	Chen, Nancy F., and Haizhou Li. "Computer-assisted pronunciation training: From pronunciation scoring towards spoken language learning." *Signal and Information Processing Association Annual Summit and Conference (APSIPA), 2016 Asia-Pacific. IEEE,* 2016

The cosine similarity $cos\,(\theta(a,b))$ shows the angle between two vectors $\{C_w(a)\}$ and $\{C_w(b)\}$, whose components are frequencies of words w in publication a and b respectively.

The results are shown in Fig. 2.

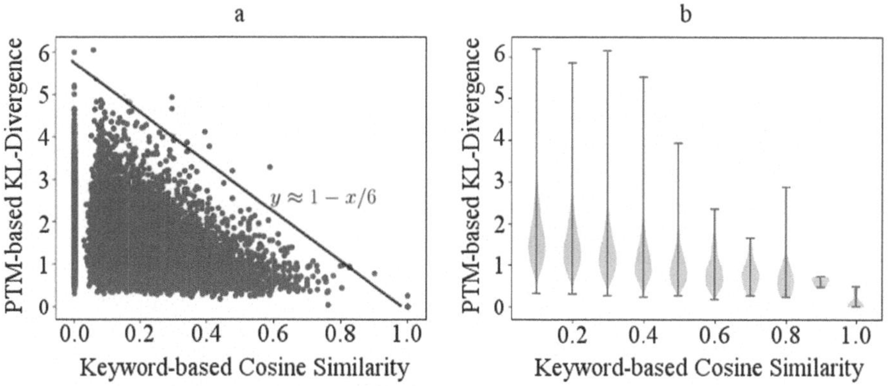

Fig. 2. Symmetric PTM-based KL-divergence and cosine similarity for the HEP abstracts: (a) scatter plot; (b) violin plot.

As we can see, the PTM-based symmetric Kullback-Leibler divergence provides upper bound for the keyword based cosine similarity:

$$cos\,(\theta(a,b)) \leq 1 - D_{KL}(a,b)/6 \tag{2}$$

which can be derived with linear regression. The reason is that the cosine similarity uses only word spelling while PTM uses R^n word embedding taking into account the meaning of terms.

The Best Age of the Seed Papers is 2–10 Years

To study the influence of a publication age on the probability of the publication citation we have attributed each edge of the PQA citation network with age calculated as difference between years of referencing publication and referenced one. The number of the edges as a function of edge age is shown on Fig. 3. We can see that the maximal number of the references is observed for the publications that are 2–8 years old. Such publications are still regarded as new ones but at the same time are old enough to be read and estimated by many researchers.

The Controlled Snowball Method Forms the Small-World Network

The distinctive features of small-world network [53] is sparse links, high average clustering coefficient and short paths connecting two nodes. In our experiments we made several variants of PQA citation networks using KL-divergence, symmetric KL-divergence and Jensen-Shannon divergence as distance measures.

Link sparsity means that node degree is much smaller than number of nodes in network (Table 2).

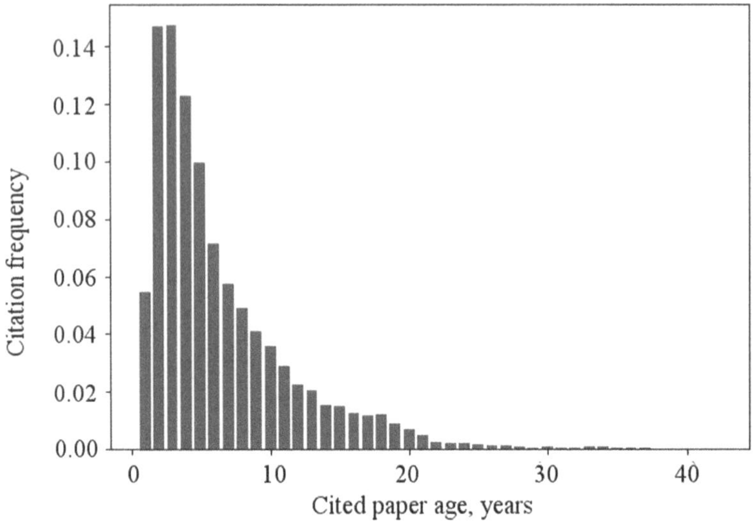

Fig. 3. Frequencies of the citation network edge ages.

Table 2. Average node degree of citation network containing 6500 nodes and built with different measures.

Difference measure	Average degree
KL-divergence	5.4363
Symmetric KL-divergence	13.3089
Jensen-Shannon divergence	11.5732
Random graph	10.0000

Average clustering coefficient is greater to the one calculated for random graph. Baseline average clustering is calculated for random graph with the same number of nodes $N = 6500$ and average node degree 10 which is baseline voluntary value close to average node degrees of other three networks (Table 3).

To ensure the possibility to reach any node from any node we need to check the existence of giant component in the graph using the Molloy-Reed criterion

$$\kappa = \left\langle k^2 \right\rangle - 2 \left\langle k \right\rangle \geq 0 \tag{3}$$

where $\langle k \rangle$ and $\left\langle k^2 \right\rangle$ are first and second moments of degree distribution respectively. Molloy and Reed [32] have proved that $\kappa = 0$ is the threshold at which a complex network will lose its giant component. So, the large positive value of κ indicates that existence of a giant component is possible.

Table 4 shows the magnitude of $\left\langle k^2 \right\rangle - 2 \langle k \rangle$ for all variants and demonstrates that all tested measures allow creation of the citation network that has giant component but symmetric KL-divergence produces more connected graph than other two divergences.

Table 3. Average clustering of citation network built with different measures. Last row contains average clustering of random graph.

Difference measure	Average clustering
KL-divergence	0.1546
Symmetric KL-divergence	0.2136
Jensen-Shannon divergence	0.2278
Random graph	0.0015

Table 4. Molloy-Reed Criterion for different measures.

Difference measure	Molloy-Reed Criterion
KL-divergence	65.84
Symmetric KL-divergence	420.15
Jensen-Shannon divergence	320.40

Next feature to check is the degree distribution. Derek de Solla Price showed in 1965 [46] that the number of references to a paper (node degree) in a citation network has a heavy-tailed distribution following the power law and thus the citation network is a small-world one. One of our initial assumptions was that the controlled snowball sampling results in a small-world networks so we need a few number of snowball iterations to achieve the high recall. Figure 4 was obtained on the base of PQA corpus built with KL-divergence difference measure and it shows that for the small node degrees the logarithm of node number has linear dependency on the logarithm of node degree and for the large node degrees the dependency has heavy tail. That means, the controlled snowball sampling produces small-world citation network as well as classical snowball. So we can be sure that a few iterations of the controlled snowball sampling allow collecting most of the relevant publications.

It is worth noting that Fig. 4 demonstrates the same degree distribution as self-organized logical P2P network [36] learned with the intelligent search strategy. The strategy includes query to adjacent nodes, query to the semantic community, update to adjacent nodes, update to the semantic community, update to query-issuing node, where updates contain information about topics of documents stored in known nodes. Intelligent P2P network constructed with numberical simulation demonstrates degree distribution with the tail much heavier than scale-free network which results in average 2 steps to answer the search query. So we can be sure that after 2 levels of snowball sampling we collect most of seminal publications.

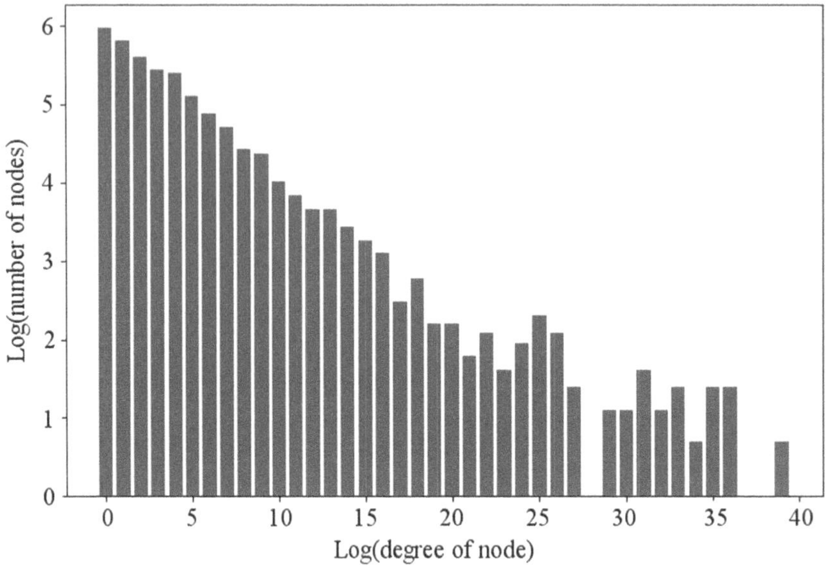

Fig. 4. Citation network node degree distribution in log-log scale.

Search Path Count Measure Provides the Most Relevant Publication Ranking

Multiple ranking criterion can be used to select the most valuable publication in the collected citation network. So we run several experiments to select the most appropriate relevance measure that correllates with scientific value in the selected domain. In our tests we consider four measures: citation index, indegree value [48], PageRank [26] and Search Path Count [6,29]. Each of the listed measures was applied to raw PQA citation network and the top 10 publication were extracted to evaluate at glance their relevance.

To estimate the quality of lists we notice that purpose of PQA is the collection of the most valued publications on "pronunciation training assessment" in computer sciences domain. Keeping in mind the purpose we can see that citation index value (Table 5) highlights the computer science textbooks and does not cover narrow target domain, PageRank measure (Table 6) lifts too old publications that passed test of time, and the most appropriate relevance measures are indegree value of nodes in citation network (Table 7) and SPC (Table 8). However the last measure provides up to time state of domain of interest so in our study we use SPC as a means of ordering the publications.

Table 5 demonstrated the known fact [8] that global measures like citation index are not suited to filter out the papers valued in the specific domain. In the list sorted by citation index, textbooks float up and seminal publications are entangled with papers studying related problems.

Table 5. Titles of publications ordered by citation index value

Year	Title
1968	The art of computer programming
1969	Data reduction and error analysis for the physical sciences
1972	Human problem solving
1966	Signal detection theory and psychophysics
1999	The unified modeling language user guide
1978	The C programming language
1950	Computing machinery and intelligence
1993	Fundamentals of speech recognition
1988	Measuring the accuracy of diagnostic systems
1988	Term-weighting approaches in automatic text retrieval

Table 6. Titles of publications ordered by PageRank value

Year	Title
1983	A maximum likelihood approach to continuous speech recognition
1976	The HARPY speech recognition system
1974	The phonological component of an automatic speech recognition system
1985	Context-dependent modeling for acoustic-phonetic recognition of continuous speech
1990	Automatic evaluation and training in English pronunciation
1970	Automatic recognition of 200 words
1979	Performance statistics of the HEAR acoustic processor
1979	Speaker-independent recognition of isolated words using clustering techniques
1980	Comparison of parametric representations for monosyllabic word recognition in continuously spoken sentences
1976	Speech recognition experiments with linear predication, bandpass filtering, and dynamic programming

Saturation of the Controlled Snowball Sampling. The core of the suggested technology is the iterative building of the short list of seminal publications so we need to know when to stop out iterations. It is known from the literature [11] that if the convergence cannot be proven the good solution is to finish the loop when the result becomes stable with respect to iterations. In our case, the stable result is the ranked list of publications that does not change when a new publication is added to the citation network.

Table 7. Titles of publications ordered by indegree value

Year	Title
2016	Measuring pronunciation improvement in users of CAPT tool TipTopTalk!
2016	On the assessment of computer-assisted pronunciation training tools
1983	A maximum likelihood approach to continuous speech recognition
1980	Comparison of parametric representations for monosyllabic word recognition in continuously spoken sentences
2000	Phone-level pronunciation scoring and assessment for interactive language learning
1993	Fundamentals of speech recognition
1996	Automatic text-independent pronunciation scoring of foreign language student speech
1997	Automatic pronunciation scoring for language instruction
1989	Speaker-independent phone recognition using hidden Markov models
2000	Automatic scoring of pronunciation quality

Table 8. Titles of publications ordered by Search Path Count value

Year	Title
2016	Large-scale characterization of non-native Mandarin Chinese spoken by speakers of European origin: Analysis on iCALL
2016	Computer-assisted pronunciation training: From pronunciation scoring towards spoken language learning
2015	Integrating acoustic and state-transition models for free phone recognition in L2 English speech using multi-distribution deep neural networks
2012	Deep neural network language models
2010	Phone recognition with the mean-covariance restricted Boltzmann machine
2015	Improved mispronunciation detection with deep neural network trained acoustic models and transfer learning based logistic regression classifiers
2014	A new neural network based logistic regression classifier for improving mispronunciation detection of L2 language learners
2007	Comparing classifiers for pronunciation error detection
2006	On the use of phonological features for pronunciation scoring
2007	Discriminative training for large-vocabulary speech recognition using minimum classification error

In our experiments, we rank the collected publications by their SPC values and then calculate the differences between two ranked lists in two ways. First, we use the Spearman's rank correlation coefficient C_S [31] between two successive lists of top N publications. However, to apply the classical formulae we need to equalize set of elements in two lists by appending to the end of each list the missing elements with equal maximal rank. Table 9 shows that Spearman's rank correlation coefficient approaches 1 demonstrating the saturation of publication list.

Second differences measure used on our study is Jaccard similarity [47], it ignores the element ordering but also confirms the saturation behaviour.

To observe the saturation we organize publications in the order they were added to the citation network and take first M ones. Then they are reordered by SPC values and top N nodes are taken as list seminal publications. Generally, the list depends on network size M and we can calculate Spearman's rank correlation coefficient C_S and Jaccard similarity J between two lists obtained for different values of M.

Table 9 shows that both difference measures are well correlated and approaching 1 showing the decreasing variation in the seminal publication lists.

Table 9. Spearman's rank correlation coefficient C_S and Jaccard similarity J between two lists of top N ($N = 100$) publications obtained from citation networks of sizes M_1 and M_2.

M_1	M_2	C_S	J
2500	3000	0.9626	0.8349
3000	3500	0.6796	0.7700
3500	4000	0.9954	0.9608
4000	4500	0.9990	0.9608
4500	5000	0.9683	0.9048
5000	5500	0.9999	0.9608
5500	6000	0.9897	0.9048
6000	6500	0.9998	0.9608

Figures 5 and 6 show the dependence of Spearman's rank correlation between two lists obtained from citation networks of sizes $M_1 = M$ and $M_2 = M - 500$ on parameter M.

Figure 5 allows comparison of the correlations obtained with different similarity measures. As we can see, the tightly connected network demonstrates worse iteration convergence. I.e. the greater is the value of Molloy–Reed criterion the worse is the saturation. The difference between KL-divergence and symmetric KL-divergence is more notable than between symmetric KL-divergence and Jensen-Shannon divergence in accordance to ratios between Molloy–Reed criterion values.

Figure 6 reveals the slower convergence for greater sizes of the required list of seminal publications. The observed behaviour can be caused by a long tail of the list where less valuable publications are shuffled during iterations.

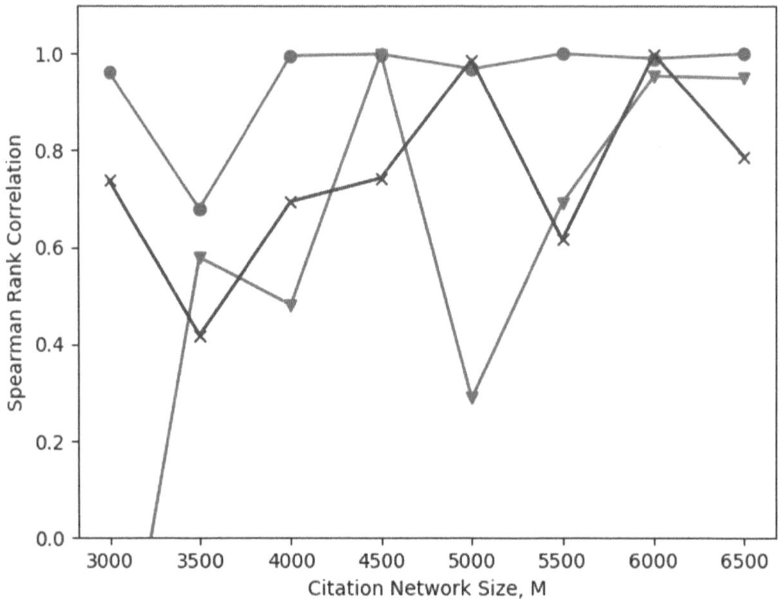

Fig. 5. Dependence of Spearman's rank correlation coefficient on size of citation networks constructed with KL-divergence (circles), symmetric KL-divergence (crosses) and Jensen-Shannon divergence (triangles). The correlation between lists derived from small citation networks constructed with the Jensen-Shannon divergence have large negative values which was not shown advisedly.

Another possible criterion of snowball iteration stop follows from the observation that, in practice, the controlled snowball sampling retrieves scientific abstracts one-by-one. Then each of retrieved abstracts is analyzed and the corresponding publication can either (a) rarely be kept in the citation network or (b) in most of cases be dropped out as irrelevant. Therefore action of keeping/dropping can be modelled as Poisson process and corresponding confidence interval can be calculated for probability of the event "when analyzing Z sequential publications at least one is accepted as part of the collected citation network". Figure 7 shows 0.95 confidence interval of Poisson distribution of paper acceptance as a coloured strip and acceptance probability threshold 0.05 as a straight line. The confidence interval was calculated on the base of 10 snowball runs for CT collection starting from random subsets of seed paper collection.

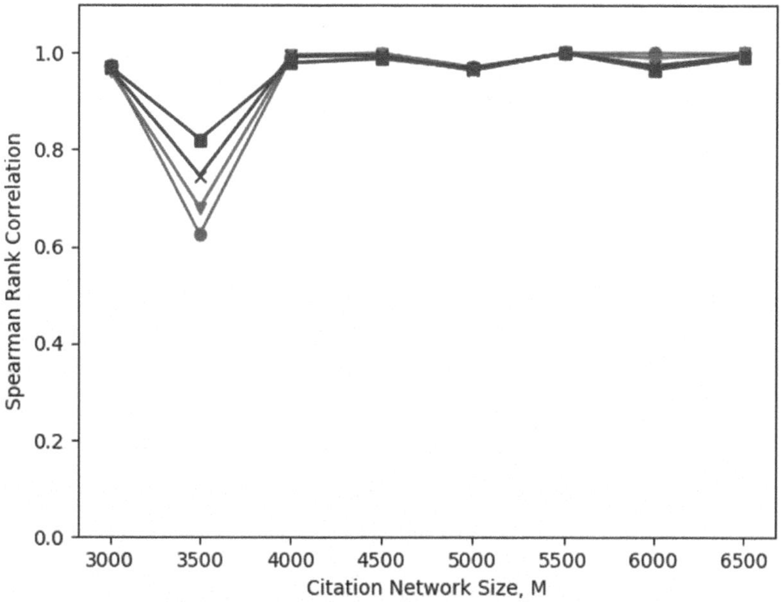

Fig. 6. Dependence of Spearman's rank correlation coefficient on size of citation networks constructed with KL-divergence for different sizes of seminal publication lists. N = 60 marked with circles, N = 100 - crosses, N = 140 - triangles, and N = 180 - squares

After some number of tested abstracts the upper bound of confidence interval becomes lower than the threshold so we conclude that the controlled snowball sampling demonstrates the saturation behaviour.

The Biased Seed Papers Can Produce Unbiased Citation Network
To estimate the stability of the controlled snowball sampling with respect to the seed papers variation we run the sampling starting from the full set of the PQA seed papers and mark the relevant papers with main path analysis. Then we run the sampling again starting from 10 random subsets of the PQA seed papers and count the number of the runs where each seminal paper occurs. In our experiments the random subsets contain 50% of the seed papers and 66% of relevant papers are detected every time, 14% – in 80% of runs, 14% – in 60% of runs, 6% – at least once (Fig. 8). So we can conclude that the PQA citation network is stable with respect to large seed paper variations being input for the controlled snowball sampling and the result of the sampling is unbiased.

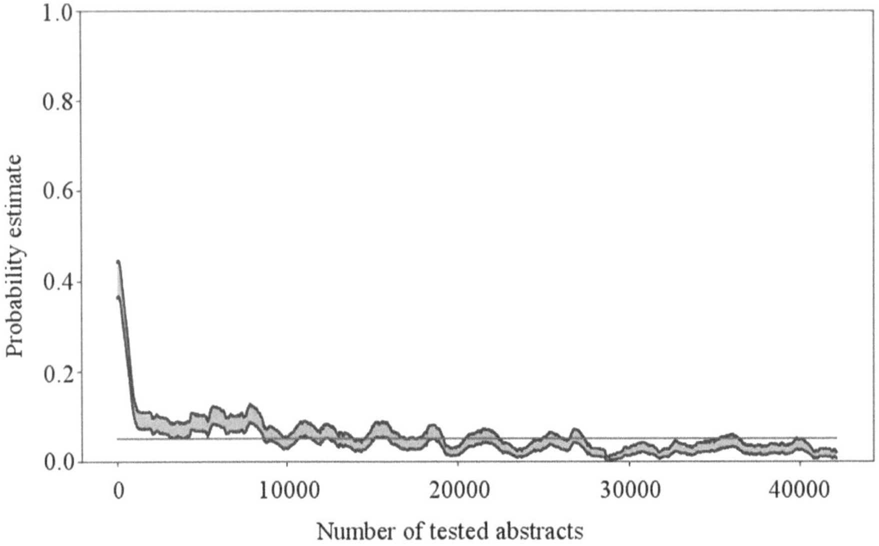

Fig. 7. 0.95 confidence interval of Poisson distribution of tagging a paper as seminal as a function of the number of already tested abstracts N and acceptance probability threshold 0.05.

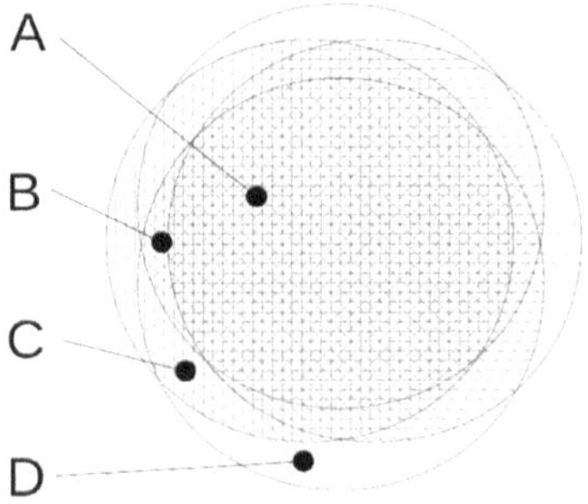

Fig. 8. Probability of the relevant paper detection: A – 66% detected every time; B – 14% detected in 80% of runs; C – 14% detected in 60% of runs; D – 6% detected at least once.

6 Conclusions and Future Studies

The objective of the current work was to present results supporting the information quality of the publications collection gathered with controlled snowball sampling method, providing reproducibility of results and authority of a domain. The performed study allows for the following recaps:

- PTM as a tool for the snowball sampling control method provides adequate semantic distance between publications.

- The collected citation network has the giant component when KL-divergence, symmetric KL-divergence and Jensen-Shannon divergence are used as difference measures.
- The KL-divergence provides better saturation than symmetric KL-divergence and Jensen-Shannon divergence.
- Search Path Count is the better ranking criterion that citation index, PageRank and indegree.
- The controlled snowball sampling guarantees the saturation.
- The maximal number of the references is observed for the publications that are 2–8 years old. Such publications are still regarded as new ones but at the same time are old enough to be read and estimated by many researchers.
- The biased seed papers produce unbiased citation network.
- The collected citation network is stable with respect to large seed paper variations being input for the controlled snowball sampling and the result of the sampling is unbiased.

The presented study was performed with a set of Python scripts which are available online[6].

The first direction of future work we envisage is the elaboration of coverage that was not addressed in the current study. The coverage measure problem turned out to be nontrivial because the evaluation methods are still incomplete, so it requires separate detailed consideration.

The second direction to follow is chasing new publications which is attractive goal as well.

Acknowledgements. The authors would like to express their gratitude to anonymous reviewers whose comments and suggestions helped improve the paper.

References

1. Ahad, A., Fayaz, M., Shah, A.S.: Navigation through citation network based on content similarity using cosine similarity algorithm. Int. J. Database Theory Appl. **9**(5), 9–20 (2016)
2. Akavipat, R., Wu, L.S., Menczer, F., Maguitman, A.G.: Emerging semantic communities in peer web search. In: Proceedings of the International Workshop on Information Retrieval in Peer-to-Peer Networks, pp. 1–8. ACM (2006)
3. Baez, M., Mirylenka, D., Parra, C.: Understanding and supporting search for scholarly knowledge. In: Proceeding of the 7th European Computer Science Summit, pp. 1–8 (2011)
4. Barabási, A.L.: Scale-free networks: a decade and beyond. Science **325**(5939), 412–413 (2009)
5. Barbosa, M.W., Costa, M.M., Almeida, J.M., Almeida, V.A.: Using locality of reference to improve performance of peer-to-peer applications. In: ACM SIGSOFT Software Engineering Notes, vol. 29, pp. 216–227. ACM (2004)
6. Batagelj, V.: Efficient algorithms for citation network analysis. arXiv preprint cs/0309023 (2003)

[6] https://github.com/gendobr/snowball.

7. Batagelj, V., Mrvar, A.: Pajek-program for large network analysis. Connections **21**(2), 47–57 (1998)
8. Beel, J., Gipp, B., Langer, S., Breitinger, C.: Paper recommender systems: a literature survey. Int. J. Digit. Librar. **17**(4), 305–338 (2016)
9. Blei, D.M., Ng, A.Y., Jordan, M.I.: Latent dirichlet allocation. J. Mach. Learn. Res. **3**(Jan), 993–1022 (2003)
10. Crespo, A., Garcia-Molina, H.: Routing indices for peer-to-peer systems. In: Proceedings 22nd International Conference on Distributed Computing Systems, pp. 23–32. IEEE (2002)
11. De Bruijn, N.G.: Asymptotic Methods in Analysis, vol. 4. Courier Corporation, Chelmsford (1981)
12. Dobrovolskyi, H., Keberle, N.: Collecting the seminal scientific abstracts with topic modelling, snowball sampling and citation analysis. In: Proceedings of the 14th International Conference on ICT in Education, Research and Industrial Applications. Integration, Harmonization and Knowledge Transfer. Volume I: Main Conference, vol. 2105, pp. 179–192. CEUR-WS (2018)
13. Dobrovolskyi, H., Keberle, N., Todoriko, O.: Probabilistic topic modelling for controlled snowball sampling in citation network collection. In: Różewski, P., Lange, C. (eds.) KESW 2017. CCIS, vol. 786, pp. 85–100. Springer, Cham (2017). https://doi.org/10.1007/978-3-319-69548-8_7
14. Dong, R., Tokarchuk, L., Ma, A.: Digging friendship: paper recommendation in social network. In: Proceedings of Networking and Electronic Commerce Research Conference, NAEC 2009, pp. 21–28 (2009)
15. Doulamis, N.D., Karamolegkos, P.N., Doulamis, A., Nikolakopoulos, I.: Exploiting semantic proximities for content search over P2P networks. Comput. Commun. **32**(5), 814–827 (2009)
16. Endres, D.M., Schindelin, J.E.: A new metric for probability distributions. IEEE Trans. Inf. Theory (2003)
17. Ermolayev, V., Batsakis, S., Keberle, N., Tatarintseva, O., Antoniou, G.: Ontologies of time: review and trends. Int. J. Comput. Sci. Appl. **11**(3) (2014)
18. Even, S.: Graph Algorithms. Cambridge University Press, Cambridge (2011)
19. Golumbic, M.C.: Algorithmic Graph Theory and Perfect Graphs, vol. 57. Elsevier, Amsterdam (2004)
20. Gori, M., Pucci, A.: Research paper recommender systems: a random-walk based approach. In: IEEE/WIC/ACM International Conference on Web Intelligence, WI 2006, pp. 778–781. IEEE (2006)
21. Hamilton, D.P., et al.: Publishing by–and for?–the numbers. Science **250**(4986), 1331–1332 (1990)
22. Huang, Z., Chung, W., Ong, T.H., Chen, H.: A graph-based recommender system for digital library. In: Proceedings of the 2nd ACM/IEEE-CS Joint Conference on Digital Libraries, pp. 65–73. ACM (2002)
23. Küçüktunç, O., Saule, E., Kaya, K., Çatalyürek, Ü.V.: Recommendation on academic networks using direction aware citation analysis. arXiv preprint arXiv:1205.1143 (2012)
24. Lao, N., Cohen, W.W.: Relational retrieval using a combination of path-constrained random walks. Mach. Learn. **81**(1), 53–67 (2010)
25. Lecy, J.D., Beatty, K.E.: Representative literature reviews using constrained snowball sampling and citation network analysis (2012)
26. Leskovec, J., Rajaraman, A., Ullman, J.D.: Mining of Massive Datasets. Cambridge University Press, Cambridge (2014)

27. Liang, Y., Li, Q., Qian, T.: Finding relevant papers based on citation relations. In: Wang, H., Li, S., Oyama, S., Hu, X., Qian, T. (eds.) WAIM 2011. LNCS, vol. 6897, pp. 403–414. Springer, Heidelberg (2011). https://doi.org/10.1007/978-3-642-23535-1_35
28. Lops, P., de Gemmis, M., Semeraro, G.: Content-based recommender systems: state of the art and trends. In: Ricci, F., Rokach, L., Shapira, B., Kantor, P.B. (eds.) Recommender Systems Handbook, pp. 73–105. Springer, Boston, MA (2011). https://doi.org/10.1007/978-0-387-85820-3_3
29. Lucio-Arias, D., Leydesdorff, L.: Main-path analysis and path-dependent transitions in histcite^{TM}-based historiograms. J. Assoc. Inf. Sci. Technol. **59**(12), 1948–1962 (2008)
30. MacKay, D.J.: Information Theory, Inference and Learning Algorithms. Cambridge University Press, Cambridge (2003)
31. Mendenhall, W.M., Sincich, T.L., Boudreau, N.S.: Statistics for Engineering and the Sciences, Student Solutions Manual. Chapman and Hall/CRC, Boca Raton (2016)
32. Molloy, M., Reed, B.: A critical point for random graphs with a given degree sequence. Random Struct. Algorithms **6**(2–3), 161–180 (1995)
33. Moya-Anegón, F., Vargas-Quesada, B., Herrero-Solana, V., Chinchilla-Rodríguez, Z., Corera-Álvarez, E., Munoz-Fernández, F.: A new technique for building maps of large scientific domains based on the cocitation of classes and categories. Scientometrics **61**(1), 129–145 (2004)
34. Newman, M.E.: The structure of scientific collaboration networks. Proc. Natl. Acad. Sci. **98**(2), 404–409 (2001)
35. Newman, M.E.: Coauthorship networks and patterns of scientific collaboration. Proc. Natl. Acad. Sci. **101**(Suppl. 1), 5200–5205 (2004)
36. Nicolini, A.L., Lorenzetti, C.M., Maguitman, A.G., Chesñevar, C.I.: Intelligent algorithms for improving communication patterns in thematic P2P search. Inf. Proces. Manag. **53**(2), 388–404 (2017)
37. Nikulin, M.S.: Hellinger distance. In: Encyclopedia of Mathematics, vol. 78 (2001)
38. Osborne, F., Motta, E.: Klink-2: integrating multiple web sources to generate semantic topic networks. In: Arenas, M., et al. (eds.) ISWC 2015. LNCS, vol. 9366, pp. 408–424. Springer, Cham (2015). https://doi.org/10.1007/978-3-319-25007-6_24
39. Pennington, J., Socher, R., Manning, C.D.: Glove: global vectors for word representation. In: EMNLP, vol. 14, pp. 1532–1543 (2014)
40. Petticrew, M., Gilbody, S.: Planning and conducting systematic reviews. Health Psychol. Pract. 150–179 (2004)
41. Pohl, S., Radlinski, F., Joachims, T.: Recommending related papers based on digital library access records. In: Proceedings of the 7th ACM/IEEE-CS Joint Conference on Digital Libraries, pp. 417–418. ACM (2007)
42. Ráez, A.M., López, L.A.U., Steinberger, R.: Adaptive selection of base classifiers in one-against-all learning for large multi-labeled collections. In: Vicedo, J.L., Martínez-Barco, P., Muńoz, R., Saiz Noeda, M. (eds.) EsTAL 2004. LNCS (LNAI), vol. 3230, pp. 1–12. Springer, Heidelberg (2004). https://doi.org/10.1007/978-3-540-30228-5_1
43. Ricci, F., Rokach, L., Shapira, B.: Recommender systems: introduction and challenges. In: Ricci, F., Rokach, L., Shapira, B. (eds.) Recommender Systems Handbook, pp. 1–34. Springer, Boston, MA (2015). https://doi.org/10.1007/978-1-4899-7637-6_1

44. Salganik, M.J., Heckathorn, D.D.: Sampling and estimation in hidden populations using respondent-driven sampling. Sociol. Methodol. **34**(1), 193–240 (2004)
45. Small, H.: Co-citation in the scientific literature: a new measure of the relationship between two documents. J. Am. Soc. Inf. Sci. **24**(4), 265–269 (1973)
46. de Solla Price, D.J.: Networks of scientific papers. Science **149**(3683), 510–515 (1965)
47. Tan, P.N., et al.: Introduction to Data Mining. Pearson Education India, Delhi (2007)
48. Trudeau, R.J.: Introduction to Graph Theory. Courier Corporation, Chelmsford (2013)
49. Valenzuela, M., Ha, V., Etzioni, O.: Identifying meaningful citations. In: AAAI Workshop: Scholarly Big Data (2015)
50. Varela, A.R., et al.: Mapping the historical development of physical activity and health research: a structured literature review and citation network analysis. Prev. Med. **111**, 466–472 (2018)
51. Vellino, A.: Usage-based vs. citation-based methods for recommending scholarly research articles. arXiv preprint arXiv:1303.7149 (2013)
52. Vorontsov, K., Potapenko, A.: Tutorial on probabilistic topic modeling: additive regularization for stochastic matrix factorization. In: Ignatov, D.I., Khachay, M.Y., Panchenko, A., Konstantinova, N., Yavorskiy, R.E. (eds.) AIST 2014. CCIS, vol. 436, pp. 29–46. Springer, Cham (2014). https://doi.org/10.1007/978-3-319-12580-0_3
53. Watts, D.J., Strogatz, S.H.: Collective dynamics of 'small-world' networks. Nature **393**(6684), 440 (1998)
54. Woodruff, A., Gossweiler, R., Pitkow, J., Chi, E.H., Card, S.K.: Enhancing a digital book with a reading recommender. In: Proceedings of the SIGCHI Conference on Human Factors in Computing Systems, pp. 153–160. ACM (2000)
55. Yan, X., Guo, J., Lan, Y., Cheng, X.: A biterm topic model for short texts. In: Proceedings of the 22nd International Conference on World Wide Web, pp. 1445–1456. ACM (2013)
56. Zeinalipour-Yazti, D., Kalogeraki, V., Gunopulos, D.: Information retrieval techniques for peer-to-peer networks. Comput. Sci. Eng. **6**(4), 20–26 (2004)
57. Zeinalipour-Yazti, D., Kalogeraki, V., Gunopulos, D.: Exploiting locality for scalable information retrieval in peer-to-peer networks. Inf. Syst. **30**(4), 277–298 (2005)
58. Zhou, D., et al.: Learning multiple graphs for document recommendations. In: Proceedings of the 17th International Conference on World Wide Web, pp. 141–150. ACM (2008)
59. Zuo, Y., Zhao, J., Xu, K.: Word network topic model: a simple but general solution for short and imbalanced texts. Knowl. Inf. Syst. **48**(2), 379–398 (2016)

Similar Terms Grouping Yields Faster Terminological Saturation

Victoria Kosa[1(✉)], David Chaves-Fraga[2], Nataliya Keberle[1], and Aliaksandr Birukou[3,4]

[1] Department of Computer Science, Zaporizhzhia National University, Zaporizhzhia, Ukraine
victoriya1402.kosa@gmail.com, nkeberle@gmail.com
[2] Ontology Engineering Group, Universidad Politécnica de Madrid, Madrid, Spain
dchaves@fi.upm.es
[3] Springer-Verlag GmbH, Heidelberg, Germany
aliaksandr.birukou@springer.com
[4] Peoples' Friendship University of Russia (RUDN University), Moscow, Russia

Abstract. This paper reports on the refinement of the algorithm for measuring terminological difference between text datasets (THD). This baseline THD algorithm, developed in the OntoElect project, used exact string matches for term comparison. In this work, it has been refined by the use of appropriately selected string similarity measures (SSM) for grouping the terms, which look similar as text strings and presumably have similar meanings. To determine rational term similarity thresholds for several chosen SSMs, the measures have been implemented as software functions and evaluated on the developed test set of term pairs in English. Further, the refined algorithm implementation has been evaluated against the baseline THD algorithm. For this evaluation, the bags of terms have been used that had been extracted from the three different document collections of scientific papers, belonging to different subject domains. The experiment revealed that the use of the refined THD algorithm, compared to the baseline, resulted in quicker terminological saturation on more compact sets of source documents, though at an expense of a noticeably higher computation time.

Keywords: Automated term extraction · OntoElect · Terminological difference · String similarity measure · Bag of terms · Terminological saturation

1 Introduction

The research presented in this paper[1] is the part of the development of the methodological and instrumental components for extracting representative (complete) sets of significant terms from the representative sub-collections of textual documents having

[1] This paper is a refined and extended version of [1].

V. Ermolayev et al. (Eds.): ICTERI 2018, CCIS 1007, pp. 43–70, 2019.
https://doi.org/10.1007/978-3-030-13929-2_3

minimal possible size. These terms are further interpreted as the required features for engineering an ontology in a particular domain of interest. Therefore, it is assumed that the documents in a collection cover a single and well-circumscribed domain. The main hypothesis, put forward in this work, is that a sub-collection can be regarded as representative to describe the domain, if it is the terminological core. It means that any additions of extra documents from the entire collection to this sub-collection do not noticeably change the terminological footprint on the domain. Such a sub-collection is further considered as complete. Therefore, a representative bag of significant terms describing its domain can be extracted from it. The approach to assess the representativeness does so by evaluating terminological saturation in a document (sub-)collection [2, 3]. Practically, the approach allows extracting statistically the same set of significant terms from a part of a collection, instead of processing the whole collection. Automated term extraction is known to be a computationally bulky process. Therefore, lowering the number and the overall volume of the input documents might substantially decrease processing times and improve scalability. For example, the terminological core of the TIME collection of 437 conference papers (Sect. 5.2), detected using the baseline THD algorithm, contains 220 papers (50.34% of the total number). We demonstrate in the paper that using the proposed THD refinement, the size of a terminological core is additionally lowered by 22 to 46% without any loss in quality.

Detecting saturation is done by measuring terminological difference (*thd*) among the pairs of the consecutive incrementally enlarged datasets, as described in Sect. 5. This measure is based on evaluating differences between individual terms. A (baseline) THD algorithm for computing *thd* [2] has been developed and implemented in the OntoElect project[2]. OntoElect develops a methodology and instrumental tool suite for refining domain ontologies. It exploits the allusion of public elections to find out what is the prevailing sentiment of the domain knowledge stakeholders. The sentiments are elicited indirectly, through terms extraction from a saturated document collection, describing the domain. Term significance scores are interpreted as the votes in favour of the corresponding ontology features. The features satisfying a simple majority of voters are represented by the terms, ordered by descending scores, with the minimal sum of the scores being higher than 1/2 of the total sum.

The baseline THD algorithm uses a simple string equivalence check for detecting similar (the same) individual terms. The objective of the research presented in this paper is to find out if it is possible to achieve better performance in measuring terminological difference by using a proper string similarity measure to compare individual terms.

The remainder of the paper is structured as follows. Section 2 reviews the related work and outlines our contributions. Section 3 presents the chosen string similarity measures and reports about the choice of the proper terms similarity thresholds for terms grouping. Section 4 sketches out the approach of OntoElect for measuring *thd* and presents our refinement of the baseline THD algorithm. Section 5 reports the set-up and results of our evaluation experiments. Our conclusions are given and plans for the future work outlined in Sect. 6.

---

[2] https://www.researchgate.net/project/OntoElect-a-Methodology-for-Domain-Ontology-Refinement.

2 Related Work

The work related to the presented research has been sought in: (i) automated term extraction (ATE) from English texts; (ii) string similarity (distance) measurement in the pairs of text strings.

2.1 Automated Term Extraction

In the majority of approaches to ATE, e.g. [4] or [5], processing is done in two consecutive phases: linguistic processing and statistical processing. Linguistic processors, like POS taggers or phrase chunkers, filter out stop words and restrict candidate terms to *n*-gram sequences: nouns or noun phrases, adjective-noun and noun-preposition-noun combinations. Statistical processing is then applied to measure the ranks of the candidate terms. These measures are [6]: either the measures of unithood, which focus on the collocation strength of units that comprise a single term; or the measures of termhood, which point to the association strength of a term to domain concepts.

For unithood, the measures are used such as mutual information [7], log likelihood [7], t-test [4, 5], modifiability and its variants [5, 8]. The measures for termhood are either term frequency-based (unsupervised approaches) or reference corpora-based (semi-supervised approaches). The most used frequency-based measures are TF/IDF (e.g. [9, 10]), weirdness [11], and domain pertinence [12]. More recently, hybrid approaches were proposed, that combine unithood and termhood measurements in a single value. A representative measure is c/nc-value [13]. C/nc-value-based approaches to ATE have received their further evolution in many works, e.g. [4, 12, 14] to mention a few.

Linguistic processing is organized and implemented in a very similar way in all ATE methods, except some of them that also include filtering out stop words. Stop words could be filtered out also at a cut-off step after statistical processing. Statistical processing is sometimes further split in two consecutive sub-phases of term candidate scoring, and ranking. For term candidates scoring, reflecting its likelihood of being a term, known methods could be distinguished by being based on measuring occurrences frequencies, including word association (c.f. [9]) or assessing occurrences contexts, using reference corpora, e.g. Wikipedia [15], or topic modelling [16, 17].

2.2 Text Similarity Measurement

In the recent surveys on text similarity measurement approaches, e.g. [18, 19], methods (or measures)[3] are grouped based on analysing: (i) characters and their sequences; (ii) tokens; (iii) terms; (iv) text corpora; or (v) synsets. In [19] hybrid measures that allow fuzzy matching between tokens are also mentioned. Brief characteristics of the groups are given immediately below. The individual methods belonging to the groups are detailed in Table 1.

[3] In this context, we do not distinguish a method and a measure. A method is understood as a way to implement the corresponding measure function.

Character- and character sequence-based measures compare characters and their sequences in strings, taking into account also the order of characters. These include the measures of common character sequences, e.g. substrings; edit distance; the number and order of the common characters between two strings.

Token-based methods model a string as a set of tokens. Individual characters, character *n*-grams, or separate words could be regarded as tokens. Quantification is done by computing the size of the overlap normalized by a measure of string length.

Term-based measures are similar to token-based measures but the tokens are different. Those are not character *n*-grams but terms, which are word *n*-grams with possibly varying *n*. Furthermore, the weights of the terms, e.g. their frequencies of occurrence, are taken into account. These measures apply more on long character strings, or documents, hence are better suited to measure document or text dataset similarity.

Corpus-based and **synset-based** (or knowledge-based) **methods** are very marginally relevant to our purposes in this paper. Corpus-based approaches determine the similarity between words based on (statistical) information gained from large text corpora. Synset-based approaches rely on semantic networks, like WordNet [20], to derive semantic similarity between words. Both approaches are therefore too bulky computationally, though may be applied to ATE – e.g. for deciding about cut-offs. Term grouping, the technique we report in this paper, is however performed after the terms have already been extracted. Hence, we omit looking at corpus- and synset-based measures.

The overview of the most popular text/string similarity measures, grouped by method types, is provided in Table 1. This overview is by far not complete as many other variants of SSM are available in the literature. Those we omit are however based on the same principles compared to the listed in Table 1, to the best of our knowledge.

Table 1. The overview of text similarity/distance measures

Name, source	Description	Specifics	Relevance	
			Term similarity	*thd*[a]
Character- and character sequence-based measures				
Longest common substring [21]	Common character sequence based measure	Returns the integer length of the longest common substring; could be normalized by the total length	Moderate	Irrelevant
Levenshtein distance [22]	Edit distance based measure	Returns an integer number of required edits	Marginal	Irrelevant
Hamming distance [23]	Edit distance based measure	Strings have to be of equal length	Marginal	Irrelevant
Monger-Elkan distance [24]	Edit distance based measure	Returns an integer number of required edits	Marginal	Irrelevant
Jaro distance [25]	Counts the minimal number of one character transforms in one string for arriving at the other string	Returns a normalized real value from [0, 1]	Good	Irrelevant

(*continued*)

Table 1. *(continued)*

Name, source	Description	Specifics	Relevance	
			Term similarity	*thd*[a]
Jaro-Winkler distance [26]	Refines Jaro measure by using a prefix scale value – prioritizes the stings that match at the beginning	Returns a normalized real value from [0, 1]	Good	Irrelevant
Token-based measures				
Sørensen-Dice coefficient [27, 28]	Counts the ratio of identical character *bi*-grams to the overall number of bi-grams in both strings	Returns a normalized real value from [0, 1]	Good	Irrelevant
Jaccard similarity [29]	Counts the ratio between the cardinalities of the intersection and union of the character sets (*uni*-grams) in the strings	Returns a normalized real value from [0, 1]	Good	Irrelevant
Cosine similarity [19]	Size of overlap in character *uni*-grams divided by the square root of the sum of the squared total numbers of unigrams in both strings	Returns a normalized positive real value	Marginal (computationally hard)	Irrelevant
Term-based measures				
Euclidian distance [30]	Measures traditional Euclidian distance in an *n*-dimensional metric space (of positive reals)	Works for documents; returns a real positive value	Irrelevant	Relevant
Cosine similarity [31]	Computes a cosine between two vectors in the term space; vectors are specified by term weights (e.g. TF of C-value)	Works for documents; returns a normalized positive real value	Irrelevant	Marginal
Pearson correlation [30]	Computes Pearson correlation for a pair of vectors in the term vector space	Works for documents; returns a normalized real value that ranges from +1 to −1; it is 1 when vectors are fully identical	Irrelevant	Marginal
Manhattan (block) distance [18]	The distance to be traveled to get from one data point to the other if a grid-like path is followed	Works for documents; resembles the *thd* measure [2]	Irrelevant	Relevant

[a]*thd* is the measure for terminological difference developed in OntoElect [2] and used in our approach – see also Sect. 4. Hence, "relevant" in this column means being appropriate for measuring terminological difference between documents of text datasets.

The authors of [32] present an expansion-based framework to measure string similarities efficiently while considering synonyms. This result is also relevant to our work as a synonym is one of the categories of term candidates that may need to be considered for grouping in our settings. In [32], it is also acknowledged that there is a rich set of string similarity measures available in the literature, including character *n*-gram similarity [33],

Levenshtein distance [22], Jaro-Winkler measure [26], Jaccard similarity [29], TF/IDF based cosine similarity [34], and Hidden Markov Model-based measure [35].

2.3 Contributions

In this work, we do not contribute any novel method for ATE. The c-value method [13] implemented in the UPM Term Extractor [36] is used as this combination of the method and implementation has been experimentally proven to be the best appropriate for detecting terminological saturation [37].

In difference and complementary to the abovementioned relevant work, we contribute several novel things. Firstly, we propose a way to rationally choose the thresholds that are used to regard string similarity as term similarity (Sect. 3). Secondly, we develop an algorithm for similar terms grouping that uses string similarity measures and term similarity thresholds (Sect. 4). Based on its use, we propose the refinement of the baseline THD algorithm [2] for measuring terminological difference between two subsequent text datasets (Sect. 4).

3 The Choice of SSMs and Terms Similarity Thresholds

From the variety of SSMs, mentioned above, due to the specifics of our task of the approximate comparison of short strings containing a few words, we filter out those: (i) that require long strings or sets of strings of a considerably big size; (ii) that are computationally hard. We also keep the representatives of all kinds of string metrics in our short list as much as possible. As a result, we form the following list of measures to be considered for further use:

- Character-based measures: Levenshtein distance [22], Hamming distance [23], Jaro similarity [25], and Jaro-Winkler similarity [26]
- Token-based measures: Jaccard similarity (*uni*-gram comparison) [29], cosine similarity (*uni*-gram comparison) [19], and Sørensen-Dice coefficient (*bi*-gram comparison) [27, 28]

Among those, Levenshtein and Hamming distances appear to be the least appropriate in our context due to their specifics. Levenshtein returns an integer number of required edits, while the rest of the measures return normalized reals. Hence, it is not clear if normalizing Levenshtein would make the result comparable to the other measures in a way to use the same term similarity threshold. Hamming measure is applicable only to the strings of equal lengths. Adding spaces to the shorter string, however, may lower the precision of measurement. Cosine similarity is based on the same principle as Jaccard, but is more computationally complex due to the presence of the square root in the denominator. Therefore, we finally choose to use Jaro, Jaro-Winkler, Jaccard[4], and Sørensen-Dice for implementation and evaluation in our work. Further, it is briefly explained how the selected measures are computed.

[4] It is expected that Cosine measure, being based on the same principle as Jaccard, is not better than Jaccard in terms of performance, though takes more time to be computed.

Jaro similarity sim_j between two strings S_1 and S_2 is computed (1) as the minimal number of one character transforms to be done to the first term (string) for getting the second string in the compared pair.

$$sim_j = \begin{cases} 0, & if\, m = 0 \\ 1/3 * \left(\frac{m}{|s_1|} + \frac{m}{|s_2|} + \frac{m-t}{m}\right) & otherwise \end{cases}, \tag{1}$$

where: $|S_1|$, $|S_2|$ are the lengths of the compared strings; m is the number of the matching characters; and t is the half of the number of transposed characters. The characters are matching if they are the same and their distance from the beginning of the string differs by no more than $\lfloor max(|S_1|, |S_2|)/2 \rfloor - 1$. The number of transposed characters is the number of matching but having different sequence order symbols.

Jaro-Winkler similarity measure sim_{j-w} refines Jaro similarity measure by using a prefix scale value p, which assigns better ratings to the strings that match from their beginnings for a prefix length l. Hence, for the two strings S_1 and S_2 it is computed as shown in (2).

$$sim_{j-w} = sim_j + l * p * (1 - sim_j), \tag{2}$$

where: l is the length of a common prefix (up to a maximum of 4 characters); p is a constant scaling factor meaning how much the similarity value is adjusted upwards for having common prefixes (up to 0.25, otherwise the measure can become larger than 1; [26] suggests that $p = 0.1$).

Sometimes Winkler's prefix bonus $l * p * (1 - sim_j)$ is given only to the pairs having Jaro similarity value higher that a particular threshold. This threshold is suggested [26] to be equal to 0.7.

Jaccard similarity index sim_{ja} is a similarity measure for finite sets, characters in our case. It is computed, for the two strings S_1 and S_2, as the ratio between the cardinalities of the intersection and union of the character sets in S_1 and S_2 as shown in (3).

$$sim_{ja} = (|S_1| \cap |S_2|)/(|S_1| \cup |S_2|) \tag{3}$$

Finally, the Sørensen-Dice coefficient is computed by counting identical character *bi*-grams in S_1 and S_2 and relating these to the overall number of *bi*-grams (4).

$$sim_{sd} = 2n_{\equiv}/(n_{S_1} + n_{S_2}), \tag{4}$$

where: $n_{\equiv}$ is the number of *bi*-grams found in S_1 and also in S_2; n_{S_1}, n_{S_2} are the numbers of all *bi*-grams in S_1 and S_2 respectively.

For the proper use of the implemented SSM functions in the context of terms comparison and grouping, it is necessary to determine what would be a reasonable threshold to distinguish between (semantically) similar and different terms. For finding that out, the following cases in string comparison need to be taken into account.

Full Positives (FP). In this case, evaluated character strings are fully the same, which clearly gives similar (the same) terms.

Full Negatives (FN). In this case, evaluated character strings are very different and the terms in these strings carry different semantics. This is also a clear situation and is characterized by low values of similarity measures.

Partial Positives (PP). In this case, evaluated character strings are partially the same and the terms in these strings carry the same or similar semantics. The terms in such strings are similar, though it may not be fully clear. The following are different categories of terms that fall into this case: the words in the terms have different endings (e.g. plural/singular forms); different delimiters are used (e.g. "-", or "–", or " - "); a symbol is missing, erroneously added, or misspelled (a typo); one term is the sub-string of the other (e.g. subsuming the second); one of the strings contains unnecessary extra characters (e.g. two or three spaces instead of one, or noise).

Partial Negatives (PN). In this case, evaluated character strings are partially the same but the terms in these strings carry different semantics. The terms in such strings are different, though it may not be fully clear. The following are the categories that fall into this case: the terms in the compared strings differ by a very few characters, but have substantially different meanings (e.g. "deprecate" versus "depreciate"); the compared multi-word terms have common word(s) but fully differ in their meanings (e.g. "affect them" versus "effect them"). These PN are the hardest case to be detected.

The test set of term pairs falling into the cases and categories described above has been manually developed[5]. For each pair of terms in this test set, all four selected string similarity measures have been computed. The extract is presented in Table 2.

Table 2. Similarity measures for different test cases

Case	Category	Terms pair	Sørensen-Dice	Jaccard	Jaro	Jaro-Winkler
Different (FN)		whirled | world	0.0	0.5	0.790	0.811
		traces | creta	0.0	0.833	0.588	0.588
		time domain | ontology lifecycle	0.0	0.428	0.445	0.445
Same (FP)		identical strings | identical strings	1.0	1.0	1.0	1.0
Similar Semantics (PP)	Extra characters	*system?problems | system problems	0.814	0.769	0.936	0.936
		sad data mining | sqr data mining	0.769	0.818	0.859	0.873
	Common parts (words)	marcov chain monte carlo methods | monte carlo methods	0.782	0.766	0.629	0.666
		data mining algorithm | data mining	0.642	0.666	0.842	0.904
		cation error | error	0.533	0.333	0.427	0.427
	Typos	fraud detection | froud dictection	0.714	0.916	0.859	0.887
		monte carlo | monte ??rlo	0.7	0.727	0.878	0.927
		data mining | data minin	0.941	0.875	0.969	0.981

(continued)

[5] The test set and computed term similarity values are publicly available at https://github.com/OntoElect/Data/blob/master/STG/Test-Set.xls.

Table 2. (*continued*)

Case	Category	Terms pair	Sørensen-Dice	Jaccard	Jaro	Jaro-Winkler
	Different delimiters	computer science | computerscience	0.896	0.916	0.979	0.987
		serial episodes | serial&&episodes	0.827	0.818	0.936	0.961
		data cube | data_cube	0.75	0.777	0.925	0.955
	Different endings	network structure | network structures	0.969	1.0	0.981	0.988
		time complexity | time complexities	0.896	0.833	0.981	0.951
		value | values	0.888	0.833	0.918	0.951
Different Semantics (PN)	Common parts (words)	database | military base	0.400	0.500	0.410	0.410
		brainstorm | stormy weather	0.363	0.428	0.509	0.509
		iron clad | iron maiden	0.444	0.636	0.804	0.882
		jellyfish | fish tank	0.352	0.307	0.614	0.614
		four delegates | delegated authority	0.451	0.666	0.557	0.557
		string theory | string format	0.583	0.571	0.812	0.887
	Very few character differences	deprecate against | depreciate against	0.909	1.0	0.903	0.941
		alternately move | alternatively move	0.933	0.916	0.9	0.94
		affect them | effect them	0.9	0.758	0.906	0.906

The average values of all four chosen similarity measures for each category have been computed using all the test set term pairs falling into this category. These values are presented in Table 3. Term similarity thresholds have to be chosen such that full and partial negatives are regarded as not similar, but full and partial positives are regarded as similar. Hence, for the case of partial positives, the thresholds have to be chosen as minimal of all the case categories, and for the partial negatives – as the maximal of all the case categories. The values of case thresholds are shown in bold in Table 3. These are further used as the margins for relevant threshold intervals in our experiments. These intervals have been evenly split by the four threshold points, as presented in Table 4. The requirements for partial positives and negatives unfortunately contradict to each other. For example, if a threshold is chosen to filter out partial negatives, also some of the partial positives will be filtered out. Therefore, subsuming that partial negatives are rare, it has been decided to use the thresholds for partial positives.

Table 3. Average string similarity measure values for different categories of term pairs from the test set

Case/Category	Items in test set	Sørensen-Dice	Jaccard	Jaro	Jaro-Winkler
Different strings (FN)	**6**	**0.03**	**0.45**	**0.55**	**0.55**
Identical strings (FP)	**3**	**1.00**	**1.00**	**1.00**	**1.00**
Similar Semantics (PP)	**32**	**0.71**	**0.72**	**0.63**	**0.70**
- Unnecessary (extra) characters	7	0.8401	0.8820	0.8714	0.8784
- Common parts (words)	6	0.7122	0.7280	0.6375	0.7043
- Typos	6	0.7797	0.8637	0.8863	0.9220
- Different delimiters	6	0.7860	0.8473	0.9125	0.9442
- Different endings	7	0.8911	0.9135	0.9410	0.9590
Different Semantics (PN)	**18**	**0.89**	**0.89**	**0.89**	**0.91**
- Common parts (words)	11	0.4336	0.5221	0.6161	0.6408
- Very few character differences	7	0.8826	0.8845	0.8914	0.9059
Total	**59**				

Table 4. Term similarity thresholds chosen for experimental evaluation

Method	Term similarity thresholds			
	Min	*Ave-1*	*Ave-2*	*Max*
Sørensen-Dice	0.71	0.76	0.83	0.89
Jaccard	0.72	0.77	0.83	0.89
Jaro	0.63	0.72	0.80	0.89
Jaro-Winkler	0.70	0.77	0.84	0.91

4 OntoElect and the Refinement of the THD Algorithm

OntoElect, as a methodology, seeks for maximizing the fitness of the developed ontology to what the domain knowledge stakeholders think about the domain. Fitness is measured as the stakeholders' "votes" – a measure that allows assessing the stakeholders' commitment to the ontology under development, reflecting how well their sentiment about the requirements is met. The more votes are collected, the higher the commitment is expected to be. If a critical mass of votes is acquired (say 50% + 1, which is a simple majority vote), it is considered that the ontology meets the requirements satisfactorily.

Unfortunately, direct acquisition of requirements from domain experts is not very realistic. The experts are expensive and not willing to do the work, which falls out of their core activity. That is why the OntoElect approach focuses on the indirect collection of the stakeholders' votes by extracting these from high quality and reasonably high impact documents authored by the stakeholders.

An important feature to be ensured for knowledge extraction from text collections is that the dataset needs to be representative to cover the opinions of the domain knowledge stakeholders satisfactorily fully. OntoElect suggests a method to measure the terminological completeness of the document collection by analysing the *saturation* of terminological footprints of the incremental slices of the document collection [2].

The approach followed in our work is finding the terminological core of a document collection by measuring terminological saturation [2, 3]. This measurement is done using our terminological difference measure (*thd*, [2]) which is a variant of a Manhattan distance measure (see e.g. [18]) or Minkovski's distance with $p = 1$ [38].

The full texts of the documents from a collection are grouped in datasets in the order of their timestamps. As pictured in Fig. 1(a), the first dataset D_1 contains the first portion of (*inc*) documents. The second dataset D_2 contains the first dataset D_1 plus the second incremental slice of (*inc*) documents. Finally, the last dataset D_n contains all the documents from the collection.

At the next step of the OntoElect workflow, the bags of multi-word terms B_1, B_2, …, B_n are extracted from the datasets D_1, D_2, …, D_n together with their significance (*c-value*) scores, using UPM Term Extractor software [36]. An example of an extracted bag of terms is shown in Fig. 1(b).

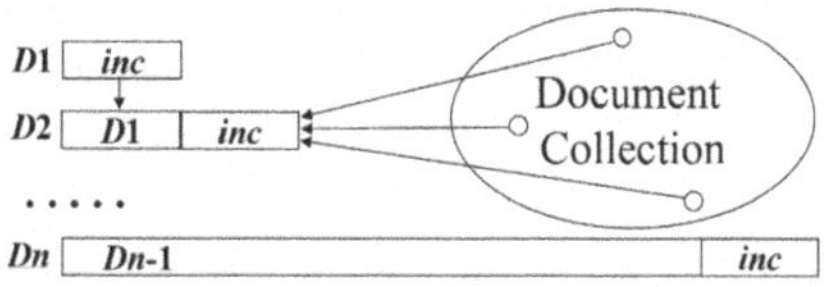

(a)

Term	c-value
temporal	6428.5
time	3950
temporal logic	2558
logic	2186.5
interval	2011
model	1957
formula	1750
relations	1613
temporal representation	1604
relation	1574.5
constraints	1557
reasoning	1549
....	

(b)

Fig. 1. (a) Incrementally enlarged datasets in OntoElect; (b) An example of a bag of terms extracted by UPM Term Extractor [36].

At the subsequent step, every extracted bag of terms B_i, $i = 1, …, n$ is processed as follows. Firstly, an individual term significance threshold (*eps*) is computed to cut off those terms that are not within the majority vote. The sum of *c-value*s with individual values above *eps* form the majority vote if this sum is higher than ½ of the sum of all *c-values*. Secondly, insignificant term candidates are cut-off at *c-value* < *eps*. Thirdly, the normalized scores are computed for each individual term: *n-score* = *c-value*/max(*c-value*). Finally, the result is saved in the bag of retained significant terms T_i. After this step only significant terms, that represent the majority vote, are considered. T_i are then evaluated for saturation by measuring pair-wise terminological difference between the subsequent bags T_i and T_{i+1}, $i = 0, …, n$-1. So far in OntoElect, it has been done by applying the baseline THD algorithm[6] [2] presented in Fig. 2.

[6] The baseline THD algorithm is implemented in Python and is publicly available at https://github.com/OntoElect/Code/tree/master/THD.

```
Algorithm THD. Compute Terminological Difference between Bags of Terms
Input:
 Ti, Ti+1 - the bags of terms with grouped similar terms.
     Each term Ti.term is accompanied with its T.n-score.
     Ti, Ti+1 are sorted in the descending order of T.n-score.
 M -  the name of the string similarity measure function to compare terms
 th - the value of the term similarity threshold from within [0,1]
Output: thd(Ti+1, Ti), thdr(Ti+1, Ti)
1.   sum := 0
2.   thd := 0
3.   for k := 1, |Ti+1|
4.     sum :=  sum + Ti+1.n-score[k]
5.     found : = .F.
6.     for m := 1, |Ti|
7.        if (Ti+1.term[k] = Ti.term[m]) | if (M(Ti+1.term[k], Ti.term[m], th))
8.           then
9.             thd += |Ti+1.n-score[k] - Ti.n-score[m]|
10.            found := .T.
11.    end for
12.    if (found = .F.) then thd += Ti+1.n-score[k]
13. end for
14. thdr := thd / sum
```

Fig. 2. Baseline THD algorithm [2] for measuring terminological difference in a pair of bags of terms and its refinement. Baseline THD uses string equalities for comparing terms (dashed rounded rectangle in line 7). The refinements are shown in solid rounded rectangles. Refined THD has two more input parameters (*M* and *th*) and uses *M* for comparing terms (line 7).

In fact, the THD algorithm accumulates the *n-score* differences, in the *thd* value for the bag T_{i+1}, if there were the same terms in T_i and T_{i+1}. If there was no the same term in T_i, it adds the *n-score* of the orphan to the *thd* value of T^{i+1}. After *thd* has been computed, the relative terminological difference *thdr* receives its value as *thd* divided by the sum of *n-scores* in T_{i+1}.

Absolute (*thd*) and relative (*thdr*) terminological differences are computed for further assessing if T_{i+1} differs from T_i more than the individual term significance threshold *eps*. If not, it implies that adding an increment of documents to D_i for producing D_{i+1} did not contribute any noticeable amount of new terminology. Hence, the subset D_{i+1} of the overall document collection may have become terminologically saturated. However, to obtain more confidence about the saturation, OntoElect suggests that more subsequent pairs of T_i and T_{i+1} are evaluated. If stable saturation is observed, then the process of looking for a minimal saturated sub-collection could be stopped.

Our task is to modify the THD algorithm in a way to allow finding not exactly the same but sufficiently similar terms by applying string similarity measures with appropriate thresholds, as explained in the previous Sect. 3. For that, the preparatory similar term grouping step has been introduced to avoid duplicate similarity detection. For each of the compared bags of terms T_i and T_{i+1} the similar term grouping (STG) algorithm is applied at this preparatory step – see Fig. 3. After term grouping is accomplished for both bags of terms, the refined THD algorithm (Fig. 2 – rounded rectangles) is performed to compute the terminological difference between T_i and T_{i+1}.

```
Algorithm STG. Group similar terms in the bag of terms
Input:
 T  - a bag of terms. Each term T.term is accompanied with its
      T.n-score. T is sorted in the descending order of T.n-score.
 M - the name of the string similarity measure function to compare
     terms
 th - the value of the term similarity threshold from within [0,1]
Output: T with grouped similar terms
    sum := 0
    for k = 1,|T|
      term :=  T.term[k]
      n-score := T.n-score[k]
      count := 1
      for m = k+1,|T|
           if M(term, T.term[m], th)
              then
                n-score += T.n-score[m]
                count += 1
                remove(T[m])
      end for
      T.n-score[k] := n-score / count
    end for
```

Fig. 3. Similar Term Grouping (STG) algorithm

5 Evaluation

This section reports on our evaluation of the refined THD algorithm against the baseline THD [2]. This evaluation is performed using the workflow of the OntoElect Requirements Elicitation Phase [3] and three document collections from different domains: TIME, DMKD-300, and DAC-cleaned. Section 5.1 outlines the set-up of our evaluation experiments. The document collections are presented in Sect. 5.2. The results of our evaluation experiments are discussed in Sect. 5.3.

5.1 The Set-Up of the Experiments

The objective of our experiments is to find out if using the refined THD algorithm yields quicker terminological saturation compared to the use of the baseline THD algorithm. We are also looking at finding out which string similarity measures best fit for measuring terminological saturation.

For making the results comparable, the same datasets, created from the document collections as described in Sect. 5.2, are fed into both the refined and baseline THD algorithms. For each document collection, we apply:

1. The refined THD – sixteen times – one per individual string similarity measure M [7] (Sect. 3) and per individual term similarity threshold th (Table 4); and
2. The baseline THD – one time as it does not depend on a term similarity threshold

[7] The functions for all the four selected SSMs have been implemented in Python 3.0 and return real values within [0, 1]. These functions are publicly available at: https://github.com/OntoElect/Code/tree/master/STG/core/methods.

The values of: (i) the number of retained terms; (ii) absolute terminological difference (*thd*); and (iii) the time taken to perform similar terms grouping by the STG algorithm (*sec*) are measured.

Finally, to verify if our SSM implementations, and hence the STG and refined THD algorithms, are correct, we check if the refined THD algorithm implementation returns the results which are satisfactorily similar to that of the baseline THD when the terms similarity threshold is set to 1.00. This threshold value straightforwardly means that only equivalent strings have to be regarded as similar terms.

All the computations are run using a Windows 7 64-bit PC with: Intel® Core™ i5 CPU, M520 @ 2.40 GHz; 8.0 Gb on-board memory; NVIDIA Geforce GT330 M GPU.

5.2 Experimental Data

The document collections used in our experiments are all composed of the papers published at the peer-reviewed international venues in three different domains:

- The TIME collection contains the full text papers of the proceedings of the Time Representation and Reasoning (TIME) Symposia series[8] published between 1994 and 2013
- The DMKD-300 collection is composed of the subset of full text articles from the Springer journal on Data Mining and Knowledge Discovery[9] published between 1997 and 2010
- The DAC-cleaned collection comprises the subset of full text papers of the Design Automation Conference[10] published between 2004 and 2006

The **chronological** order of adding documents is chosen for generating experimental datasets from the documents of all the three collections using our Dataset Generator [37]. The characteristics of all the document collections and generated datasets are summarized in Table 5.

Table 5. The characteristics of the used document collections and datasets

Document collection	Paper type and layout	No doc	Noise	Processing	Inc	No datasets
TIME	Conference, IEEE 2-column	437	Manually cleaned	Manual conversion to plain text, automated dataset generation	20 papers	22
DMKD-300	Journal, Springer 1-column	300	Not cleaned, moderately noisy	Automated [37]	20 papers	15
DAC-cleaned	Conference, IEEE 2-column	506	Quite noisy	Automated, stop terms removal [37]	20 papers	26

[8] http://time.di.unimi.it/TIME_Home.html.

[9] https://link.springer.com/journal/10618.

[10] http://dac.com/.

5.3 Results and Discussion

The measurements, taken in our experiments for different collections and terms similarity threshold points, are not presented in the paper in a tabular form due to page limits. Instead, the results are presented diagrammatically in figures below and made available in full, including values, publicly online[11].

The results of our measurements of terminological saturation (*thd*) are pictured in Figs. 4, 5 and 6. Saturation (*thd*) measurements reveal that the refined THD algorithm detects terminological saturation faster than the baseline THD algorithm, no matter what the chosen term similarity measure (*M*) or similarity threshold (*th*) is. If the results for different measures are compared, then it may be noted that the respective saturation curves behave differently, depending on the similarity threshold point.

Overall, as one could see in Figs. 4, 5 and 6(a)–(d), the use of the Sørensen-Dice measure demonstrates the least volatile behaviour along the terms similarity threshold points. Sørensen-Dice also results in making the refined THD algorithm to detect saturation slower than the three other measures for *Min*, *Ave*-1, and *Ave*-2. For *Max*, it is as fast as Jaro and slightly slower than Jaccard and Jaro-Winker.

One more observation is that, integrally, all the implemented term similarity measures coped well with retaining significant terms from all the three document collections. This is indicated by the co-locations of terminology contribution peaks at the diagrams (a)–(d) in Figs. 4, 5 and 6. One can see in Figs. 4, 5 and 6(d), for the *Max* threshold point, that all the string similarity methods curves follow the shape of the

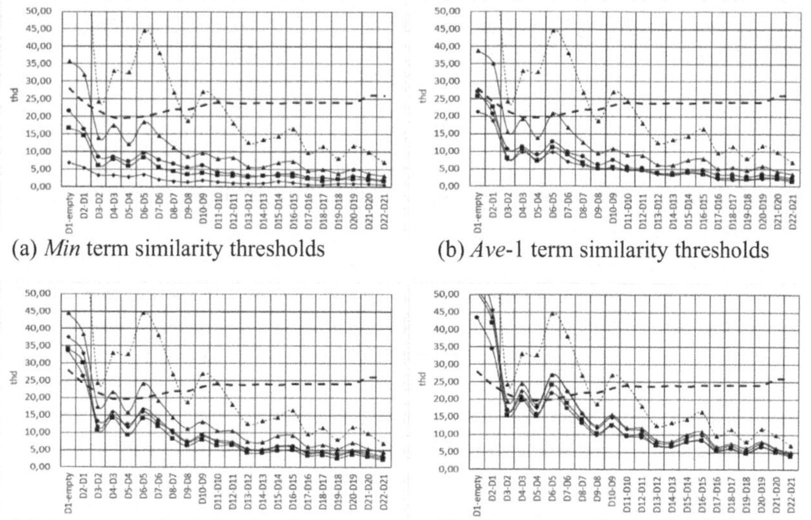

(a) *Min* term similarity thresholds (b) *Ave*-1 term similarity thresholds

(c) *Ave*-2 term similarity thresholds (d) *Max* term similarity thresholds

Legend: Sorensen-Dice — Jaccard — Jaro — Jaro-Winkler — Baseline — eps

Fig. 4. Terminological saturation measurements on TIME for different similarity threshold points

[11] https://github.com/OntoElect/Experiments/tree/master/STG. File names are {TIME, DMKD-300, DAC-cleaned}-Results-Alltogether-{min, ave, ave2, max, 1}-th.xlsx.

baseline THD curve quite closely. Hence, they have the peaks exactly at the same *thd* measurement points where the baseline has, pointing at more new significant terms. The most sensitive to terminology contribution peaks was Sørensen-Dice.

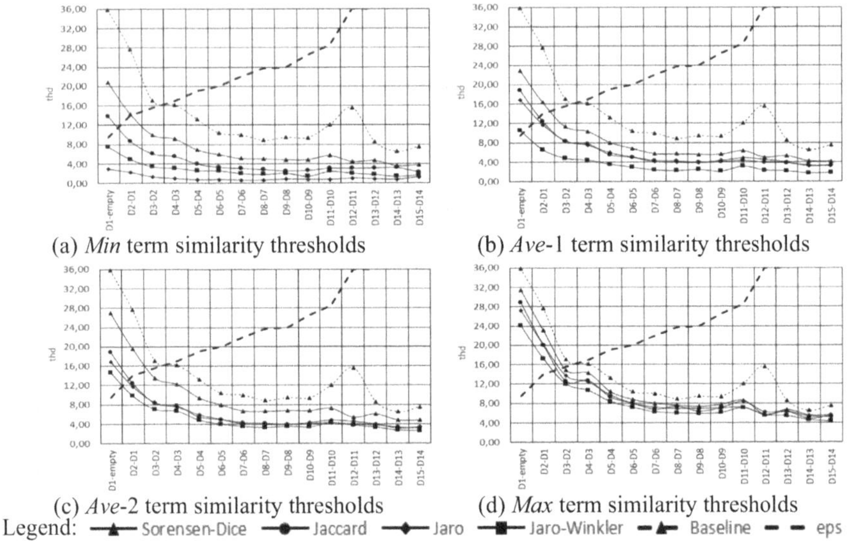

(a) *Min* term similarity thresholds (b) *Ave*-1 term similarity thresholds (c) *Ave*-2 term similarity thresholds (d) *Max* term similarity thresholds

Fig. 5. Terminological saturation measurements on DMKD-300 for different similarity threshold points

The diagrams in Figs. 7, 8 and 9 show the times spent by the STG algorithm to detect and group similar terms for different chosen term similarity thresholds. One particular diagram corresponds to a particular terms similarity threshold point (*Min*, *Ave*-1, *Ave*-2, and *Max*).

It needs to be mentioned that the introduction of string similarity measures in the computation of terminological difference (THD algorithm) increases the computational complexity quite substantially. As it could be noticed in Figs. 7, 8 and 9(a)–(d), the times grow with the value of the terms similarity threshold (*th*) and reach thousands of seconds for *Max* threshold values. It is worth acknowledging that Sørensen-Dice and Jaccard are substantially more stable to the increase of *th* than Jaro and Jaro-Winkler. Sørensen-Dice takes, however, times more time than Jaccard. From the other hand, Jaccard is not very sensitive to terminological peaks and retains significantly less terms than Sørensen-Dice.

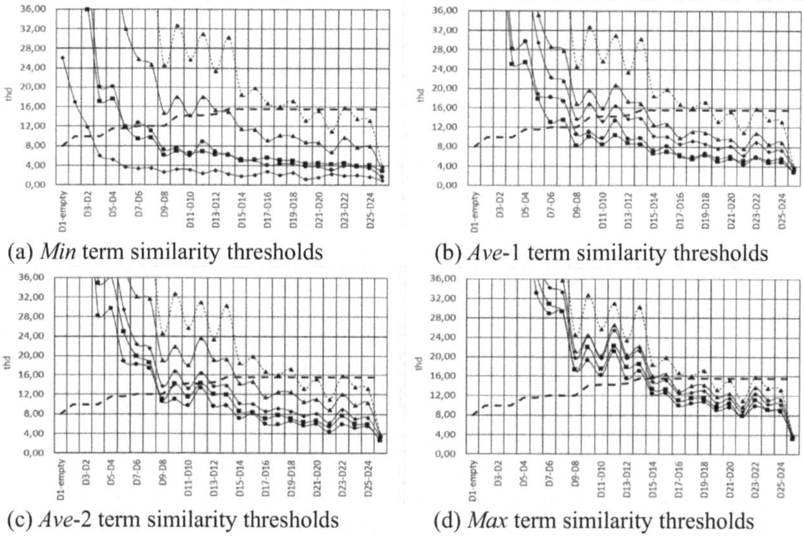

(a) *Min* term similarity thresholds (b) *Ave*-1 term similarity thresholds

(c) *Ave*-2 term similarity thresholds (d) *Max* term similarity thresholds

Legend: Sorensen-Dice Jaccard Jaro Jaro-Winkler Baseline eps

Fig. 6. Terminological saturation measurements on DAC-cleaned for different similarity threshold points

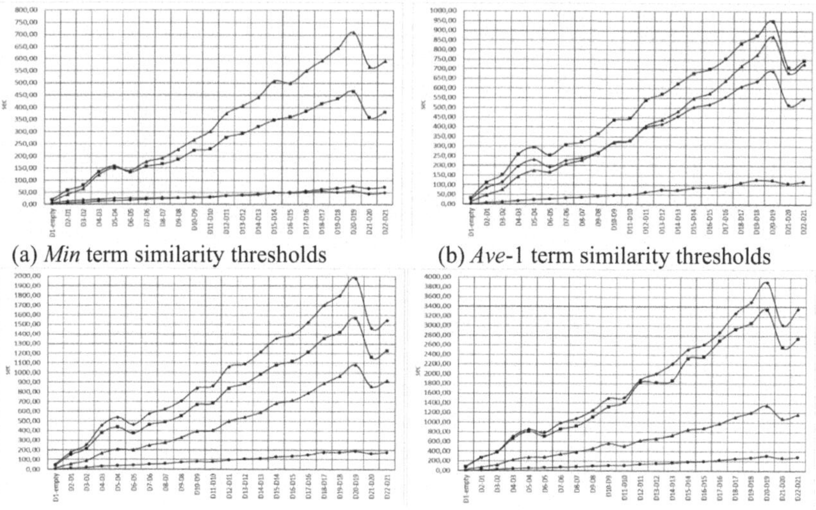

(a) *Min* term similarity thresholds (b) *Ave*-1 term similarity thresholds

(c) *Ave*-2 term similarity thresholds (d) *Max* term similarity thresholds

Legend: Sorensen-Dice Jaccard Jaro Jaro-Winkler

Fig. 7. Time (sec) spent by the STG algorithm for grouping similar terms on TIME bags of terms

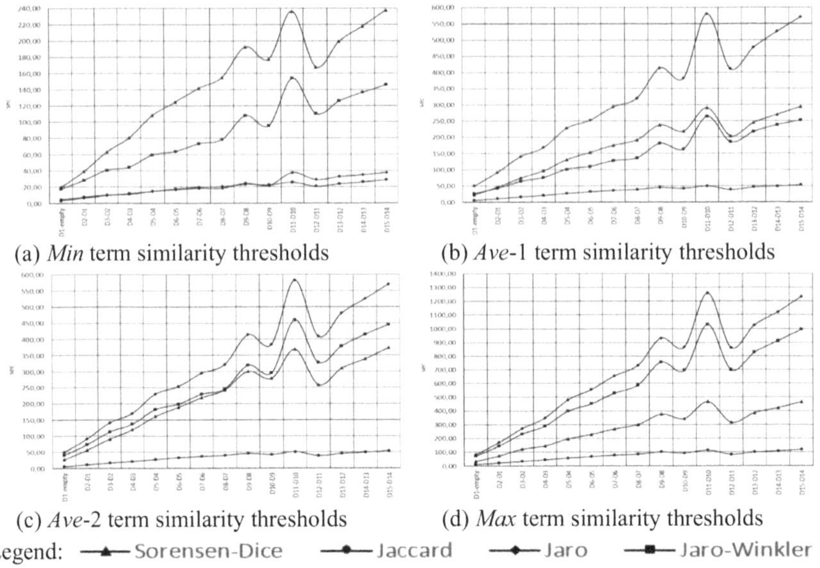

(a) *Min* term similarity thresholds (b) *Ave*-1 term similarity thresholds

(c) *Ave*-2 term similarity thresholds (d) *Max* term similarity thresholds

Fig. 8. Time (sec) spent by the STG algorithm for grouping similar terms on DMKD-300 bags of terms

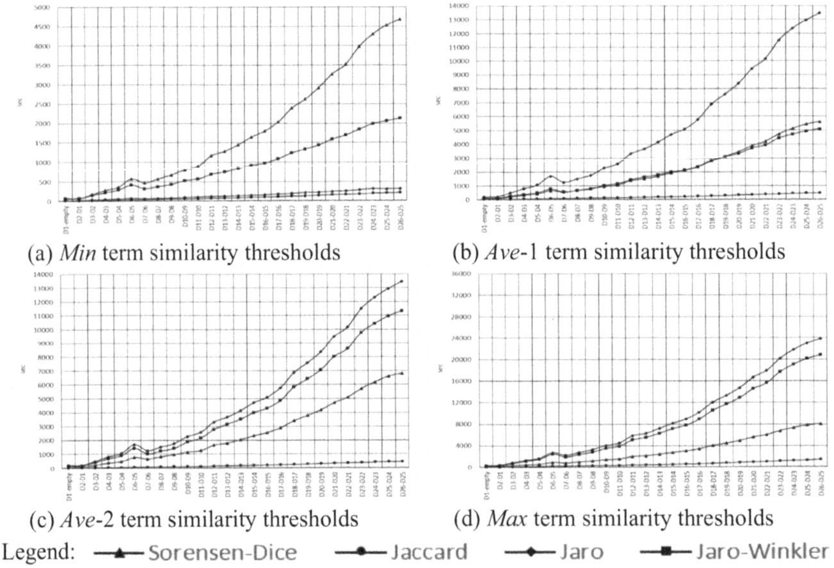

(a) *Min* term similarity thresholds (b) *Ave*-1 term similarity thresholds

(c) *Ave*-2 term similarity thresholds (d) *Max* term similarity thresholds

Fig. 9. Time (sec) spent by the STG algorithm for grouping similar terms on DAC-cleaned bags of terms

Figure 10 pictures the proportions of the retained to all extracted terms when saturation has been detected, computed at different terms similarity threshold points, for the bags of terms extracted from our three document collections. It is clear from Fig. 10 that Sørensen-Dice yields the second highest proportions for all the collections and used term similarity thresholds, after the baseline, which does not group terms.

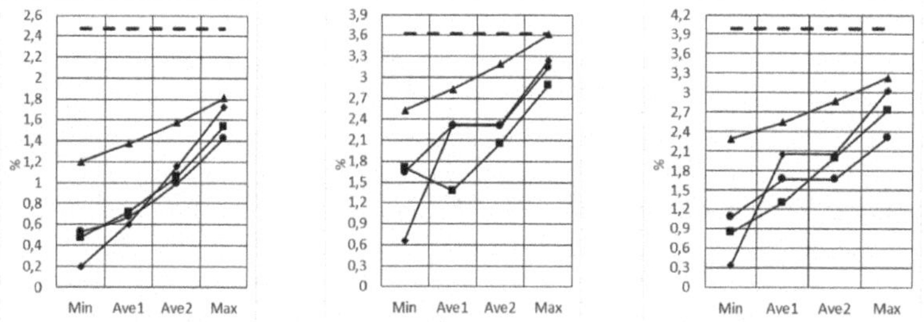

(a) TIME collection (b) DMKD-300 collection (c) DAC-cleaned collection

Legend: Sorensen-Dice Jaccard Jaro Jaro-Winkler Baseline

Fig. 10. The proportions of retained to all extracted terms for different term similarity measures per document collections

Finally, the terms similarity threshold is set to 1.00 and the refined THD implementation is evaluated for all three collections for the pairs of the bags of terms in a few pair vicinity of the saturation points. The task is to check if the refined THD with similar terms grouping: (i) detects terminological saturation at the same point as the baseline THD, therefore, *thd* values are measured; and (ii) retains the same number of significant terms as the baseline THD, therefore, the numbers of retained terms are measured. We are also interested in comparing the time taken to accomplish term grouping (STG).

The results for the DMKD-300 collection are presented graphically in Fig. 11. The results for the TIME and DAC-cleaned collections[12] are very much similar to these for DMKD-300 and do not change our conclusion and recommendation.

It may be seen in Fig. 11(a) and (b) that Jaro and Jaro-Winkler implementations fully repeat the baseline THD results, both in the measured *thd* values and numbers of retained significant terms. Sørensen-Dice behaves similarly to Jaro and Jaro-Winkler up to the saturation point. After that, it returns slightly lower *thd* and retains slightly less significant terms. This behaviour is acceptable as the measurements after the saturation point are of marginal interest. Jaccard implementation however appears to return significantly lower *thd* values and significantly less retained terms at all measurement points – before and after detecting saturation. Jaccard also detects saturation one measurement point earlier than the rest of the SSMs, which is not correct for this threshold (1.00).

[12] These results could be accessed at https://github.com/OntoElect/Experiments/tree/master/STG. File names are {TIME, DMKD-300, DAC-cleaned}-Results-Alltogether-1-th.xlsx.

Figure 11(c) reveals that, for being accurate in measurements at the very high threshold of 1.00, Jaro and Jaro-Winkler take too much of a computational overhead. Sørensen-Dice and Jaccard however remain more stable to the increase of the *th*, similarly as it was before for *Ave1*, *Ave2*, and *Max* threshold points.

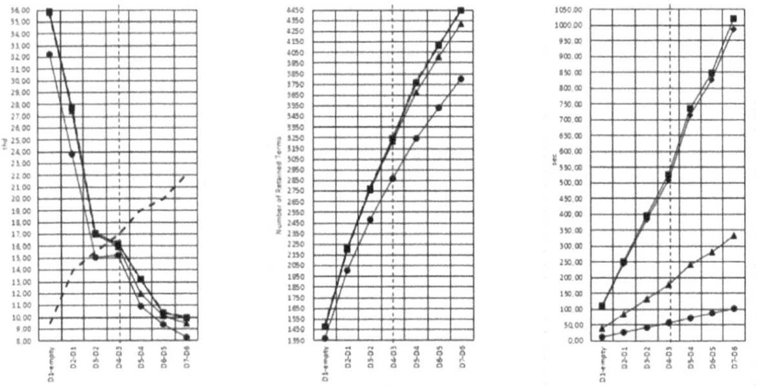

(a) *thd* values (b) numbers of retained terms (c) time taken by STG

Legend: Sorensen-Dice, Jaccard, Jaro, Jaro-Winkler, Baseline, eps

Fig. 11. Evaluation of the refined THD implementation at *th* = 1.00 on DMKD-300 bags of terms. Vertical dashed lines mark terminological saturation point.

The summary of our experimental findings is collected in Table 6 in the form of the rankings. We rank the performance of all the evaluated SSMs and the baseline THD on a scale from 1 (the best) to 5 (the worst) for every document collection and every terms similarity threshold point (*Min*, *Ave*1, *Ave*2, *Max*) within each collection. We also look at the average rankings for all four thresholds points.

The aspects we look at in this ranking are: (i) the fastness of detecting terminological saturation, the faster – the better (Figs. 4, 5 and 6); (ii) the number of retained significant terms, the more – the better (Fig. 10); and (iii) the time taken by the method to accomplish the computation, the less – the better (Figs. 7, 8 and 9).

Table 7 contains the values of performance indices for different SSMs and the baseline THD regarding the four terms similarity thresholds points and their average values. This is done for two cases: (a) taking into account the execution time criterion (Less Time Taken in Table 6); and (b) not taking the execution time criterion into account. It has been done to analyse the value of using an SSM if the computational overhead is not important. The values were calculated by summing all the ranks for different collections and criteria taken from the corresponding threshold point rows of Table 6.

Table 6. The ranking of the evaluated SSMs

Criterion	String similarity threshold	Rank (1–5)				
		Baseline THD	Sørensen-Dice	Jaccard	Jaro	Jaro-Winkler
TIME collection						
Faster detection of saturation	Min	5	4	1	1	1
	Ave1	5	4	1	1	1
	Ave2	5	4	1	1	1
	Max	5	3	1	3	1
	Average	5	3.75	1	1.5	1
More significant terms retained	Min	1	2	3	5	4
	Ave1	1	2	4	5	3
	Ave2	1	2	5	3	4
	Max	1	2	5	3	4
	Average	1	2	4.25	4	3.75
Less time taken	Min	1	5	3	2	4
	Ave1	1	4	2	3	5
	Ave2	1	3	2	5	4
	Max	1	3	2	5	4
	Average	1	3.75	2.25	3.75	4.25
DMKD-300 collection						
Faster detection of saturation	Min	5	4	3	1	1
	Ave1	5	4	1	1	1
	Ave2	5	4	1	1	1
	Max	5	1	1	1	1
	Average	5	3.25	1.5	1	1
More significant terms retained	Min	1	2	3	5	4
	Ave1	1	2	4	5	3
	Ave2	1	2	5	3	4
	Max	1	2	5	3	4
	Average	1	2	4.25	4	3.75
Less time taken	Min	1	5	2	3	4
	Ave1	1	4	2	5	3
	Ave2	1	3	2	5	4
	Max	1	3	2	5	4
	Average	1	3.75	2	4.5	3.75
DAC-cleaned collection						
Faster detection of saturation	Min	5	4	3	1	2
	Ave1	5	4	1	3	1
	Ave2	5	4	1	3	1
	Max	5	4	1	1	1
	Average	5	4	1.5	2	1.25

(continued)

Table 6. *(continued)*

Criterion	String similarity threshold	Rank (1–5)				
		Baseline THD	Sørensen-Dice	Jaccard	Jaro	Jaro-Winkler
More significant terms retained	Min	1	2	3	5	4
	Ave1	1	2	4	3	5
	Ave2	1	2	5	3	4
	Max	1	2	5	3	4
	Average	1	2	4.25	3.5	4.25
Less time taken	Min	1	5	2	3	4
	Ave1	1	4	2	5	3
	Ave2	1	3	2	5	4
	Max	1	3	2	5	4
	Average	1	3.75	2	4.5	3.75

These sums have further been subtracted from the highest rank value[13] in order to revert to "the higher – the better" scale in Table 7. Performance indices are also pictured in Fig. 12.

Table 7. The performance indices of the evaluated SSMs with and without accounting for taken execution time

Threshold	Baseline THD	Sørensen-Dice	Jaccard	Jaro	Jaro-Winkler
(a) Execution time criterion is taken into account					
Min	12	0	10	7	5
Ave1	12	3	12	2	8
Ave2	12	6	9	4	6
Max	12	10	9	4	6
Average (Table 6)	12	4.75	10	4.25	6.25
(b) Execution time criterion is not taken into account					
Min	0	0	2	0	2
Ave1	0	0	3	0	4
Ave2	0	0	0	4	3
Max	0	4	0	4	3
Average (Table 6)	0	1	1.25	2	3

[13] The rank value is the sum of all ranks for a method within a particular threshold in Table 6. The highest rank value indicates the lowest performance. For case (a) it equals to 12, which is for Sørensen-Dice at *Min* threshold. For case (b) it equals to 4.

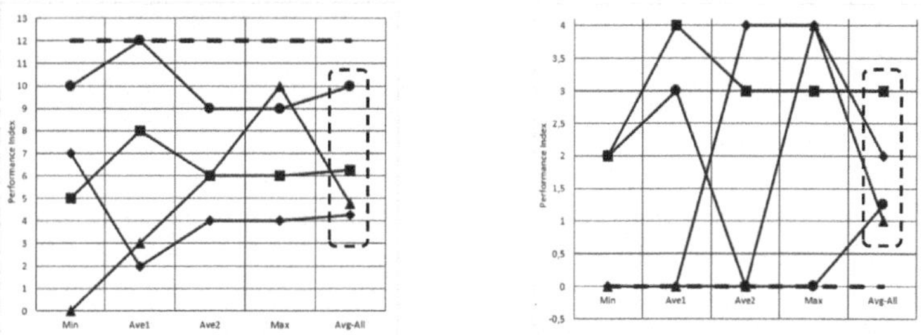

(a) Accounting for execution time (b) Without accounting for execution time

Legend: Sorensen-Dice Jaccard Jaro Jaro-Winkler Baseline

Fig. 12. Performance indices of the evaluated SSMs per terms similarity thresholds with (a) and without (b) taking their execution time ranks into account. The points in the rounded dashed rectangles represent the averages for all the thresholds.

Regarding the evaluation of correctness at *th* = 1.00, the SSM, that behaves both correctly, up to the saturation point, and computationally efficiently, is Sørensen-Dice. Jaro and Jaro-Winkler, though are correct, take too much of the computational overhead at this high *th* value. Jaccard is not fully correct.

Let us now summarize the comparative analysis of the performance of all the SSMs in the two cases, (a) and (b), presented in Table 7 and Fig. 12.

In case (a), when the computation time is taken into account in the comparative rating, the summary is the following. Probably surprisingly, Jaccard, which is computationally the most lightweight SSM (Figs. 7, 8 and 9), demonstrates the best overall performance. In this case, it still does not outperform the baseline THD because: (i) it takes extra time for STG; and (ii) it retains less significant terms[14]. Jaccard is the best balanced on all evaluation criteria, compared to the rest of the evaluated SSMs. One important drawback of Jaccard is that it does not perform fully correctly at *th* = 1.00. Therefore, the use of Jaccard may be recommended in the cases of low terms similarity thresholds (like *Min* or *Ave*1) and hard constraints on the time of computation. Performance indices are also good for Sørensen-Dice and Jaro-Winkler which both work acceptably correctly at *th* = 1.00. These two SSMs appear to be mutually complementary in the terms that: (i) Jaro-Winkler is better than Sørensen-Dice at lower terms similarity thresholds, like *Min* or *Ave*1; (ii) Sørensen-Dice outperforms Jaro-Winkler at higher terms similarity thresholds, like *Ave*2 or *Max*. Jaro in case (a) is a clear negative outlier and is not recommended for use.

In case (b), when the computation time is not taken into account in the comparative rating, the summary is different. As it is clearly seen in Fig. 12(b), all the SSMs outperform the baseline THD on average and at *Max* threshold. Jaro-Winkler is the best performing for *Min* and *Ave*1 thresholds, but gives up to Jaro at *Ave*2 and *Max*. It is also outperformed by Sørensen-Dice at *Max*. However, Jaro-Winkler appears to be most balanced in performance regarding all the four thresholds, which is highlighted by

[14] Which should be so as the baseline THD does not group terms. Hence, any alternative method that does similar terms grouping retains less significant terms.

the *Avg-All* value. Jaccard in case (b) is a clear negative outlier and therefore cannot be recommended for use.

If the assessments for the cases (a) and (b) are combined, the following recommendation could be given. At an expense of a substantially higher execution time, the THD algorithm refined by Jaro-Winkler (at all thresholds except *Max*) or Sørensen-Dice (at *Max* threshold) are our recommended choices for measuring terminological saturation. Jaro-Winkler is the first choice, because it is the most balanced in performance for all the evaluated thresholds.

6 Conclusions and Future Work

In this paper, we investigated if a simple string equivalence measure, used in the baseline THD algorithm, could be outperformed if a carefully chosen string similarity measure is used instead.

Overall, we found out that the use of STG, even at high terms similarity thresholds, rewards quite substantially in reducing the volume of processed data. The numbers of these gains are provided in the Terminological Core part of Table 8. Depending on

Table 8. The gains of the use of STG and refined THD

	Saturation point	Terminological core			Terms			
		No papers	Volume, Mb	% baseline	Extracted terms	Retained terms	% baseline	*eps*
	TIME (Max)							
Baseline	D11	220	6.55	**100.00**	287887	7110	**100.00**	23.77
Jaccard	D6	120	3.53	**53.89**	190263	2717	**38.21**	21.00
Sorensen-Dice	D7	140	4.17	**63.66**	200176	3629	**51.04**	22.00
Jaro-Winkler	D6	120	3.53	**53.89**	190263	2717	**38.21**	21.00
	DMKD-300 (Max)							
Baseline	D3	60	3.14	**100.00**	89617	7110	**100.00**	17.00
Jaccard	D2	45	2.46	**78.34**	67913	2135	**30.03**	15.50
Sorensen-Dice	D2	45	2.46	**78.34**	67913	2453	**34.50**	15.50
Jaro-Winkler	D2	45	2.46	**78.34**	67913	1963	**27.61**	15.50
	DAC-cleaned (Max)							
Baseline	D23	460	12.40	**100.00**	514364	20558	**100.00**	15.51
Jaccard	D14	280	7.46	**60.16**	320473	7406	**36.02**	15.51
Sorensen-Dice	D16	320	8.54	**68.87**	356749	11528	**56.08**	15.51
Jaro-Winkler	D14	280	7.46	**60.16**	320473	8736	**42.49**	15.51

how fast saturation is achievable in different collections, the use of STG allowed lowering the size of a terminological core by 22 to 46%.

It is also remarkable that, in general, the numbers of retained significant terms, due to their grouping, were also decreased substantially, by 44 to 72% depending on the collection. At the same time, the individual term significance thresholds (*eps*) were very slightly changed. This hints that the use of STG did not result in a noticeable loss of significant terms.

Because of applying our THD algorithm refinement, using all four evaluated SSMs, terminological saturation has been detected faster. Hence, in that sense, the refined THD with STG outperformed the baseline method. Three of the SSMs gave also acceptably correct results at th = 1.00. A somewhat discouraging result was, however, that the use of SSMs for STG causes a substantial computational overhead. Therefore, none of the methods involving STG outperformed the baseline THD integrally if execution time is an important criterion for assessing performance – case (a) in Table 7 and Fig. 12. If execution time is not very important and may be disregarded, the result is substantially different – case (b) in Table 7 and Fig. 12. Overall, putting together the findings in these two cases, the recommendation was made to use the THD algorithm refined by Jaro-Winkler (at all thresholds except *Max*) or Sørensen-Dice (at *Max* threshold) for measuring terminological saturation. Jaro-Winkler was recommended as the first choice, because it is the most balanced in performance for all the evaluated thresholds.

The plans for our future work are implied by the presented results. Firstly, we would like to admit that the test set of term pairs (Table 2) is not big enough to consider the choice of the thresholds fully reliable. Therefore, we will extend the test set in short term and apply a variation of a clustering technique to check our thresholds. Secondly, we would like to explore the ways to improve the performance of the Sørensen-Dice and Jaro-Winkler measures implementations, as their high computational complexity is the only obstacle to outperform the rest of the evaluated SSMs and, possibly, the baseline. To put it more generally, we plan to explore the ways to improve the performance of similar terms grouping, as the times taken by the STG algorithm are too long. Thirdly, we are interested in finding out if a similar terms grouping algorithm, using Sørensen-Dice or Jaro-Winkler, would be plausible for grouping features while building feature taxonomies. This task is on the agenda for the second (Conceptualization) phase of OntoElect [3, 39].

Acknowledgements. The research leading to this publication has been performed in part in cooperation between the Department of Computer Science of Zaporizhzhia National University, the Ontology Engineering Group of the Universidad Politécnica de Madrid, the Applied Probability and Informatics Department at the RUDN University, and Springer-Verlag GmbH. The first author is funded by a PhD grant awarded by Zaporizhzhia National University and the Ministry of Education and Science of Ukraine. The second author is supported by the FPI grant (BES-2017-082511) under the DATOS 4.0: RETOS Y SOLUCIONES - UPM project (TIN2016-78011-C4-4-R) funded by Ministerio de Economía, Industria y Competitividad of Spanish government and EU FEDER funds. The fourth author acknowledges the support of the "RUDN University Program 5-100". The authors would like to acknowledge the contributions by Alyona Chugunenko and Rodion Popov for their research contributions leading to this publication. In particular, they helped develop the approach for term grouping and implement the software for it. The collection of full text Springer journal papers dealing with Knowledge Management, including DMKD-300, has been provided by Springer-Verlag. The authors would also like to express their gratitude to anonymous reviewers whose comments and suggestions helped improve the paper.

References

1. Chugunenko, A., Kosa, V., Popov, R., Chaves-Fraga, D., Ermolayev, V.: Refining terminological saturation using string similarity measures. In: Ermolayev, V., et al. (eds.) Proceedings of the ICTERI 2018. Volume I: Main Conference, Kyiv, Ukraine, 14–17 May 2018, vol. 2105, pp. 3–18. CEUR-WS, online
2. Tatarintseva, O., Ermolayev, V., Keller, B., Matzke, W.-E.: Quantifying ontology fitness in ontoelect using saturation- and vote-based metrics. In: Ermolayev, V., Mayr, H.C., Nikitchenko, M., Spivakovsky, A., Zholtkevych, G. (eds.) ICTERI 2013. CCIS, vol. 412, pp. 136–162. Springer, Cham (2013). https://doi.org/10.1007/978-3-319-03998-5_8
3. Ermolayev, V.: OntoElecting requirements for domain ontologies. The case of time domain. EMISA Int. J. Concept. Model. **13**(Sp. Issue), 86–109 (2018)
4. Fahmi, I., Bouma, G., van der Plas, L.: Improving statistical method using known terms for automatic term extraction. In: Computational Linguistics in the Netherlands, CLIN 17 (2007)
5. Wermter, J., Hahn, U.: Finding new terminology in very large corpora. In: Clark, P., Schreiber, G. (eds.) Proceedings of the 3rd International Conference on Knowledge Capture, K-CAP 2005, pp. 137–144. ACM, Banff (2005)
6. Zhang, Z., Iria, J., Brewster, C., Ciravegna, F.: A comparative evaluation of term recognition algorithms. In: Proceedings of the 6th International Conference on Language Resources and Evaluation, LREC 2008, Marrakech, Morocco (2008)
7. Daille, B.: Study and implementation of combined techniques for automatic extraction of terminology. In: Klavans, J., Resnik, P. (eds.) The Balancing Act: Combining Symbolic and Statistical Approaches to Language, pp. 49–66. The MIT Press, Cambridge (1996)
8. Caraballo, S.A., Charniak, E.: Determining the specificity of nouns from text. In: Proceedings of the 1999 Joint SIGDAT Conference on Empirical Methods in Natural Language Processing and Very Large Corpora, pp. 63–70 (1999)
9. Astrakhantsev, N.: ATR4S: toolkit with state-of-the-art automatic terms recognition methods in scala. arXiv preprint arXiv:1611.07804 (2016)
10. Medelyan, O., Witten, I.H.: Thesaurus based automatic keyphrase indexing. In: Marchionini, G., Nelson, M.L., Marshall, C.C. (eds.) Proceedings of the ACM/IEEE Joint Conf on Digital Libraries, JCDL 2006, pp. 296–297. ACM, Chapel Hill (2006)
11. Ahmad, K., Gillam, L., Tostevin, L.: University of surrey participation in TREC8: weirdness indexing for logical document extrapolation and retrieval (WILDER). In: Proceedings of the 8th Text Retrieval Conference, TREC-8 (1999)
12. Sclano, F., Velardi, P.: TermExtractor: a web application to learn the common terminology of interest groups and research communities. In: Proceedings of the 9th Conference on Terminology and Artificial Intelligence, TIA 2007, Sophia Antipolis, France (2007)
13. Frantzi, K.T., Ananiadou, S.: The C/NC value domain independent method for multi-word term extraction. J. Nat. Lang. Proc. **6**(3), 145–180 (1999)

14. Kozakov, L., Park, Y., Fin, T., Drissi, Y., Doganata, Y., Cofino, T.: Glossary extraction and utilization in the information search and delivery system for IBM Technical Support. IBM Syst. J. **43**(3), 546–563 (2004)
15. Astrakhantsev, N.: Methods and software for terminology extraction from domain-specific text collection. Ph.D. thesis, Institute for System Programming of Russian Academy of Sciences (2015)
16. Bordea, G., Buitelaar, P., Polajnar, T.: Domain-independent term extraction through domain modelling. In: Proceedings of the 10th International Conference on Terminology and Artificial Intelligence, TIA 2013, Paris, France (2013)
17. Badenes-Olmedo, C., Redondo-García, J.L., Corcho, O.: Efficient clustering from distributions over topics. In: Proceedings of the K-CAP 2017, Article 17, 8 p. ACM, New York (2017)
18. Gomaa, W.H., Fahmy, A.A.: A survey of text similarity approaches. Int. J. Comput. Appl. **68**(13), 13–18 (2013)
19. Yu, M., Li, G., Deng, D., Feng, J.: String similarity search and join: a survey. Front. Comput. Sci. **10**(3), 399–417 (2016)
20. Miller, G.A., Beckwith, R., Fellbaum, C.D., Gross, D., Miller, K.: WordNet: an online lexical database. Int. J. Lexicograph. **3**(4), 235–244 (1990)
21. Arnold, M., Ohlebusch, E.: Linear time algorithms for generalizations of the longest common substring problem. Algorithmica **60**(4), 806–818 (2011)
22. Levenshtein, V.I.: Binary codes capable of correcting deletions, insertions, and reversals. Sov. Phys. Dokl. **10**(8), 707–710 (1966)
23. Hamming, R.W.: Error detecting and error correcting codes. Bell Syst. Tech. J. **29**(2), 147–160 (1950)
24. Monger, A., Elkan, C.: The field-matching problem: algorithm and applications. In: Proceedings of the 2nd International Conference on Knowledge Discovery and Data Mining, pp. 267–270. AAAI Press (1996)
25. Jaro, M.A.: Advances in record-linkage methodology as applied to matching the 1985 census of Tampa, Florida. J. Am. Stat. Assoc. **84**(406), 414–420 (1989)
26. Winkler, W.E.: String comparator metrics and enhanced decision rules in the Fellegi-Sunter model of record linkage. In: Proceedings of the Section on Survey Research Methods. ASA, pp. 354–359 (1990)
27. Dice, L.R.: Measures of the amount of ecologic association between species. Ecology **26**(3), 297–302 (1945)
28. Sørensen, T.: A method of establishing groups of equal amplitude in plant sociology based on similarity of species and its application to analyses of the vegetation on Danish commons. Kongelige Danske Videnskabernes Selskab **5**(4), 1–34 (1948)
29. Jaccard, P.: The distribution of the flora in the alpine zone. New Phytol. **11**, 37–50 (1912)
30. Huang, A.: Similarity measures for text document clustering. In: Proceedings of the 6th New Zealand Computer Science Research Student Conference (NZCSRSC2008), Christchurch, New Zealand, pp. 49–56 (2008)
31. Singhal, A.: Modern information retrieval: a brief overview. Bull. the IEEE Comput. Soc. Tech. Comm. Data Eng. **24**(4), 35–43 (2001)
32. Lu, J., Lin, C., Wang, W., Li, C., Wang, H.: String similarity measures and joins with synonyms. In: Proceedings of the 2013 ACM SIGMOD International Conference on the Management of Data, pp. 373–384 (2013)
33. Lee, H., Ng, R.T., Shim, K.: Power-law based estimation of set similarity join size. Proc. VLDB Endow. **2**(1), 658–669 (2009)

34. Tsuruoka, Y., McNaught, J., Tsujii, J., Ananiadou, S.: Learning string similarity measures for gene/protein name dictionary look-up using logistic regression. Bioinformatics **23**(20), 2768–2774 (2007)
35. Qin, J., Wang, W., Lu, Y., Xiao, C., Lin, X.: Efficient exact edit similarity query processing with the asymmetric signature scheme. In: Proceedings of the 2011 ACM SIGMOD International Conference on Management of Data, pp. 1033–1044. ACM, New York (2011)
36. Corcho, O., Gonzalez, R., Badenes, C., Dong, F.: Repository of indexed ROs. Deliverable No. 5.4. Dr Inventor project (2015)
37. Kosa, V., et al.: Cross-evaluation of automated term extraction tools by measuring terminological saturation. In: Bassiliades, N., et al. (eds.) ICTERI 2017. CCIS, vol. 826, pp. 135–163. Springer, Cham (2018). https://doi.org/10.1007/978-3-319-76168-8_7
38. Minkowski, H.: Geometrie der Zahlen. Bibliotheca Mathematica Teubneriana, Band 40 Johnson Reprint Corp., New York-London, 256 pp. (1968). (in German)
39. Moiseenko, S., Ermolayev, V.: Conceptualizing and formalizing requirements for ontology engineering. In: Antoniou, G., Zholtkevych, G. (eds.) Proceedings of the ICTERI 2018 Ph. D. Symposium, Kyiv, Ukraine, 14–17 May, vol. 2122, pp. 35–44. CEUR-WS (2018, online)

Inference Rules for the Partial Floyd-Hoare Logic Based on Composition of Predicate Complement

Ievgen Ivanov and Mykola Nikitchenko(✉)

Taras Shevchenko National University of Kyiv, 64/13, Volodymyrska Street, Kyiv 01601, Ukraine

ivanov.eugen@gmail.com, nikitchenko@unicyb.kiev.ua

Abstract. Classical Floyd-Hoare logic is sound when total pre- and post-conditions are considered. In the case of partial conditions (predicates) the logic becomes unsound. This situation may be corrected by introducing additional constraints to the rules of the logic. But such constraints, in particular, for the sequence and while rules, are rather complicated. In this paper we propose new simpler rules formulated in a program algebra extended with the composition of predicate complement. The obtained logic is called the Complemented Partial Floyd-Hoare Logic (CPFHL). The predicate component of this logic is related to three-valued logic. We prove the soundness theorem for CPFHL and discuss further investigations of the problem. The obtained results can be useful for software verification.

Keywords: Formal methods · Software verification · Partial predicate · Floyd-Hoare logic

1 Introduction

The research presented in this paper[1] is devoted to generalization of the classical Floyd-Hoare logic [2–4] to the case of partial pre- and post-conditions. This logic is a useful tool for proving partial correctness of sequential programs. It is based on properties of a Floyd-Hoare triple (assertion) of the form $\{p\}f\{q\}$ which consists of pre- and post-conditions p, q (predicates) and a program f. This assertion is treated in the following way: when the program's input data d satisfies the pre-condition ($p(d)$ is true), and the program terminates on d, the program's output (the value $f(d)$) satisfies the post-condition ($q(f(d))$ is true).

Although in the classical Floyd-Hoare logic it is assumed that a program may not terminate and its output may be undefined on a particular input d ($f(d)$ is undefined), it is also assumed that the pre- and post-conditions p, q are predicates which are defined on all possible data, i.e. that they are total predicates.

[1] This paper is a refined and extended version of [1].

V. Ermolayev et al. (Eds.): ICTERI 2018, CCIS 1007, pp. 71–88, 2019.
https://doi.org/10.1007/978-3-030-13929-2_4

Partiality of these predicates can arise naturally, if they are expressed using operations which can cause errors, nontermination, or non-well-defined behavior. For example, one may want to consider the Floyd-Hoare triple written in Octave syntax [5]

{`true`} `a(1)=0` {`a(1)==0`}

where `a(1)=0` denotes the operation of assignment of the value 0 to the 1-st element of the array (or a more sophisticated key-value map) `a`, and `a(1)==0` is a predicate stating that the 1-st element of `a` is zero. Logically, it makes sense to consider the latter predicate (the truth value of which depends on the content of `a`) as partial and assume that it is defined only when `a` has an element with index/key 1 (e.g. it is undefined when `a` has no content). Note that depending on the rules of the programming language, the assignment operation may be always well-defined (e.g. if the assignment `a(1)=0` automatically allocates the space for the new value, if it is not allocated), but the post-condition predicate may be partial, because extraction from `a` is not always defined in the sense that it causes an error or another type of abnormal behavior. In particular, a situation of this kind occurs in Matlab [6] and Octave [5] languages for element insertion and extraction operations for vectors (these languages and the corresponding environments are widely used in scientific computing).

In the literature [7] one finds discussions of the ways of emulating partial predicates and functions in software specifications using total predicates and/or functions, however, almost all approaches which try to avoid partiality have some drawbacks which are described in [7–9].

To deal with this issue, in the previous works [10–13] extensions of the classical Floyd-Hoare logic which allowed one to reason about partial correctness of sequential programs using Floyd-Hoare triples with partial programs and partial pre- and post-conditions were investigated.

Here a Floyd-Hoare triple $\{p\}f\{q\}$ means that when the program's input data d satisfies the pre-condition ($p(d)$ is defined and true), and the program terminates on d, and the post-condition is defined on the program's output (the value $q(f(d))$ is defined), then the program's output satisfies the post-condition ($q(f(d))$ is true). We call this interpretation of a triple with partial pre- and post-conditions a *weak Floyd-Hoare triple* (the reason is that it does not require the post-condition to be defined, if the pre-condition is defined; one alternative is to require that the post-condition is defined whenever the pre-condition is defined which we call the *strong Floyd-Hoare triple*, but which we do not consider in this paper). The logic itself will be called Partial Floyd-Hoare Logic (PFHL).

An important fact is that the classical inference system for the Floyd-Hoare logic for the language WHILE [14] is not sound in the presence of partial pre- and post-conditions [10]. One reason of this is unsoundness of the *sequence rule* when p, q, r are partial predicates:

$$R\_SEQ\ \frac{\{p\}\ f\ \{q\}, \{q\}\ g\ \{r\}}{\{p\}\ f \bullet g\ \{r\}}$$

where $f \bullet g$ denotes the sequential composition of programs f and g (i.e. g runs after f on the result of f).

This can be explained on the following simple example in Octave syntax. Let `n` be an integer-valued variable. The expression `zeros(n,1)` evaluates to a `n` $\times$ 1 vector of zeros. If the variable `a` contains a vector, the expression `length(a)` evaluates to the length of `a`. The `i`-th ($i = 1,2,\ldots$) component can be accessed using the expression `a(i)` which raises an error, if the value of `i` is not a valid index, e.g. if the length of `a` is less than the value of `i`. Assignment is denoted as `=`, equality test is denoted as `==`, and comparisons are denoted as `>=`, `>`. Then we can assume that the following assertions are valid (in the sense of weak Floyd-Hoare triples):

```
{n>=0} a=zeros(n,1) {a(i)==0},
{a(i)==0} m=length(a) {m>0}
```

We assume that in the first triple the post-condition is undefined, if the length of `a` is zero (`a` is empty), which happens if `n` is zero. However,

```
{n>=0} a=zeros(n,1); m=length(a) {m>0}
```

is not a valid assertion (in sense of weak Floyd-Hoare triples), since if `n` is zero, then `a` is a zero-length vector (is empty) and `m` is zero.

Because of unsoundness of some rules of the classical inference system in the presence of partial predicates, new inference systems for Partial Floyd-Hoare Logic should be constructed. In our previous works we have considered two ideas of such construction. The first idea consists in restricting the class of assertions to T-increasing assertions [11]; in this case the rules preserve their validity and no other changes are required. The second idea consists in introducing constraints which restrict applicability of the rules of logic. Such a system with constrained rules was proposed in [10]. In this system the regular sequence rule was replaced with the following *constrained sequence rule*:

$$R\_SEQ' \; \frac{\{p\}\; f\; \{q\}, \{q\}\; g\; \{r\}}{\{p\}\; f \bullet g\; \{r\}}, p \models PC(f \bullet g, r)$$

where

- $f \bullet g$ denotes the function $d \mapsto g(f(d))$ which is the result of sequential composition of f and g;
- $p \models q$ means that each interpretation of the formula $\neg p \vee q$ (i.e. $p \to q$) never takes the false value (i.e. it is always either true or undefined);
- PC is the *Predicate transformer composition* [11] such that $PC(f, q)$ is a partial predicate r such that for any data d, $r(d) = q(f(d))$, if $f(d)$ (i.e. program value) and $q(f(d))$ (i.e. the value of the predicate q on the result of f on data d) are defined, and $r(d)$ is undefined otherwise (i.e. if $f(d)$ or $q(f(d))$ are undefined).

In a similar way some other rules were modified.

However, the presence of complicated constraints makes application of such rules difficult in all but the most trivial cases. Therefore in [11] such constraints were even called trifling constraints.

Thus, the above mentioned ideas lead to PFHL limitation (T-increasing assertions) or do not lead to practical inference systems (constrained rules). This implies that further investigation of PFHL is necessary and the inference rules have to be based on some other ideas.

In this paper we propose an extension of the predicate language with a new composition called predicate complement. Introduction of this composition permits us to modify the rules in such a way that they become sound and no constraints are required. The obtained inference system for PFHL based on extended program algebra can be useful for software verification.

2 Notation

The symbol $\tilde{\rightarrow}$ will denote partial functions and $\rightarrow$ will denote total functions (e.g. $f : A \tilde{\rightarrow} B$ means that f is a partial function on a set A with values in a set B; $f : A \rightarrow B$ means that f is a total function from A to B). We will use the symbol $\xrightarrow{n}$ for partial functions with finite graph. For $f : D \tilde{\rightarrow} D'$:

- $f(d) \downarrow$ denotes that f is defined on $d \in D$;
- $f(d) \downarrow = d'$ denotes that f is defined on $d \in D$ and has the value $d' \in D'$;
- $f(d) \uparrow$ denotes that f is undefined on $d \in D$;
- $\mathrm{dom}(f) = \{d \in D \mid f(d) \downarrow\}$ is the domain of a function (note that there are different definitions of the domain of a partial function in different branches of mathematics); we use the convention from recursion theory).

The notation $f_1(d_1) \cong f_2(d_2)$ means the *strong equality*, i.e. that $f_1(d_1) \downarrow$ if and only if $f_2(d_2) \downarrow$, and if $f_1(d_1) \downarrow$, then $f_1(d_1) = f_2(d_2)$.

We use the following notations for predicate $p : D \tilde{\rightarrow} Bool$:

- $p^T = \{d \mid p(d) \downarrow = T\}$ is the truth domain of a predicate p;
- $p^F = \{d \mid p(d) \downarrow = F\}$ is the falsity domain of p.

3 Program Algebras with Predicate Complement

In accordance with the composition-nominative approach [15–18] we consider program logics which are based on program algebras. Such algebras are constructed in the following way:

– first, a set D of data processed by programs is defined; in our case we treat D as hierarchical nominative data [19];

– then, classes of predicates $Pr = D \tilde{\rightarrow} Bool$ and functions $Fn = D \tilde{\rightarrow} D$ are defined;

– at last, operations (compositions) over Pr and Fn are specified.

This scheme leads to two-sorted program algebras. In our previous works [10,11] we considered program algebras with traditional compositions. But the problem of defining unconstrained rules requires new compositions. In this paper we introduce a program algebra extended with a composition of predicate complement.

3.1 Hierarchical Nominative Data

Informally speaking, simple hierarchical nominative data are constructed inductively over sets of basic names V and values A and can be presented as trees, the arcs of which are labeled by elements of V and leafs are labeled by elements of A. Here we consider a more complex case in which names can be presented as sequences of elements of V of the form $v_1v_2...v_n \in V^+$. Introduction of complex names from V^+ requires additional restrictions induced by the *principle of unambiguous associative naming.* This principle demands that for any hierarchical nominative data d with complex names for any two paths $(u_1, u_2, \ldots, u_k)$ and $(v_1, v_2, \ldots, v_l)$ in d, neither of which is a prefix of another, the words $u_1u_2 \ldots u_k$ and $v_1v_2 \ldots v_l$ are incomparable in the sense of the prefix relation. Such data are also called *complex-named data* [20]. An example of such data is $[uv \mapsto 1, w \mapsto [uw \mapsto 3]]$, $u, v, w \in V$. We denote this class of data as $ND_{CC}(V, A)$. Formal definitions can be found in [19,21].

3.2 Basic Operations on Nominative Data

The basic operations on nominative data are the operations of

- *denaming* (taking the value of a name),
- *naming* (assigning a new value to a name),
- *overlapping.*

We define these operations for the class $ND_{CC}(V, A)$ of nominative data with complex names and values ($d \in ND_{CC}(V, A)$).

Definition 1 ([19], Denaming). *The (associative) denaming is an operation $v \Rightarrow_a$ with a parameter $v \in V^+$ defined by induction on the length of v:*

- *if $v \in V$, then* $v \Rightarrow_a (d) \cong \begin{cases} d(v), & \text{if } d(v)\downarrow; \\ d/v, & \text{if } d(v)\uparrow \text{ and } d/v \neq \emptyset; \\ \text{undefined}, & \text{if } d(v)\uparrow \text{ and } d/v = \emptyset, \end{cases}$
  where $d/u = [v_1 \mapsto d(v) \mid d(v)\downarrow, v = uv_1, v_1 \in V^+]$;
- *if $v \in V^+ \backslash V$, then $v \Rightarrow_a (d) \cong v_2 \Rightarrow_a (v_1 \Rightarrow_a (d))$, where v_1 is the first symbol of v and v_2 is the suffix, i.e. v_1, v_2 are (unique) words such that $v = v_1v_2$ and $v_1 \in V$.*

Definition 2 ([19], Naming). *Naming is an unary operation $\Rightarrow v$ with a parameter $v \in V^+$ such that $\Rightarrow v(d) = [v \mapsto d]$.*

Overlapping is a kind of updating operation which updates the values of names in its first argument with the values of names in its second argument. For different types of nominative data different overlapping operations can be considered. Here we will define two kinds of overlapping: global and local overlapping. Global (associative or structural) overlapping ∇_a updates several values in the first argument while the local one ∇_a^v (with a parameter $v \in V^+$) updates only one value which is associated with the name v.

Global overlapping can be used, e.g. for formalizing procedures calls, while the local overlapping can be used as a formalization of the assignment operator in programming languages.

Definition 3 ([19], Global overlapping). *This is a binary operation ∇_a defined by induction on the rank of the first argument as follows.*

Let $ND_{CC_k}(V, A)$ be the class of data with the rank less of equal to k.

Induction base of the definition. If $d_1 \in ND_{CC_0}(V, A)$, then

$$d_1 \nabla_a d_2 \cong \begin{cases} d_2, & \text{if } d_1 = \emptyset \text{ and } d_2 \in ND_{CC}(V, A) \backslash A; \\ \text{undefined}, & \text{if } d_1 \in A \text{ or } d_2 \in A. \end{cases}$$

Induction step of the definition. Assume that the value $d_1 \nabla_a d_2$ is already defined for all d_1, d_2 such that $d_1 \in ND_{CC_k}(V, A)$. Let

$$d_1 \in ND_{CC_{k+1}}(V, A) \backslash ND_{CC_k}(V, A).$$

Then $d_1 \nabla_a d_2 = d$, where d is defined for each name $u \in V^+$ as follows:

(1) $d(u) = d_2(u)$, if $u \in dom(d_2)$ and u does not have a proper prefix which belongs to $dom(d_1)$;
(2) $d(u) = d_1(u) \nabla_a (d_2/u)$, if $d_1(u)$ is defined and does not belong to A and u is a proper prefix of some element of $dom(d_2)$, where
$$d_2/u = [v_1 \mapsto d_2(v) \mid d_2(v) \downarrow, v = uv_1, v_1 \in V^+];$$
(3) $d(u) = d_2/u$, if $d_1(u)$ is defined and belongs to A and u is a proper prefix of some element of $dom(d_2)$;
(4) $d(u) = d_1(u)$, if $d_1(u)$ is defined and u is not comparable (in the sense of the prefix relation) with any element of $dom(d_2)$;
(5) $d(u) \uparrow$, otherwise.

Definition 4 ([19], Local overlapping). *This is a binary operation ∇_a^v with a parameter $v \in V^+$ defined as follows: $d_1 \nabla_a^v d_2 \cong d_1 \nabla_a (\Rightarrow v(d_2))$.*

3.3 Compositions

Now we define compositions of program algebras over nominative data with complex names and values.

Let V and A be fixed sets of basic names and values. Denote
$Pr_{CC}(V, A) = ND_{CC}(V, A) \xrightarrow{\sim} \{T, F\}$,
$Fn_{CC}(V, A) = ND_{CC}(V, A) \xrightarrow{\sim} ND_{CC}(V, A)$.

We will assume that T and F do not belong to $ND_{CC}(V, A)$.

We will call the elements of $Pr_{CC}(V, A)$ (*partial nominative*) *predicates* and the elements of $Fn_{CC}(V, A)$ *(partial nominative) functions.*

Let us denote by $\bar{U}$ the set of all tuples $(u_1, u_2, \ldots, u_n)$, $n \geq 1$ of complex names from V^+ such that whenever $i \neq j$, u_i and u_j are incomparable in the sense of the prefix relation.

In the following definitions $d \in ND_{CC}(V, A)$, $f, g, f_1, \ldots, f_n \in Fn_{CC}(V, A)$, $p, p_1, p_2 \in Pr_{CC}(V, A)$.

Definition 5 (Compositions).

- *Composition of superposition into a predicate*

$$S_{P_{CC}}^{u_1,\ldots,u_n} : Pr_{CC}(V, A) \times (Fn_{CC}(V, A))^n \to Pr_{CC}(V, A)$$

  with a parameter $(u_1, \ldots, u_n) \in \bar{U}$ *is defined as follows:*

$$S_P^{u_1,\ldots,u_n}(p, f_1, \ldots, f_n)(d) \cong p(\ldots(d\nabla_a^{u_1} f_1(d)) \ldots \nabla_a^{u_n} f_n(d)).$$

  We will also use the following notation for this composition: $S_P^{\bar{u}}$.

- *Composition of superposition into a function*

$$S_F^{u_1,\ldots,u_n} : Fn_{CC}(V, A) \times (Fn_{CC}(V, A))^n \to Fn_{CC}(V, A)$$

  with a parameter $(u_1, \ldots, u_n) \in \bar{U}$ *is defined as follows:*

$$S_F^{u_1,\ldots,u_n}(f, f_1, \ldots, f_n)(d) \cong f(\ldots(d\nabla_a^{u_1} f_1(d))\ldots\nabla_a^{u_n} f_n(d)).$$

  We will also use the following notation for this composition: $S_F^{\bar{u}}$.

- *Assignment composition* $AS^u : Fn_{CC}(V, A) \to Fn_{CC}(V, A)$ *with a parameter* $u \in V^+$ *is defined as follows:*

$$(AS^u(f))(d) \cong d\nabla_a^u f(d).$$

- *Sequential composition of functions (denoted using the infix notation)* $\bullet$: $Fn_{CC}(V, A) \times Fn_{CC}(V, A) \to Fn_{CC}(V, A)$ *is defined as follows:*

$$(f \bullet g)(d) \cong g(f(d)).$$

- *Branching composition* $IF : Pr_{CC}(V, A) \times Fn_{CC}(V, A) \times Fn_{CC}(V, A) \to Fn_{CC}(V, A)$ *is defined as follows:*

$$IF(p, f, g)(d) = \begin{cases} f(d), & \text{if } p(d) \downarrow= T \text{ and } f(d) \downarrow; \\ g(d), & \text{if } p(d) \downarrow= F \text{ and } g(d) \downarrow; \\ \text{undefined} & \text{in other cases.} \end{cases}$$

- *Cycle composition* $WH : Pr_{CC}(V, A) \times Fn_{CC}(V, A) \to Fn_{CC}(V, A)$ *is defined as follows:*
  $WH(p, f)(d) \downarrow= f^{(n)}(d)$, *if there exists* $n \geq 0$ *such that* $p(f^{(i)})(d) \downarrow= T$ *for all* $i \in \{0, 1, \ldots, n-1\}$ *and* $p(f^{(n)}p)(d) \downarrow= F$, *where* $f^{(n)}$ *is a n-times sequential composition of* f *with itself* ($f^{(0)}$ *is the identity function), and* $WH(p, f)(d)$ *is undefined otherwise.*

- *Identity null-ary composition (function)* $id : Fn_{CC}(V, A)$ ($id : ND_{CC}(V, A) \to ND_{CC}(V, A)$) *is defined as follows:*
$$id(d) = d.$$
- *Composition of disjunction* $\vee : Pr_{CC}(V, A) \times Pr_{CC}(V, A) \to Pr_{CC}(V, A)$ *is a composition defined as follows:*

$$(p_1 \vee p_2)(d) = \begin{cases} T, & \text{if } p_1(d) \downarrow= T \text{ or } p_2(d) \downarrow= T; \\ F, & \text{if } p_1(d) \downarrow= F \text{ and } p_2(d) \downarrow= F; \\ \text{undefined} & \text{in other cases.} \end{cases}$$

- *Composition of negation* $\neg : Pr_{CC}(V, A) \to Pr_{CC}(V, A)$ *is a composition such that*
$$(\neg p)(d) = \begin{cases} T, & \text{if } p(d) \downarrow= F; \\ F, & \text{if } p(d) \downarrow= T; \\ \text{undefined} & \text{in other cases.} \end{cases}$$

- *Composition of existential quantification over hierarchical data is a unary composition* $Pr_{CC}(V, A) \to Pr_{CC}(V, A)$ *with a parameter* $x \in V^+$ *such that*

$$(\exists x\, p)(d) = \begin{cases} T, \text{ if } p(d \nabla_a^x d') \downarrow= T \text{ for some } d' \in ND_{CC}(V, A), \\ F, \text{ if } p(d \nabla_a^x d') \downarrow= F \text{ for all } d' \in ND_{CC}(V, A), \\ \text{undefined in other cases.} \end{cases}$$

- *Composition of predicate complement is a composition* $\sim: Pr_{CC}(V, A) \to Pr_{CC}(V, A)$ *such that*

$$(\sim p)(d) = \begin{cases} T, & \text{if } p(d) \text{ is undefined;} \\ \text{undefined} & \text{in other cases.} \end{cases}.$$

Having compositions, we can define a special program algebra investigated in this paper.

Definition 6. *A complemented program algebra over hierarchical nominative data with complex names and values* $CPAND_{CC}(V, A)$ *is a two-sorted algebra*
$$CPAND_{CC}(V, A) = (Pr_{CC}(V, A), Fn_{CC}(V, A);$$
$$AS^u, id, \bullet, IF, WH, S_F^{\bar{u}}, S_P^{\bar{u}}, \Rightarrow v, v \Rightarrow_a, \vee, \neg, \exists x, \sim),$$
where $v, u, x \in V^+, \bar{u} \in \bar{U}$.

Derived compositions like conjunction $\wedge$ and universal quantification $\forall x$ are defined in a traditional way.

Let us discuss briefly predicate compositions of this algebra. Operations (compositions) $\vee$, $\neg$, and $\exists x$ are defined according to the truth tables of Kleene's strong logic of indeterminacy [22]. Please note that $\vee$, $\wedge$ and $\neg$ on the set of all partial predicates form a Kleene algebra (a De Morgan algebra which satisfies the normality axiom) [23].

In contrast to these compositions, the composition of predicate complement is more complicated. First, it does not have the monotonicity property, second, it does not have nice distributivity properties. Even more, it is not computable in the sense that the set of partial recursive predicates is not closed under this composition.

Introduction of the composition of predicate complement makes investigation of the corresponding logics more difficult. In this case methods developed for three-valued logics can be used.

Nevertheless, the algebra $(Pr_{CC}(V,A); \vee, \wedge, \neg, \sim)$ has certain properties which make it useful in program partial correctness proofs:

- it can be proven that the algebra $(Pr_{CC}(V,A); \vee, \wedge, \neg, \sim)$ has the same (up to the names of operations) set of identities as the algebra

$$(\{-1,0,1\}; \max(\cdot,\cdot), \min(\cdot,\cdot), x \mapsto -x, x \mapsto 1-|x|);$$

- all scalar functions of n variables expressible in the latter algebra are non-expanding maps from $\{-1,0,1\}^n \to \{-1,0,1\}$ with respect to Chebyshev distance $dist((x_1,\ldots,x_n),(y_1,\ldots,y_n)) = \max_{i=1}^n |x_i - y_i|$.

Definition 7. *A semantic weak Floyd-Hoare triple is a tuple* (p,f,q)*, where* $f: D \tilde{\to} D'$, $p: D \tilde{\to} Bool$, $q: D' \tilde{\to} Bool$ *for some* D, D' *such that for each* $d \in D$*, if* $p(d) \downarrow = T$ *and* $f(d)\downarrow$ *and* $q(f(d))\downarrow$*, then* $q(f(d)) = T$.

We will use the following notation:

- $\{p\}f\{q\}$ means that (p,f,q) is a semantic weak Floyd-Hoare triple.

Please note that a semantic weak Floyd-Hoare triple induces ternary Floyd-Hoare composition $FH : Pr_{CC}(V,A) \times Fn_{CC}(V,A) \times Pr_{CC}(V,A) \to Pr_{CC}(V,A)$ [10,11], but in this paper we do not include it into program algebras in order to not make them overcomplicated.

We start with new inference rules for the sequential composition. These rules are valid for any set of data D.

4 New Inference Rules for Sequential Composition

Theorem 1. *Assume that* $\{p\}f\{q\}$, $\{q\}g\{r_1\}$*, and* $\{\sim q\}g\{r_2\}$.
Then $\{p\}f \bullet g\{r_1 \vee r_2\}$.

Proof. Let $d \in D$. Assume that $p(d) \downarrow = T$, $(f \bullet g)(d)\downarrow$, and $(r_1 \vee r_2)((f \bullet g)(d))\downarrow$.
Then $f(d)\downarrow$ and $g(f(d))\downarrow$. Denote $d' = f(d)$ and $d'' = (f \bullet g)(d) = g(f(d))$.
Let us show that $(r_1 \vee r_2)(d'') = T$.

Suppose that $(r_1 \vee r_2)(d'') \downarrow = F$. Then $r_1(d'') \downarrow = F$ and $r_2(d'') \downarrow = F$. We have that either $q(d')\downarrow$, or $q(d')\uparrow$.

Consider the case when $q(d')\downarrow$. Then $q(f(d))\downarrow$, and since $p(d)\downarrow = T$ and $\{p\}f\{q\}$, we have $q(f(d)) = q(d') = T$. Then since $r_1(g(d')) \cong r_1(d'')\downarrow$ and

$\{q\}g\{r_1\}$, we have $r_1(g(d')) = r_1(d'') = T$, but this contradicts the above mentioned statement $r_1(d'') = F$.

Consider the case when $q(d') \uparrow$. Then $(\sim q)(d') \downarrow= T$. Then since $r_2(g(d')) \cong r_2(d'') \downarrow$ and $\{\sim q\}g\{r_2\}$, we have $r_2(g(d')) = r_2(d'') = T$, but this contradicts the above mentioned statement $r_2(d'') = F$.

In both cases we have a contradiction, so $(r_1 \vee r_2)(d'') \downarrow= F$ is impossible. But $(r_1 \vee r_2)(d'') \cong (r_1 \vee r_2)((f \bullet g)(d)) \downarrow$ by the assumption, so $(r_1 \vee r_2)(d'') = T$. □

This result can be used as a semantic foundation of the following unconstrained inference rule for sequential composition for the inference system for Floyd-Hoare logic with partial pre- and post-conditions (which involves the $\sim$ operation on predicates):

$$R\_USEQ \frac{\{p\}\ f\ \{q\}, \{q\}\ g\ \{r_1\}, \{\sim q\}g\{r_2\}}{\{p\}\ f \bullet g\ \{r_1 \vee r_2\}}$$

In the special case of coinciding r_1 and r_2, it can be rewritten as:

$$R\_SSEQ \frac{\{p\}\ f\ \{q\}, \{q\}\ g\ \{r\}, \{\sim q\}g\{r\}}{\{p\}\ f \bullet g\ \{r\}}$$

Theorem 1 implies that addition of the rules R_USEQ and/or R_SSEQ to the inference system AC proposed in [10,11] (with the proper extension of syntax of pre- and post-condition predicate formulas to accommodate the symbol of $\sim$ operation) does not change its soundness.

As an informal example, consider how these rules can be applied in the case mentioned in the Introduction:

```
{n>=0} a=zeros(n,1) {a(1)==0},
{a(1)==0} m=length(a) {m>0}.
```

These two assumptions alone are not sufficient to establish the triple concerning the sequential composition. A missing piece of information is the triple describing the behavior of the instruction `m=length(a)` when the predicate `(a(1)==0)` is undefined. Under the interpretation assumed in the Introduction, this undefinedness means that an attempt of evaluation of `(a(1)==0)` leads to an abnormal/error state. If `a` is a defined vector, this happens exactly when `a` has zero length (is empty). If `a` is undefined, then an attempt of evaluation of `length(a)` causes an error. Thus we can state that

```
{~(a(1)==0)} m=length(a) {m=0},
```

where $\sim$ is not a part of Octave syntax, but just a notation to represent the statement that the expression `(a(1)==0)` is undefined (causes an abnormal/error state). Thus by the R_USEQ rule:

```
{n>=0} a=zeros(n,1); m=length(a) {m>0 ∨ m=0}.
```

Again, here $\vee$ is not a part of Octave syntax, but a notation to represent the disjunction of two predicates.

5 New Inference Rule for Cycle Composition

Let r be a partial predicate on D and $f : D \tilde{\to} D$.

Recall that the cycle composition $WH(r, f)$ is defined as follows: it returns a function in $D \tilde{\to} D$ such that for each $d \in D$,

$$WH(r, f)(d) \downarrow = f^{(n)}(d),$$

if there exists an integer $n \geq 0$ such that $r(f^{(i)}(d)) \downarrow = T$ for all $i = 0, 1, \ldots, n-1$ and $r(f^{(n)}(d)) \downarrow = F$, where $f^{(i)}$ denotes $\underbrace{f \bullet f \bullet \ldots \bullet f}_{i}$ and $f^{(0)}$ is the identity function on D (i.e. $f^{(0)}(d') = d'$ for all $d' \in D$);
and $WH(r, f)(d) \uparrow$, otherwise.

WH is intended to capture the semantics of the loop of the form `while_do_` in imperative programming languages which support structured programming. Here r represents the semantics of the loop condition and f represents the semantics of the loop body.

In terms of WH the loop rule for the classical inference system for the Floyd-Hoare logic with total pre- and post-conditions [14] can be reformulated as follows [24, p. 15]:

$$R\_WH \ \frac{\{r \wedge p\} \ f \ \{p\}}{\{p\} \ WH(r, f) \ \{\neg r \wedge p\}}$$

Here p represents the loop invariant.

This rule, generally, is not valid in the case of partial pre- and post-conditions, but can be replaced with a constrained rule [24, p. 16] (R_WH') in this case.

Applying the approach which we used in the statement and proof of Theorem 1, we can propose an alternative unconstrained rule for the while loop which is more convenient to apply.

Theorem 2. *Assume* $\{r \wedge p\} f \{p\}$, $\{r \wedge (\sim p)\} f \{p\}$. *Then* $\{p\} WH(r, f) \{\neg r \wedge p\}$.

Proof. Let $d \in D$. Assume that

$p(d) \downarrow = T$, $WH(r, f)(d) \downarrow = d'$, and $(\neg r \wedge p)(WH(r, f)(d)) \downarrow$.

Let us show that $(\neg r \wedge p)(WH(r, f)(d)) = (\neg r \wedge p)(d') = T$.

From the definition on WH it follows that there exists an integer $n \geq 0$ such that $r(f^{(i)}(d)) \downarrow = T$ for all $i = 0, 1, \ldots, n-1$, and $r(f^{(n)}(d)) \downarrow = F$, and

$$d' = WH(r, f)(d) = f^{(n)}(d).$$

If $n = 0$, then $d' = d$, so $r(d') \downarrow = F$ and $p(d') \downarrow = T$, whence $(\neg r \wedge p)(d') = T$.

Now we will assume that $n \geq 1$.

Let us show by induction on $i \in \{0, 1, \ldots\}$ that if $i \in \{0, 1, \ldots, n-1\}$, then $p(f^{(i)}(d)) \downarrow = T$ or $p(f^{(i)}(d)) \uparrow$.

Base of induction: $p(f^{(0)}(d)) \cong p(d) \downarrow = T$, so the statement holds.

Inductive step. Assume that $i \in \{0, 1, \ldots, n-1\}$. Assume that $p(f^{(i)}(d)) \downarrow= T$ or $p(f^{(i)}(d)) \uparrow$. Assume that $i+1 \in \{0, 1, \ldots, n-1\}$. Note that $r(f^{(i)}(d)) \downarrow= T$ because $i < n$. Denote $d_1 = f^{(i)}(d)$.

Consider the case $p(f^{(i)}(d)) \downarrow= T$. Then since $r(d_1) \downarrow= T$ and $p(d_1) \downarrow= T$, we have $(r \wedge p)(d_1) \downarrow= T$. If $p(f(d_1)) \downarrow$, then since $\{r \wedge p\} f \{p\}$, we have $p(f^{(i+1)}(d)) \cong p(f(d_1)) \downarrow= T$. Otherwise, $p(f(d_1)) \uparrow$. In either case, either $p(f^{(i+1)}(d)) \downarrow= T$, or $p(f^{(i+1)}(d)) \uparrow$ holds.

Consider the case $p(f^{(i)}(d)) \uparrow$. Then $p(d_1) \uparrow$ and $r(d_1) \downarrow= T$. Further, $(\sim p)(d_1) \downarrow= T$, so $(r \wedge (\sim p))(d_1) \downarrow= T$. Since $\{r \wedge (\sim p)\} f \{p\}$, we have either $p(f^{(i+1)}(d)) \cong p(f(d_1)) \downarrow= T$, or $p(f^{(i+1)}(d)) \cong p(f(d_1)) \uparrow$.

In both cases $p(f^{(i+1)}(d)) \downarrow= T$ or $p(f^{(i+1)}(d)) \uparrow$, so the inductive step is completed.

Since $n \geq 1$ by our assumption, the proven statement implies that either $p(f^{(n-1)}(d)) \downarrow= T$, or $p(f^{(n-1)}(d)) \uparrow$. We have

$$(\neg r \wedge p)(f^{(n)}(d)) \cong (\neg r \wedge p)(WH(r, f)(d)) \downarrow.$$

Moreover, $r(f^{(n)}(d)) \downarrow= F$, so $(\neg r)(f^{(n)}(d)) \downarrow= T$, whence $p(f^{(n)}(d)) \downarrow$.

Consider the case $p(f^{(n-1)}(d)) \downarrow= T$. We have $r(f^{(n-1)}(d)) \downarrow= T$, therefore $(r \wedge p)(f^{(n-1)}(d)) \downarrow= T$. Then since $\{r \wedge p\} f \{p\}$ and $p(f(f^{(n-1)}(d))) \cong p(f^{(n)}(d)) \downarrow$, we have $p(f^{(n)}(d)) = T$.

Consider the case $p(f^{(n-1)}(d)) \uparrow$. Then because $f^{(n-1)}(d) \downarrow$, we have $(\sim p)(f^{(n-1)}(d)) \downarrow= T$. Moreover, we have $r(f^{(n-1)}(d)) \downarrow= T$, whence $(r \wedge (\sim p))(f^{(n-1)}(d)) \downarrow= T$. Then because $\{r \wedge (\sim p)\} f \{p\}$ and, moreover, $p(f(f^{(n-1)}(d))) \cong p(f^{(n)}(d)) \downarrow$, we have $p(f^{(n)}(d)) = T$.

In both cases we have $p(f^{(n)}(d)) = T$. Since $r(f^{(n)}(d)) \downarrow= F$, we have $(\neg r \wedge p)(WH(r, f)(d)) \cong (\neg r \wedge p)(f^{(n)}(d)) \downarrow= T$. □

This result can be used as a semantic foundation of the following unconstrained inference rule (for the case of partial pre- and post-conditions):

$$R\_UWH \frac{\{r \wedge p\}\ f\ \{p\}, \{r \wedge (\sim p)\} f \{p\}}{\{p\}\ WH(r, f)\ \{\neg r \wedge p\}}$$

6 Syntax and Interpretation of Complemented Partial Floyd-Hoare Logic

Algebra $CPAND_{CC}(V, A)$ has strong expressive power that is not required for our goal: to construct a special program logic. Therefore we restrict syntactically the class of terms of this algebra. The idea is to consider programs as special nominative functions (program functions) constructed with the help of compositions AS^x, id, $\bullet$, IF, WH, $S_F^{\bar{x}}$. Functions of other types can be represented by functional expressions. Formulas represent partial predicates over nominative data. The signature of the constructed logic is $\Sigma = (V, Ps, FEs, Prgs)$ where $Ps, FEs, Prgs$ are sets of predicate, function, and program symbols respectively.

Let us give definitions of the sets of formulas Fr^{Σ}, functional expressions FEx^{Σ}, program texts Pt^{Σ}, and Floyd-Hoare assertions $FHFr^{\Sigma}$.

The sets Fr^{Σ}, FE^{Σ}, and Pt^{Σ} are defined inductively (here we use the symbols of compositions, predicates and functions in the purely syntactic sense, i.e. they are currently not associated with semantics):

1. if $ps \in Ps$, then $ps \in Fr^{\Sigma}$;
2. if $fes \in FEs$, then $fes \in FE^{\Sigma}$;
3. if $prgs \in Prgs$, then $prgs \in Pt^{\Sigma}$;
4. if $\Phi, \Psi \in Fr^{\Sigma}$, then $\Phi \vee \Psi$, $\neg\Phi$, $\sim \Phi$, $\exists x\Phi \in Fr^{\Sigma}$;
5. $\Rightarrow v$, $v \Rightarrow_a \in FE^{\Sigma}$;
6. if $n \geq 1$, $\Phi \in Fr^{\Sigma}$, $fe_1, \ldots, fe_n \in FE^{\Sigma}$, and $\bar{x} \in \bar{U}$, then $S_P^{\bar{x}}(\Phi, fe_1, \ldots, fe_n) \in Fr^{\Sigma}$;
7. if $n \geq 1$, fe, $fe_1, \ldots, fe_n \in FE^{\Sigma}$, and $\bar{x} \in \bar{U}$, then $S_F^{\bar{x}}(fe, fe_1, \ldots, fe_n) \in FE^{\Sigma}$;
8. if $n \geq 1$, $prg \in Pt^{\Sigma}$, $fe_1, \ldots, fe_n \in FE^{\Sigma}$, and $\bar{x} \in \bar{U}$, then $S_F^{\bar{x}}(prg, fe_1, \ldots, fe_n) \in Pt^{\Sigma}$;
9. if $x \in V^+$ and $fe \in FE^{\Sigma}$, then $AS^x(fe) \in Pt^{\Sigma}$;
10. $id \in Pt^{\Sigma}$;
11. if $prg_1, prg_2 \in Pt^{\Sigma}$, then $pr_1 \bullet pr_2 \in Pt^{\Sigma}$;
12. if $\Phi \in Fr^{\Sigma}$ and $prg_1, prg_2 \in Pt^{\Sigma}$, then $IF(\Phi, prg_1, prg_2) \in Pt^{\Sigma}$;
13. if $\Phi \in Fr^{\Sigma}$ and $prg \in Pt^{\Sigma}$, then $WH(\Phi, prg) \in Pt^{\Sigma}$.

To avoid syntactical nondeterminism, parentheses can be used.

The set $FHFr^{\Sigma}$ is the set of all formulas of the form $\{p\}f\{q\}$, where p, $q \in Fr^{\Sigma}$ and $f \in Pt^{\Sigma}$.

Please note that we often use the same notation both for predicates and for formulas, e.g. depending on the context, p can be treated as a predicate or as a function; the same concerns functions and functional expressions.

Definition 8. *Let $\Sigma = (V, Ps, FEs, Prgs)$ be a logic signature and A be a set. Then an interpretation J is a tuple $(CPAND_{CC}(V, A), I_{Ps}, I_{FEs}, I_{Prgs})$, where $I_{Ps} : Ps \rightarrow Pr_{CC}(V, A)$ is an interpretation mapping for predicate symbols, $I_{FEs} : FEs \rightarrow Fn_{CC}(V, A)$ and $I_{Prs} : Prs \rightarrow Fn_{CC}(V, A)$ are interpretation mappings for function and program symbols, respectively.*

For any interpretation $J = (CPAND_{CC}(V, A), I_{Ps}, I_{FEs}, I_{Prgs})$ we denote by J_{Fr}, J_{FE}, and J_{Pt} the *formula, function, and program text interpretation mappings*

$$J_{Fr} : Fr^{\Sigma} \rightarrow Pr_{CC}(V, A),$$

$$J_{FE} : FE^{\Sigma} \rightarrow Fn_{CC}(V, A),$$

$$J_{Pt} : Pt^{\Sigma} \rightarrow Fn_{CC}(V, A),$$

which are the standard extensions of I_{Ps}, I_{FEs}, and I_{Prgs} to Fr^{Σ}, FE^{Σ}, and Pt^{Σ} respectively (defined by structural induction). Also, we denote by J_{FHFr} the *interpretation mapping of Floyd-Hoare assertions* $J_{FHFr} : FHFr^{\Sigma} \longrightarrow Pr_{CC}(V, A)$ defined as follows:

$$J_{FHFr}(\{p\}f\{q\}) = FH(J_{Fr}(p), J_{Pt}(f), J_{Fr}(q)).$$

Here we will not define interpretations explicitly expecting that they are clear from the context. For any $P \in Fr^{\Sigma}$ or $P \in FHFr^{\Sigma}$ we will denote by P_J or $(P)_J$ the predicate that corresponds to P under interpretation J. We will omit the index J when it is clear from the context.

Definition 9. *A formula $P \in Fr^{\Sigma}$ or a Floyd-Hoare assertion $P \in FHFr^{\Sigma}$ is valid (irrefutable) in an interpretation J (denoted as $J \models P$), if $P_J^F = \emptyset$.*

Definition 10. *A formula $P \in Fr^{\Sigma}$ or a Floyd-Hoare assertion $P \in FHFr^{\Sigma}$ is logically valid (denoted as $\models P$), if it is valid in every interpretation.*

We will also need *special logical truth-consequence and falsity-consequence relations* [10] $\models_T, \models_F \subseteq Fr^{\Sigma} \times Fr^{\Sigma}$.

Definition 11. *A formula $Q \in Fr^{\Sigma}$ is a truth-consequence of a formula $P \in Fr^{\Sigma}$ in an interpretation J (denoted as $P_J \models_T Q$), if $P_J^T \subseteq Q_J^T$. A formula $Q \in Fr^{\Sigma}$ is a logical truth-consequence of a formula $P \in Fr^{\Sigma}$ (denoted as $P \models_T Q$), if $P_J \models_T Q$ for every interpretation J.*

Definition 12. *A formula $Q \in Fr^{\Sigma}$ is a falsity-consequence of a formula $P \in Fr^{\Sigma}$ in an interpretation J (denoted as $P_J \models_F Q$), if $P_J^F \supseteq Q_J^F$. A formula $Q \in Fr^{\Sigma}$ is a logical falsity-consequence of a formula $P \in Fr^{\Sigma}$ (denoted as $P \models_F Q$), if $P_J \models_F Q$ for every interpretation J.*

7 Inference System for a Complemented Partial Floyd-Hoare Logic

To make the program logic CPFHL which we have defined applicable to software verification problems it is necessary to present an inference system. Such an inference system could be based on the inference system for the classical Floyd-Hoare logic with total predicates for the language WHILE [14], but it is known to be unsound in the case of partial predicates [11] which is considered in the paper. For this reason we present new inference rules based on program algebras with the composition of predicate complement. Obtained system will be sound.

We will write $\vdash_X p$ to denote that a formula p is *derived* in some inference system X. An inference system X is *sound*, if $\vdash_X p \Rightarrow \ \models p$ for each formula p, and is *complete*, if $\models p \Rightarrow \vdash_X p$ for each p.

Taking into consideration the obtained results we write the following inference rules ($v, x \in V^+, \bar{x} \in \bar{U}$, $p, p', q, q', r \in Fr^{\Sigma}$, $h, g_1, \ldots, g_n \in FE^{\Sigma}$, $f, g \in Pt^{\Sigma}$):

$$R\_AS \ \frac{}{\{S_P^x(p,h)\}\ AS^x(h)\ \{p\}}$$

$$R\_SKIP \ \frac{}{\{p\}\ id\ \{p\}}$$

$$R\_SSEQ \ \frac{\{p\}\ f\ \{q\}, \{q\}\ g\ \{r\}, \{\sim q\} g \{r\}}{\{p\}\ f \bullet g\ \{r\}}$$

$$R\_IF \ \frac{\{r \wedge p\}\ f\ \{q\}, \{\neg r \wedge p\}\ g\ \{q\}}{\{p\}\ IF(r, f, g)\ \{q\}}$$

$$R\_UWH \ \frac{\{r \wedge p\}\ f\ \{p\}, \{r \wedge (\sim p)\} f \{p\}}{\{p\}\ WH(r, f)\ \{\neg r \wedge p\}}$$

$$R\_SFID \ \frac{}{\{S_P^{\bar{x}}(p, g_1, \ldots, g_n)\} S_F^{\bar{x}}(id, g_1, \ldots, g_n)\{p\}}$$

$$R\_SF \ \frac{\{p\} S_F^{\bar{x}}(id, g_1, \ldots, g_n) \bullet f\{q\}}{\{p\} S_F^{\bar{x}}(f, g_1, \ldots, g_n)\{q\}}$$

$$R\_CONS_{TF} \ \frac{\{p'\}\ f\ \{q'\}}{\{p\}\ f\ \{q\}}, p \models_T p', q' \models_F q$$

Let us make the following comments to these rules:

- rules R_AS, R_SFID and R_SF are the only rules oriented on the class of nominative data with complex names and values; other rules can be considered for any class of data D;
- rules R_SKIP and R_IF are traditional rules for Floyd-Hoare logics; they do not require any changes;
- rules R_SSEQ and R_UWH were proposed and investigated in the previous sections;
- rules R_SFID and R_SF specify procedure calls;
- consequence rules can be formulated in different forms; here we use the rule R_CONS_{TF} based on special consequence relations $\models_T$ and $\models_F$. In the case of total predicates this rule will be equivalent to traditional consequence rule.

We denote the inference system presented by the above rules as RCN.

Theorem 3. *The inference system RCN is sound, i.e. for any Floyd-Hoare assertion* $P \in FHFr^{\Sigma}$ *we have that*

$$\vdash_{RCN} P \Rightarrow \models P.$$

Proof. We prove the theorem by induction on the length of inference of P. Let $J = (CPAND_{CC}(V, A), I_{Ps}, I_{FEs}, I_{Prgs})$.

- Consider the case when P has the form $\{S_P^x(p, h)\}\ AS^x(h)\ \{p\}$ and P is inferred in RCN, i.e. $\vdash_{RCN} P$ $(q, p \in Fr^{\Sigma}, h \in FE^{\Sigma})$ by rule R_AS.
  Given an interpretation J we should prove that $J \models \{S_P^x(p, h)\}\ AS^x(h)\ \{p\}$. This means (by the definition of weak Floyd-Hoare triple) that we should prove the following statement: for any $d, d' \in ND_{CC}(V, A)$ if $S_P^x(p, h)_J(d) \downarrow= T$, $AS^x(h)_J(d) \downarrow= d'$, and $p_J(d') \downarrow$ then $p_J(d') = T$.
  By definition of the composition of superposition into a predicate we have that $S_P^x(p, h)_J(d) = p_J(d \nabla_a^x h_J(d))$. By definition of assignment composition we have that $AS^x(h)_J(d) = d \nabla_a^x h_J(d)$. Since $p_J(d \nabla_a^x h_J(d)) = T$ and $d \nabla_a^x h_J(d) = d'$ we obtain that $p_J(d') = T$.
- The case when P has the form $\{p\}\ id\ \{p\}$ i.e. the rule R_SKIP is used to infer P is trivial.

- The case when P is obtained by rule R_SSEQ has been proved in Sect. 4 of this paper.
- The case when P is obtained by rule R_IF is a traditional one.
- The case when P is obtained by rule R_UWH has been proved in Sect. 5 of this paper.
- Consider the case when P is obtained by rule R_SFID. It means that P has the form $\{S_P^{\bar{x}}(p, g_1, \ldots, g_n)\}S_F^{\bar{x}}(id, g_1, \ldots, g_n)\{p\}$ $(\bar{x} = (x_1, \ldots, x_n)$.
  Given an interpretation J we should prove that
  $$J \models \{S_P^{\bar{x}}(p, g_1, \ldots, g_n)\}S_F^{\bar{x}}(id, g_1, \ldots, g_n)\{p\}.$$
  This means (by the definition of weak Floyd-Hoare triple) that we should prove the following statement:
  for any $d, d' \in ND_{CC}(V, A)$ if $S_P^{\bar{x}}(p, g_1, \ldots, g_n)_J(d) \downarrow = T$, $S_F^{\bar{x}}(id, g_1, \ldots, g_n)_J(d) \downarrow = d'$, and $p_J(d') \downarrow$ then $p_J(d') = T$.
  By the definition of the composition of superposition into a predicate we have that $S_P^{x_1,\ldots,x_n}(p, g_1, \ldots, g_n)_J(d) = p_J(\ldots(d\nabla_a^{x_1} g_{1J}(d))\ldots\nabla_a^{x_n} g_{nJ}(d))$. Obtained value is equal to T.
  By the definition of the composition of superposition into a function we have that $S_F^{x_1,\ldots,x_n}(id, g_1, \ldots, g_n)_J(d) = id_J(\ldots(d\nabla_a^{x_1} g_{1J}(d))\ldots\nabla_a^{x_n} g_{nJ}(d)) = d'$. Therefore, $p_J(d') \downarrow = T$.
- Consider the case when P is obtained by rule R_SF. In this case P has the form $\{p\}S_F^{\bar{x}}(f, g_1, \ldots, g_n)\{q\}$. Given an interpretation J we should prove that $J \models \{p\}S_F^{\bar{x}}(f, g_1, \ldots, g_n)\{q\}$ under assumption $J \models \{p\}S_F^{\bar{x}}(id, g_1, \ldots, g_n) \bullet f\{q\}$. It means that we should prove the following statement:
  for any $d, d' \in ND_{CC}(V, A)$ if $p_J(d) \downarrow = T$, $S_F^{\bar{x}}(f, g_1, \ldots, g_n)_J(d) \downarrow = d'$, and $q_J(d') \downarrow$ then $q_J(d') = T$ using inductive hypothesis.
  First, let us prove that $(S_F^{\bar{x}}(id, g_1, \ldots, g_n) \bullet f)_J(d) \cong S_F^{\bar{x}}(f, g_1, \ldots, g_n)_J(d)$ for any $d \in ND_{CC}(V, A)$.
  Indeed, $(S_F^{\bar{x}}(id, g_1, \ldots, g_n) \bullet f)_J(d) \cong f_J(S_F^{\bar{x}}(id, g_1, \ldots, g_n)_J(d)) \cong$
  $\cong f_J(id_J(\ldots(d\nabla_a^{u_1} g_{1J}(d))\ldots\nabla_a^{u_n} g_{nJ}(d))) \cong$
  $\cong f_J(\ldots(d\nabla_a^{u_1} g_{1J}(d))\ldots\nabla_a^{u_n} g_{nJ}(d)) \cong S_F^{\bar{x}}(f, g_1, \ldots, g_n)_J(d)$.
  Let $p_J(d) \downarrow = T$, $S_F^{\bar{x}}(f, g_1, \ldots, g_n)_J(d) \downarrow = d'$, and $q_J(d') \downarrow$.
  Since $(S_F^{\bar{x}}(id, g_1, \ldots, g_n) \bullet f)_J(d) \cong S_F^{\bar{x}}(f, g_1, \ldots, g_n)_J(d)$ we have that $(S_F^{\bar{x}}(id, g_1, \ldots, g_n) \bullet f)_J(d) \downarrow = d'$.
  Then, by induction hypothesis for $\{p\}S_F^{\bar{x}}(id, g_1, \ldots, g_n) \bullet f\{q\}$ we obtain the required property $q_J(d') \downarrow = T$.
- Consider the case when P is obtained by rule R_CONS_{TF}. It means that P has the form $\{p\}f\{q\}$. Given an interpretation J we should prove $J \models \{p\}f\{q\}$ under assumptions $J \models \{p'\}f\{q'\}$, $p_J \models_T p'$ and $q'_J \models_F q$.
  Indeed, let $d \in ND_{CC}(V, A)$. Assume that $p_J(d) \downarrow = T$, $f_J(d) \downarrow = d'$ and $q_J(d') \downarrow$ for some $d' \in ND_{CC}(V, A)$. Since $p_J \models_T p'$ we have $p'_J(d) \downarrow = T$. Since $f_J(d) \downarrow = d'$ and $J \models \{p'\}f\{q'\}$ we have $q'_J(d') = T$ when $q'_J(d') \downarrow$. Assume that $q_J(d') = F$. Since $q'_J \models_F q$ we should have $q'_J(d') = F$. We have a contradiction with $q'_J(d') = T$. Therefore, $q_J(d') = T$.

□

In the system RCN new unconventional consequence relations $\models_T$ and $\models_F$ were used. Their main semantic properties were studied in [25]. Further investigation will permit one to substitute these consequence relations by the corresponding inference relations $\vdash_T$ and $\vdash_F$. A detailed investigation of inference methods for CPFHL is planned for the forthcoming publications.

8 Conclusion

We have proposed a modified inference system for an extended Floyd-Hoare logic for partial pre- and post-conditions and partial programs studied in [10, 11, 26]. The modifications primarily concern the sequence and while rules and have been formulated in program algebras extended with the composition of predicate complement. The addition of these rules does not change the soundness of the system. Moreover, the new rules have no semantic constraints. The obtained results can be useful for verification of programs with respect to specifications which can contain partial operations.

In the future we plan to make detailed comparison of inference systems and propose new modifications to improve their efficiency.

References

1. Ivanov, I., Nikitchenko, M.: On the sequence rule for the Floyd-Hoare logic with partial pre-and post-conditions. In: CEUR Workshop Proceedings. Proceedings of the 14th International Conference on ICT in Education, Research and Industrial Applications. Integration, Harmonization and Knowledge Transfer. Volume II: Workshops, Kyiv, Ukraine, 14–17 May 2018, vol. 2104, pp. 716–724 (2018)
2. Floyd, R.: Assigning meanings to programs. In: Mathematical Aspects of Computer Science, vol. 19, pp. 19–32 (1967)
3. Hoare, C.: An axiomatic basis for computer programming. Commun. ACM **12**(10), 576–580 (1969)
4. Apt, K.: Ten years of Hoare's logic: a survey - part I. ACM Trans. Program. Lang. Syst. **3**(4), 431–483 (1981)
5. GNU: Octave. https://www.gnu.org/software/octave/
6. MathWorks: MATLAB. https://www.mathworks.com/products/matlab.html
7. Jones, C.: Reasoning about partial functions in the formal development of programs. Electron. Notes Theor. Comput. Sci. **145**, 3–25 (2006). Proceedings of AVoCS 2005. Elsevier
8. Hähnle, R.: Many-valued logic, partiality, and abstraction in formal specification languages. Logic J. IGPL **13**(4), 415–433 (2005)
9. Gries, D., Schneider, F.: Avoiding the undefined by underspecification. Technical report, Ithaca, NY, USA (1995)
10. Nikitchenko, M., Kryvolap, A.: Properties of inference systems for Floyd-Hoare logic with partial predicates. Acta Electrotechnica et Informatica **13**(4), 70–78 (2013)
11. Kryvolap, A., Nikitchenko, M., Schreiner, W.: Extending Floyd-Hoare logic for partial pre- and postconditions. In: Ermolayev, V., Mayr, H.C., Nikitchenko, M., Spivakovsky, A., Zholtkevych, G. (eds.) ICTERI 2013. CCIS, vol. 412, pp. 355–378. Springer, Cham (2013). https://doi.org/10.1007/978-3-319-03998-5_18

12. Kornilowicz, A., Kryvolap, A., Nikitchenko, M., Ivanov, I.: Formalization of the algebra of nominative data in Mizar. In: Ganzha, M., Maciaszek, L.A., Paprzycki, M. (eds.) Proceedings of the 2017 Federated Conference on Computer Science and Information Systems, FedCSIS 2017, Prague, Czech Republic, 3–6 September 2017, pp. 237–244 (2017)
13. Korniłowicz, A., Kryvolap, A., Nikitchenko, M., Ivanov, I.: Formalization of the nominative algorithmic algebra in Mizar. In: Świątek, J., Borzemski, L., Wilimowska, Z. (eds.) ISAT 2017. AISC, vol. 656, pp. 176–186. Springer, Cham (2018). https://doi.org/10.1007/978-3-319-67229-8_16
14. Nielson, H., Nielson, F.: Semantics with Applications - a Formal Introduction. Wiley, Hoboken (1992). Wiley professional computing
15. Nikitchenko, N.S.: A composition nominative approach to program semantics. Technical report, IT-TR 1998–020, Technical University of Denmark (1998)
16. Skobelev, V., Ivanov, I., Nikitchenko, M.: Nominative data with ordered set of names. Comput. Sci. J. Moldova **25**, 195–216 (2017)
17. Ivanov, I.: On representations of abstract systems with partial inputs and outputs. In: Gopal, T.V., Agrawal, M., Li, A., Cooper, S.B. (eds.) TAMC 2014. LNCS, vol. 8402, pp. 104–123. Springer, Cham (2014). https://doi.org/10.1007/978-3-319-06089-7_8
18. Ivanov, I.: An abstract block formalism for engineering systems. In: Ermolayev, V., et al. (eds.) Proceedings of the 9th International Conference on ICT in Education, Research and Industrial Applications: Integration, Harmonization and Knowledge Transfer, CEUR Workshop Proceedings, Kherson, Ukraine, 19–22 June 2013, vol. 1000, pp. 448–463. CEUR-WS.org (2013)
19. Skobelev, V., Nikitchenko, M., Ivanov, I.: On algebraic properties of nominative data and functions. In: Ermolayev, V., Mayr, H., Nikitchenko, M., Spivakovsky, A., Zholtkevych, G. (eds.) ICTERI 2014. CCIS, vol. 469, pp. 117–138. Springer, Cham (2014). https://doi.org/10.1007/978-3-319-13206-8_6
20. Nikitchenko, M., Ivanov, I.: Composition-nominative languages of programs with associative denaming. Visnyk (Bull.) Lviv Univ. Ser. Appl. Math. Inform. **16**, 124–139 (2010)
21. Nikitchenko, M., Ivanov, I., Skobelev, V.: Proving properties of programs on hierarchical nominative data. Comput. Sci. J. Moldova **24**(3(72)), 371–398 (2016)
22. Kleene, S.: Introduction to Metamathematics. North-Holland Publishing Co. and P. Noordhoff, Amsterdam and Groningen (1952)
23. Korniłowicz, A., Ivanov, I., Nikitchenko, M.: Kleene algebra of partial predicates. Formalized Math. **26**, 11–20 (2018)
24. Korniłowicz, A., Kryvolap, A., Nikitchenko, M., Ivanov, I.: An approach to formalization of an extension of Floyd-Hoare logic. In: Proceedings of the 13th International Conference on ICT in Education, Research and Industrial Applications. Integration, Harmonization and Knowledge Transfer, Kyiv, Ukraine, 15–18 May 2017, pp. 504–523 (2017)
25. Nikitchenko, M., Shkilniak, S.: Semantic properties of T-consequence relation in logics of quasiary predicates. Comput. Sci. J. Moldova **23**(2(68)), 102–122 (2015)
26. Nikitchenko, M., Ivanov, I., Korniłowicz, A., Kryvolap, A.: Extended Floyd-Hoare Logic over relational nominative data. In: Bassiliades, N., et al. (eds.) ICTERI 2017. CCIS, vol. 826, pp. 41–64. Springer, Cham (2018). https://doi.org/10.1007/978-3-319-76168-8_3

Category Methods for Modelling Logical Time Based on the Concept of Clocks

Grygoriy Zholtkevych(✉), Lyudmyla Polyakova, and Hassan Khalil El Zein

Mathematics and Computer Science School, V.N. Karazin Kharkiv National University, 4, Svobody Sqr., Kharkiv 61022, Ukraine
{g.zholtkevych,l.yu.polyakova}@karazin.ua, dr.hassanelzein@icloud.com

Abstract. This paper continues a series of articles presenting authors' ideas and results relating to development of category-theoretic methods for specifying and analysing models of logical time in distributed systems including cyber-physical systems. Results of this paper generalise results of the previous articles by the way of generalising the concepts of a clock structure and a schedule. The paper shows that all the main results obtained earlier remain valid for the generalisation under consideration. In addition, the proposed generalisation gives a tool for identifying in category-theoretic terms the concepts of global clocks and synchronisation processes.

Keywords: Category · Morphism · Functor · Equivalence of categories · Logical time · Logical clock · Schedule · Clock structure · Clock morphism · Schedule morphism · Global clock · Synchronising morphism

1 Introduction

The current state of the information technology development is characterised by two principle trends (detail analysis see in [8,9]). The first of these is the explosive expansion of computational intelligence application, and the second is the stable increasing use of hybrid (cyber-physical) systems that combine both software and physical components. The example of carrying out these trends is Internet-of-Things, which use grows permanently. Systems representing both of these trends require effective using distributed computations. This requirement is a technological answer to the practical achievement of the upper bound of processor performance on the one side and the development of communication tools on the other.

We need to stress that a distributed system is more complex in comparison with a single processor system. Increasing system complexity is caused by increasing uncertainty of the system behaviour related to the absence of the uniform mechanism measuring time. This problem was firstly marked by Lamport in [3]. He explained also that using the physical time for specifying causality relationships between events generated different components of a distributed system can lead to logical contradictions. This was a reason for introducing the concept of logical time and formulating the first mathematical model of logical time based on logical clocks. The detailed illustration of problems related to physical time is given in [9, Subsect. 2.2].

V. Ermolayev et al. (Eds.): ICTERI 2018, CCIS 1007, pp. 89–101, 2019.
https://doi.org/10.1007/978-3-030-13929-2_5

The time model based on logical clocks gained momentum at the following papers [2,5,6]. Despite the fact that the language CCSL built on the basis of a logical clock model is used in the embedded system design standard UML/MARTE (see, [1]), its semantics does not have a complete formal definition.

Category-theoretic methods, as it is well-known, give adequate conceptual tools to construct semantic models for formal languages. Our hypothesis is that the formulation of category-theoretic models for different aspects of logical time and studying them with using technique of Category Theory would provide different semantic views for CCSL and its generalisations and these views would give the complete characterising of the corresponding languages semantic meaning.

This paper is focused on the problem of modelling logical time in distributed systems, in particular on the model based on the concept of logical time, and is a continuation of authors' results presented in [8–10].

In contrast to models presented in the conference ICTERI 2018 proceeding volume, here we refuse of fixing the set of clocks for a clock model. Such a refuse gives a chance to construct a compositional theory of clock structures unlike the presumption that the set of clocks is predefined for all of them.

As known, there are two approaches to model logical time basing on the concept of logical clocks. The first one based on the denotational style, which requires to consider some set corresponding to an Universe of events equipped with the causality relation and the second one based on the operational style, which introduces criteria for distinguishing correct and incorrect individual behaviours represented by sequences of simultaneous occurrences of events. In this context, to understand the interrelationships between the mentioned two approaches is a very important problem.

Authors' studying this problem led them to the hypothesis that Category Theory [4,7] is an adequate mathematical language for such research.

This paper has the following structure:

- in Sect. 2, we construct and discuss the category of clock structures. Subsection 2.1 recalls the notion of a quasi-ordered set; in Subsect. 2.2 the categorical properties of clock structures and linear clock structures are given; Subsect. 2.3 gives some technique of slices in clock structures; in Subsect. 2.4 we discuss the so called global clock – a weak zero object of the category of clock structures; in Subsect. 2.5 we prove the analogue of Szpilrajn Extension Theorem for clock structures;
- in Sect. 3, we construct the category of schedules and prove the Main Theorem about the equivalence of the linear clock structures category and the category of schedules.
- Section 4 concludes the paper.

2 Category of Clock Structures

In [9], the denotational approach to model cyber-physical systems (CPS) in terms of clock structures with the fixed set of clocks was proposed and the corresponding category of clock structures was built. Here, we extend the notion of clock structure by including into the component list of this structure the clock set. This allows us to obtain more capacious category and some additional categorical constructions, such as finite co-products, weakly terminal and initial objects.

2.1 Quasi-Ordered Sets

Using the denotational approach of CPS modelling one can order the event occurrences not in physical but in logical time by specifying the causality relations between them.

Consideration of causality relationship as a quasi-order (i.e. a reflexive and transitive relation) on the set of event occurrences is a generally accepted approach. Following this approach, we understand an Universe of events as a set $\mathcal{I}$ of event occurrences equipped with a quasi-order "$\preccurlyeq$". Usually, we consider along with the relation "$\preccurlyeq$" the relations $\equiv$ (synchronisation), $\prec$ (precedence), # (mutually exclusion), and $\|$ (independence). These relations are defined by the causality relation as in Table 1 (see also [9]).

Table 1. Relations derived from the causality relation

Notation	Meaning
$i \equiv j$	both $i \preccurlyeq j$ and $j \preccurlyeq i$ are fulfilled
$i \prec j$	$i \preccurlyeq j$ is fulfilled but $j \preccurlyeq i$ is not fulfilled
$i \# j$	either $i \prec j$ is fulfilled or $j \prec i$ is fulfilled
$i \parallel j$	neither $i \preccurlyeq j$ is fulfilled nor $j \preccurlyeq i$ is fulfilled

2.2 Clock Structures

Now we give the definition of a clock structure as an Universe of events with additional structures and properties.

Definition 1. *A clock structure is a quadruple $\mathcal{S} = (C, \mathcal{I}, \gamma, \preccurlyeq)$ where*

- *C is a finite set of clocks, whose each element is interpreted as a reference to the source,*
- *$\mathcal{I}$ is the set of instants corresponding to the occurrences of events,*
- *$\gamma\colon \mathcal{I} \to C$ is a surjective mapping that associates the clock that is the source of an instant with this instant,*
- *"$\preccurlyeq$" is a quasi-order on $\mathcal{I}$ that models the causality relation between instants.*

This quadruple should also meet the following axioms
the axiom of unbounded liveness: *the set $\mathcal{I}$ is infinite;*
the axiom of finite causality:
for any $i \in \mathcal{I}$, the corresponding principal ideal $(\,i\,]$ is finite;
the axiom of total ordering for clock time-lines:
for each $c \in C$, the c-time-line $\mathcal{I}_c = \gamma^{-1}(c)$ is linearly ordered by "$\prec$".

Note 1. For a fixed set of clocks C, clock structures $\mathcal{S} = (C, \mathcal{I}, \gamma, \preccurlyeq)$ are C-structures, which have been considered in [9, 10].

The notion of a morphism of clock-structures is modified as follows

Definition 2. *Let* $\mathcal{S}' = (C', \mathcal{I}', \gamma', \preccurlyeq)$ *and* $\mathcal{S}'' = (C'', \mathcal{I}'', \gamma'', \preccurlyeq)$ *be the clock structures then a pair* (f, σ) *of mappings* $f\colon \mathcal{I}' \to \mathcal{I}''$ *and* $\sigma\colon C' \to C''$ *is called a morphism of clock structures if the following holds*

- *for any* $i \in \mathcal{I}'$*, the equation* $\gamma''(f\,i) = \sigma(\gamma'\,i)$ *is fulfilled,*
- *for any* $i \in \mathcal{I}'$ *and* $j \in \mathcal{I}'$*,* $f\,i \preccurlyeq f\,j$ *whenever* $i \preccurlyeq j$*,*
- *for any* $i \in \mathcal{I}'$ *and* $j \in \mathcal{I}'$*,* $f\,i \,\#\, f\,j$ *whenever* $i \,\#\, j$*.*

We get the notion of the morphism of C-structures (C-morphism) from [10] if $C' = C''$ and $\sigma = id_{C'}$. We write C-morphism simply as f without σ.

The following proposition is evident.

Proposition 1. *The class of all structures equipped with morphisms of structures forms a category denoted below by* **Struct***.*
The class of all C*-structures equipped with morphisms of* C*-structures forms a category denoted below by* $\mathbf{Struct}_C$*.*

Note that $\mathbf{Struct}_C$ is a subcategory of **Struct**, but it is not full since other morphisms besides C-morphisms are possible for C-structures (see the example of global clock below).

An important special class of clock structures is formed by so-called linear clock structures, defined in [9].

Definition 3. *A structure* $\mathcal{L} = (C, \mathcal{I}, \gamma, \preccurlyeq)$ *is called a linear structure if* $i \parallel j$ *is false for all* $i, j \in \mathcal{I}$*.*

The validity of the following sentence is obvious.

Proposition 2. *The class of all linear structures equipped with morphisms of structures forms a category denoted below by* **LinStruct***.*
LinStruct *is a full subcategory of* **Struct***.*

Construction (co-Product of Clock Structures). Let us assume that a finite family $\{\mathcal{S}_\alpha\}_{\alpha \in A}$ of clock structures $\mathcal{S}_\alpha = (C_\alpha, \mathcal{I}_\alpha, \gamma_\alpha, \preccurlyeq)$ where $\alpha \in A$ is given. Then let us define the quadruple $\mathcal{S} = (C, \mathcal{I}, \gamma, \preccurlyeq)$ in the following manner

$$C = \sum_{\alpha \in A} C_\alpha, \qquad \mathcal{I} = \sum_{\alpha \in A} \mathcal{I}_\alpha, \qquad \gamma(i) = \gamma_\alpha(i) \quad \text{if } i \in \mathcal{I}_\alpha$$
$$i \preccurlyeq j \text{ in } \mathcal{I} \quad \text{if } i, j \in \mathcal{I}_\alpha \text{ and } i \preccurlyeq j \text{ in } \mathcal{I}_\alpha \text{ for some } \alpha \in A.$$

Further, denote the embeddings of $\mathcal{I}_\alpha$ into $\mathcal{I}$ and C_α into C by f_α and σ_α respectively.

Proposition 3. *The quadruple* $\mathcal{S}$ *defined in the previous construction is a clock structure.*
For each $\alpha \in A$*, the pairs* $(f_\alpha, \sigma_\alpha)$ *are morphisms from* $\mathcal{S}_\alpha$ *into* $\mathcal{S}$*.*
The clock structure $\mathcal{S} = (C, \mathcal{I}, \gamma, \preccurlyeq)$ *with the morphism family* $\{(f_\alpha, \sigma_\alpha) \mid \alpha \in A\}$ *is the co-product of the family* $\{\mathcal{S}_\alpha \mid \alpha \in A\}$ *in category* **Struct***.*

Proof. The first two items of the proposition are trivially true.
Let us prove the last item. For arbitrary $\mathcal{S}' = (C', \mathcal{I}', \gamma', \preccurlyeq)$ and morphisms $(f'_\alpha, \sigma'_\alpha) : \mathcal{S}_\alpha \to \mathcal{S}'$, define $f : \mathcal{I} \to \mathcal{I}'$ as $f(i) = j$ if $i \in \mathcal{I}_\alpha$ and $f_\alpha(i) = j$ and similarly $\sigma : C \to C'$ as $\sigma(c) = c'$ if $c \in C_\alpha$ and $\sigma'_\alpha(c) = c'$. One can check directly that $(f, \sigma) : \mathcal{S} \to \mathcal{S}'$ is a morphism of clock structures and the conditions $f'_\alpha = f \circ f_\alpha$, $\sigma'_\alpha = \sigma \circ \sigma_\alpha$ hold for every α that allows to consider $\mathcal{S}'$ as a co-product by definition (see for instance [4]). □

2.3 Slices

Now let us recall some results obtained in [9, 10] concerning the arrangement of clock structures. The original properties were proved for C-structures, but they can be extended to general case without any or with some slight modifications. We give them here for the completeness of presentation.

Definition 4. *Let $\mathcal{S} = (C, \mathcal{I}, \gamma, \preccurlyeq)$ be a clock structure, $A \subset \mathcal{I}$ and $i \in A$ then i is called a minimal instant in A if the statement $j \prec i$ is false for any $j \in A$.*

To refer to the subset of minimal instants of A we use the denotation $\min A$. We associate the sequence of slices $\mathcal{I}[0], \mathcal{I}[1], \ldots, \mathcal{I}[n], \ldots$ with any structure $\mathcal{S}$ in the following manner

$$\begin{aligned} \mathcal{I}[0] &= \min \mathcal{I}; \\ \mathcal{I}[n] &= \min \left(\mathcal{I} \setminus \bigcup_{k=0}^{n-1} \mathcal{I}[k] \right) \qquad \text{for } n \in \mathbb{N}_+ \end{aligned} \tag{1}$$

where $\mathcal{I}$ is the instant set of $\mathcal{S}$.

Proposition 4. *For a clock structure $\mathcal{S} = (C, \mathcal{I}, \gamma, \preccurlyeq)$ the following properties hold*

1. $|\mathcal{I}[n]| \leq |C|$ *for each* $n \in \mathbb{N}$;
2. *if* $i, j \in \mathcal{I}[n]$ *then either* $i \parallel j$ *or* $i \equiv j$ *for some* $n \in \mathbb{N}$;
3. *the sequence of slices is a covering of the set of instants;*
4. *if* $i \in \mathcal{I}[n]$, $j \in \mathcal{I}$, *and* $j \equiv i$ *then* $j \in \mathcal{I}[n]$ *for some* $n \in \mathbb{N}$;
5. *if* $i \in \mathcal{I}[n+1]$ *then there exists* $j \in \mathcal{I}[n]$ *such that* $j \prec i$ *for some* $n \in \mathbb{N}$.

Corollary 1. *If $\mathcal{L}$ with an instant set $\mathcal{I}$ is a linear structure then $i, j \in \mathcal{I}[n]$ implies $i \equiv j$.*

Corollary 2. *The restriction γ on $\mathcal{I}[n]$ is injective for all $n \in \mathbb{N}$.*

Proof. If $\gamma i = \gamma j$ for some $i, j \in \mathcal{I}[n]$ then the axiom of total ordering for the clock time-lines of Definition 1 makes impossible the case $i \parallel j$ in Item 2 of Proposition 4. Further $i \equiv j$ together with the mentioned axiom implies $i = j$. □

Proposition 5. *Let $\mathcal{S} = (C, \mathcal{I}, \gamma, \preccurlyeq)$ be a clock structure and for some $m, n \in \mathbb{N}$, $i \in \mathcal{I}[m]$ and $j \in \mathcal{I}[n]$ then the following properties hold*

1. *if* $i \preccurlyeq j$ *then* $m \leq n$, *while* $i \prec j$ *implies* $m < n$;
2. *if* $\mathcal{S}$ *is linear and* $m \leq n$ *then* $i \preccurlyeq j$, *while* $m < n$ *implies* $i \prec j$.

Proof. To prove Item 1 suppose $m > n$. Then Item 5 of Proposition 4 implies the existence of $j' \in \mathcal{I}[n]$ such that $j' \prec i$. But $j' \prec i \preccurlyeq j$ contradicts by Item 2 of Proposition 4 the fact $j', j \in \mathcal{I}[n]$.
To prove Item 2 note the following.
If $m = n$ we directly obtain the statement by Corollary 1.
If $m < n$ then Item 5 of Proposition 4 implies the existence of $i' \in \mathcal{I}[m]$ such that $i' \prec j$ then $i' \equiv i$ and from $i \preccurlyeq i' \prec j$ we conclude $i \prec j$. □

The established properties of the clock structures make it possible to establish the properties of their morphisms.

Proposition 6. *For any morphism $(f, \sigma)\colon \mathcal{S}' \to \mathcal{S}''$ from the clock structure $\mathcal{S}' = (C', \mathcal{I}', \gamma', \preccurlyeq)$ into the clock structure $\mathcal{S}'' = (C'', \mathcal{I}'', \gamma'', \preccurlyeq)$ and $i \in \mathcal{I}'[m]$, $f\,i \in \mathcal{I}''[n]$ for some $m, n \in \mathbb{N}$ then the following properties hold*

1. *the mapping $f\colon \mathcal{I}' \to \mathcal{I}''$ preserves relations "$\preccurlyeq$", "$\equiv$", "$\prec$", and "$\#$";*
2. *if for some $j \in \mathcal{I}'$, we have $i \equiv j$ then $f\,j \in \mathcal{I}''[n]$;*
3. *$n \geq m$;*
4. *if f is an isomorphism then $m = n$.*

Proof. To prove Item 1 note that f preserves relations "$\preccurlyeq$" and "$\#$" by Definition 2. The statement $i \equiv j$ is equivalent to $i \preccurlyeq j$ and $j \preccurlyeq i$ and, therefore, f preserves "$\equiv$". Further taking into account that $i \prec j$ if and only if $i \preccurlyeq j$ and $i \# j$ one can immediately obtain that f preserves "$\prec$".
To prove Item 2 let us use proven Item 1. Really, $i \equiv j$ ensures, as Item 1 claims, $f\,i \equiv f\,j$. Now use of Item 4 of Proposition 4 leads to the required statement.
To prove Item 3 we use induction in m. For $m = 0$ it is true that $n \geq m$. Suppose the statement holds for $m \leq k$ and let $m = k + 1$, $i \in \mathcal{I}'[m]$, $f\,i \in \mathcal{I}''[n]$. Suppose $n < k + 1$. Then by Item 5 of Proposition 4 there exists $j \in \mathcal{I}'[n]$ such that $j \prec i$. As above shown $fj \prec f\,i$. If $fj \in \mathcal{I}''[r]$ then by induction hypothesis $r \geq n$. Taking into account $j \prec i$ and Item 1 of Proposition 5 one can derive that $r < n$. Thus, we have obtained two mutually excluding inequalities $r \geq n$ and $r < n$ and conclude that our supposition is incorrect i.e. $n \geq m$.
Item 4 follows immediately from Item 3. □

Morphisms of linear clock structures hold additional properties.

Proposition 7. *For any morphism $(f, \sigma)\colon \mathcal{L}' \to \mathcal{L}''$ from the linear structure $\mathcal{L}' = (C', \mathcal{I}', \gamma', \preccurlyeq)$ into the linear structure $\mathcal{L}'' = (C'', \mathcal{I}'', \gamma'', \preccurlyeq)$ and $i, j \in \mathcal{I}'$ the following properties hold*

1. *if $f\,i \equiv f\,j$ then $i \equiv j$;*
2. *if $i \in \mathcal{I}'[m]$, $j \in \mathcal{I}'$, $f\,i, f\,j \in \mathcal{I}''[n]$ for some $m, n \in \mathbb{N}$ then $j \in \mathcal{I}'[m]$;*
3. *$f\,\mathcal{I}'[m] \subset \mathcal{I}''[n]$ for some n such that $m \leq n$.*

Proof. To prove Item 1 note that $i \not\equiv j$ is equivalent to $i \# j$ for linear clock structures. Item 1 of Proposition 6 ensures $f\,i \# f\,j$ but this contradicts to $f\,i \equiv f\,j$.
To prove Item 2 note $f\,i, f\,j \in \mathcal{I}''[n]$ ensures $f\,i \equiv f\,j$ for a linear clock structure. Taking into account the previous item one can conclude that $i \equiv j$. Now we need to use Item 4 of Proposition 4 to obtain the required statement.
Item 3 follows directly from Corollary 1 and Item 2 of Proposition 6. □

2.4 Global Clock

The possibility of defining global clocks is a significant advantage of the considered model compared with the model with the fixed clock set.

Definition 5. *A global clock is a clock structure $\mathcal{GC} = (C, \mathcal{I}, \gamma, \preccurlyeq)$ where C is a singleton $\{gc\}$, $\mathcal{I} = \mathbb{N}$, $\gamma(n) = gc$ for every $n \in \mathbb{N}$, and $n \preccurlyeq m$ iff $n \leq m$.*

According to [4, X.2] a weakly terminal object of category admits a morphism from every object into itself, while for (strongly) terminal object such a morphism should be unique. In dual way a weakly initial object admits a morphism into every object from itself. Both (weakly) terminal and (weakly) initial object is called (weakly) zero object.

Proposition 8. *The global clock is weakly zero object of* **Struct**.

Proof. Let $\mathcal{S} = (C, \mathcal{I}, \gamma, \preccurlyeq)$. One can consider the synchronising morphism $(s, \iota) : \mathcal{S} \to \mathcal{GC}$ with $\iota(c) = gc$ for every $c \in C$ and $s(i) = n$ if $i \in \mathcal{I}[n]$. Then Item 1 of Proposition 5 implies that (s, ι) is correctly defined and so $\mathcal{GC}$ is weak terminal.
To construct a morphism $(f, \sigma) : \mathcal{GC} \to \mathcal{S}$ fix the clock $c \in C$, consider the only time line $\mathcal{I}_c = \gamma^{-1}(c)$ and put $\sigma(gc) = c$ and $f(n) = i$ if $|\{j \in \mathcal{I}_c \mid j \prec i\}| = n$. One can directly check that (f, σ) is a morphism and so $\mathcal{GC}$ is weak initial. □

For every clock structure $\mathcal{S}$ the synchronising morphism $(s, \iota) : \mathcal{S} \to \mathcal{GC}$ (see the proof of Proposition 8) can be extended through the synchronising morphism of some linear structure. More precisely, for $\mathcal{S} = (C, \mathcal{I}, \gamma, \preccurlyeq)$, one can construct the linear clock structure $\mathcal{L} = (C^*, \mathcal{I}^*, \gamma^*, \preccurlyeq)$ (so called linearization of $\mathcal{S}$) as follows:

$$C^* = C, \qquad \mathcal{I}^* = \mathcal{I} \times \mathbf{1} \text{ where } \mathbf{1} = \{*\} \text{ is a singleton,}$$
$$\gamma^*(i, *) = \gamma(i) \text{ for } i \in \mathcal{I},$$
$$(i, *) \preccurlyeq (j, *) \text{ iff } i \in \mathcal{I}[n],\ j \in \mathcal{I}[m], \text{ and } n \leq m \text{ for some } n, m \in \mathbb{N}.$$

Considering the mapping $l : \mathcal{I} \to \mathcal{I}^*$ defined by the formula $l(i) = (i, *)$ we have that (l, id_C) is a morphism from $\mathcal{S}$ onto $\mathcal{L}$ due to construction of $\mathcal{L}$ and Item 1 of Proposition 5.

Proposition 9. *Let $\mathcal{S}$ be a clock structure and $\mathcal{L}$ its linearization then the diagram*

$$\begin{array}{ccc} \mathcal{S} & \xrightarrow{(l,\, \mathrm{id}_C)} & \mathcal{L} \\ & \underset{(s,\, \iota)}{\searrow} & \downarrow{\scriptstyle (s_L,\, \iota)} \\ & & \mathcal{GC} \end{array}$$

with synchronising morphisms (s, ι) and (s_L, ι) is commutative.

Proof. To prove the commutativity of the diagram we show that if $i \in \mathcal{I}[n]$ then $(i, *) \in \mathcal{I}^*[n]$ by induction on n.
It is evident that for $i \in \mathcal{I}[0]$ and any $j \in \mathcal{I}$ the condition $(i, *) \preccurlyeq (j, *)$ holds, therefore $(i, *) \in \mathcal{I}^*[0]$.

Let for any $n \leq k$ the statement be true and $i \in \mathcal{I}[k+1]$. Suppose $(i,*) \in \mathcal{I}^*[m]$ and $m \neq k+1$. Item 3 of Proposition 6 implies $m > k+1$. Then by Item 5 of Proposition 4 there exists $j \in \mathcal{I}$ such as $(j,*) \prec (i,*)$ and $(j,*) \in \mathcal{I}^*[k+1]$. Let $j \in \mathcal{I}[t]$ for some $t \in \mathbb{N}$. Then Item 3 of Proposition 6 implies $t \leq k+1$ and induction hypothesis implies $t \geq k+1$. Therefore $t = k+1$, and the fact $i, j \in \mathcal{I}[k+1]$ contradicts the fact $(j,*) \prec (i,*)$, which proves the statement. □

Note that the global clock is not (strongly) terminal and (strongly) initial since all monotonous functions on $\mathbb{N}$ (and only them) give the morphisms of $\mathcal{GC}$ into itself. Nevertheless the global clock has some kind of the universal property.

Proposition 10. *Let* $(s,\iota) : \mathcal{L} \to \mathcal{GC}$ *be a synchronising morphism of linear structure* $\mathcal{L}$ *and* $(f,\sigma) : \mathcal{L} \to \mathcal{GC}$ *be another morphism then there exists the unique morphism* $(u, \mathrm{id}_{\{gc\}})$ *such as* $(f,\sigma) = (u, id_{\{gc\}}) \circ (s,\iota)$, *i.e. the diagram*

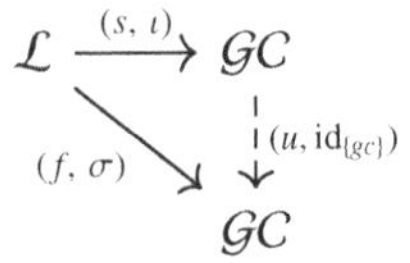

is commutative.

Proof. We should check that the mapping $(u, id_{\{gc\}})$, with $u(n) = f(i)$ iff $i \in \mathcal{I}[n]$ is correctly defined and is a morphism.
Correctness follows from Item 3 of Proposition 7. If $n \leq m$ and $u(n) = f(i)$ for some $i \in \mathcal{I}[n]$, $u(m) = f(j)$ for some $j \in \mathcal{I}[m]$ then Item 2 of Proposition 7 implies $i \preccurlyeq j$, so $u(n) = f(i) \preccurlyeq f(j) = u(m)$, while by the same reason $n < m$ implies $u(n) < u(m)$. That is $(u, id_{\{gc\}})$ is a morphism. □

We have seen in Proposition 8 that the weak terminality of global clock is just a categorical way to formulate Item 1 of Proposition 5. The categorical interpretation of Item 3 of Proposition 7 can be given as

Proposition 11. *Let* $(f, \sigma) : \mathcal{L}' \to \mathcal{L}''$ *be a morphism of linear structures and* $(s', \iota') : \mathcal{L}' \to \mathcal{GC}$, $(s'', \iota'') : \mathcal{L}'' \to \mathcal{GC}$ *be synchronising morphisms then there exists the unique morphism* $(u, \mathrm{id}_{\{gc\}}) : \mathcal{GC} \to \mathcal{GC}$ *such as the diagram*

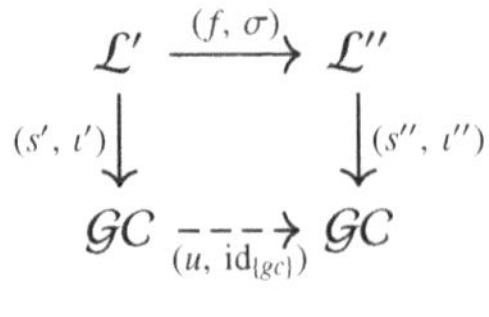

is commutative.

Proof. Put $(s'', \iota'') \circ (f, \sigma)$ as (f, σ) in Proposition 10 and obtain directly. □

2.5 Analogue of Szpilrajn Extension Theorem

The Szpilrajn Extension Theorem is an important fact in the theory of ordered sets. As it turned out, this fact has an analogue for clock structures, which was proved in [8]. Here we use the linearization constructed above to prove it.

Definition 6 *([8]). Let $\mathcal{S} = (C, \mathcal{I}, \gamma, \preccurlyeq)$ be a clock structure then a C-morphism $e: \mathcal{S} \to \mathcal{S}'$ for some C-structure $\mathcal{S}'$ with surjective mapping $e : \mathcal{I} \to \mathcal{I}'$ is called an extension of $\mathcal{S}$.*
Whenever $\mathcal{S}'$ is a linear C-structure we say that e is a linear extension of S.

Theorem 1 (about Linear Extension). *Let $\mathcal{S} = (C, \mathcal{I}, \gamma, \preccurlyeq)$ be a clock structure and i_*, j_* be some pair of independent instants belonging to $\mathcal{I}$ then there exists a linear extension $e : \mathcal{S} \to \mathcal{L}$ such that the condition $e(i_*) \prec e(j_*)$ holds.*

We need the following lemma:

Lemma 1. *Let $\mathcal{S} = (C, \mathcal{I}, \gamma, \preccurlyeq)$ be a clock structure and i_*, j_* be two independent instants in $\mathcal{I}$ then there exists an extension $e : \mathcal{S} \to \mathcal{S}^*$ such that $e(i_*) \prec e(j_*)$.*

Proof. Let $\mathbf{1} = \{*\}$ be some singleton then one can define

$$\mathcal{I}^* = \mathcal{I} \times \mathbf{1}, \qquad \gamma^*(i, *) = \gamma(i) \text{ for any } i \in \mathcal{I},$$
$$(i, *) \preccurlyeq (j, *) \text{ iff either } i \preccurlyeq j \text{ or } i \preccurlyeq i_* \text{ and } j_* \preccurlyeq j \text{ for any } i, j \in \mathcal{I},$$
$$e(i) = (i, *) \text{ for any } i \in \mathcal{I}.$$

Let us check that so defined relation "$\preccurlyeq$" on $\mathcal{I}^*$ is a quasi-order. Indeed, it is evident that this relation is reflexive. If now we have that $(i, *) \preccurlyeq (j, *)$ and $(j, *) \preccurlyeq (k, *)$ for some $i, j, k \in \mathcal{I}$ then the next variants are only possible:
1. $i \preccurlyeq j$ and $j \preccurlyeq k$; these conditions ensure $i \preccurlyeq k$ and, hence, $(i, *) \preccurlyeq (k, *)$;
2. $i \preccurlyeq i_*$, $j_* \preccurlyeq j$, and $j \preccurlyeq k$; these conditions ensure $i \preccurlyeq i_*$ and $j_* \preccurlyeq k$, but this means that $(i, *) \preccurlyeq (k, *)$;
3. $i \preccurlyeq j$, $j \preccurlyeq i_*$, $j_* \preccurlyeq k$; these conditions ensure $(i, *) \preccurlyeq (k, *)$ that is checked similarly to above.
Thus "$\preccurlyeq$" is a quasi-order on $\mathcal{I}^*$.
One can easily see that $(i_*, *) \preccurlyeq (j_*, *)$, but $(j_*, *) \not\preccurlyeq (i_*, *)$, i. e. $e(i_*) \prec e(j_*)$.
The construction ensures evidently that $i \# j$ implies $e(i) \# e(j)$.
And, finally, it is evident that if $(j, *) \in ((i, *)]$, then $j \in (i] \cup (i_*]$. Therefore, the set $((i, *)]$ is finite for any $(i, *) \in \mathcal{I}^*$.
Thus, $S^* = (C, \mathcal{I}^*, \gamma^*, \preccurlyeq)$ is a clock structure, $e : \mathcal{S} \to \mathcal{S}^*$ is a morphism with surjective $e : \mathcal{I} \to \mathcal{I}^*$ and $e(i_*) \prec e(j_*)$. □

Proof (of Theorem about Linear Extension). Let $e' : \mathcal{S} \to \mathcal{S}'$ be an extension of S constructed in accordance with Lemma 1 and $l : \mathcal{S}' \to \mathcal{L}$ be a linearization morphism constructed as in Proposition 9. Then $e = l \circ e'$ is a linear extension of $\mathcal{S}$. Since $e'(i_*) \prec e'(j_*)$ and l preserves the "$\prec$" relation, we get $e(i_*) \prec e(j_*)$. □

3 Category of Schedules

The operational approach to describe logical time dependencies in distributed systems (including cyber-physical systems) is based on observing streams of system messages. Below this approach is described with the language of the category theory.

Definition 7. *Let C be a finite set of logical clocks then any non-empty subset of C is called a clock message.*

Informally, a clock message contains the information about which clocks ticked at the same time-point.

To refer to the set of clock messages associated with a clock set C we use below the denotation M_C.

We do not assume now that the clock set C is fixed (as we did it in [10]) and construct more capacious category for schedules as it has been already done for clock structures.

Definition 8. *A schedule (or more precisely a C-schedule) is an infinite sequence $\boldsymbol{\pi} = (\pi[0], \pi[1], \ldots, \pi[n], \ldots)$ of clock messages from M_C.*
For a C'-schedule $\boldsymbol{\pi}'$ and C''-schedule $\boldsymbol{\pi}''$, a morphism from $\boldsymbol{\pi}'$ into $\boldsymbol{\pi}''$ is a quadruple $\langle \boldsymbol{\pi}', k, \sigma, \boldsymbol{\pi}'' \rangle$ where

1. *$\sigma : C' \to C''$ is a mapping;*
2. *$k : \mathbb{N} \to \mathbb{N}$ is a strictly monotonous mapping;*
3. *$\sigma(\pi'[n]) \subset \pi''[k(n)]$ for any $n \in \mathbb{N}$.*

Note that if for a given schedule $\boldsymbol{\pi}$, the clock set C is not indicated one can assume that $C = \bigcup_{n \in \mathbb{N}} \pi[n]$ is finite.
Now we use the formula

$$\langle \boldsymbol{\pi}'', k_2, \sigma_2, \boldsymbol{\pi}''' \rangle \circ \langle \boldsymbol{\pi}', k_1, \sigma_1, \boldsymbol{\pi}'' \rangle = \langle \boldsymbol{\pi}', k_2 \circ k_1, \sigma_2 \circ \sigma_1, \boldsymbol{\pi}''' \rangle$$

for defining the composition of morphisms $\langle \boldsymbol{\pi}', k_1, \sigma_1, \boldsymbol{\pi}'' \rangle$ and $\langle \boldsymbol{\pi}'', k_2, \sigma_2, \boldsymbol{\pi}''' \rangle$. It is evident that

1. morphisms of the form $\langle \boldsymbol{\pi}, 1_{\mathbb{N}}, 1_C, \boldsymbol{\pi} \rangle$ are units of this composition (where $\boldsymbol{\pi}$ is a C-schedule).
2. the associative law is fulfilled for this composition.

Thus, the following statement is true.

Proposition 12. *The set of all schedules equipped with morphisms of schedules forms a category denoted below by* **Sched**.

It was shown in [10] that categories $\mathbf{LinStruct}_C$ and $\mathbf{Sched}_C$ are equivalent. We generalise this statement as follows.

Main Theorem. *The categories* **LinStruct** *and* **Sched** *are equivalent i.e. there exists a pair of functors*

$$F\colon \mathbf{Sched} \to \mathbf{LinStruct} \quad \text{and} \quad G\colon \mathbf{LinStruct} \to \mathbf{Sched}$$

such that $\boldsymbol{F}\,\boldsymbol{G}$ *is naturally isomorphic to the identity endofunctor of* **LinStruct** *and* $\boldsymbol{G}\,\boldsymbol{F}$ *is naturally isomorphic to the identity endofunctor of* **Sched**.

Note 2. All necessary definitions and facts about natural transformations and natural isomorphisms can be found in [4].

We use the following theorem as the main tool to establish the validity of Main Theorem.

Theorem 2. *Categories* **C** *and* **D** *are equivalent if and only if there exists a functor* $\boldsymbol{F}\colon \mathbf{C} \to \mathbf{D}$ *such that*

1. *for any object* $\boldsymbol{d}$ *in* **D**, *there exists an object* $\boldsymbol{c}$ *in* **C** *such that* $\boldsymbol{F\,c}$ *and* $\boldsymbol{d}$ *are isomorphic;*
2. *for any objects* $\boldsymbol{c}'$ *and* $\boldsymbol{c}''$ *in* **C**, *the mapping* $\boldsymbol{F}\colon \mathbf{C}(\boldsymbol{c}', \boldsymbol{c}'') \to \mathbf{D}(\boldsymbol{F\,c}', \boldsymbol{F\,c}'')$ *is a bijection.*

The proof of this theorem one can find in [7, Theorem 7.1], the proof of the more general statement is given in [4, IV.4, Theorem 1].

Following [9] we associate the linear C-structure $\mathcal{L}^{\pi} = (C, \mathcal{I}^{\pi}, \gamma, \preccurlyeq\ )$ with a C-schedule $\boldsymbol{\pi}$ in the following manner

$$\mathcal{I}^{\pi} = \{\langle c, n\rangle \in C \times \mathbb{N} \mid c \in \pi[n]\}, \qquad \gamma\langle c, n\rangle = c \text{ for } \langle c, n\rangle \in \mathcal{I}^{\pi},$$
$$\langle c', n'\rangle \preccurlyeq \langle c'', n''\rangle \text{ if and only if } n' \le n'' \text{ for } \langle c', n'\rangle,\ \langle c'', n''\rangle \in \mathcal{I}^{\pi}.$$

Here we generalise this association by its extension up to a functor $\boldsymbol{F}$ from the category **Sched** into the category **LinStruct**.
To do this we assign the required correspondences as follows

$$\boldsymbol{F\,\pi} = \mathcal{L}^{\pi} \quad \text{for } \boldsymbol{\pi} \in \textbf{Sched} \text{ with clock set } C; \tag{2a}$$
$$\boldsymbol{F}\,\langle \boldsymbol{\pi}', k, \sigma, \boldsymbol{\pi}''\rangle = (f, \sigma) \in \textbf{LinStruct}(\mathcal{L}^{\pi'}, \mathcal{L}^{\pi''}) \tag{2b}$$
$$\text{where } f\,\langle c, n\rangle = \langle \sigma(c), k(n)\rangle.$$

Note that σ is the same clock mapping in the schedule morphism $\langle \boldsymbol{\pi}', k, \sigma, \boldsymbol{\pi}''\rangle$ and in the clock structure $(f, \sigma) : \mathcal{L}^{\pi'} \to \mathcal{L}^{\pi''}$ and that (f, σ) is correctly defined morphism of $\mathcal{L}^{\pi'}$ and $\mathcal{L}^{\pi''}$ since k is monotonous.

Now we need to check the validity of some statements. The corresponding checks are gathered in the following proposition.

Proposition 13. *Let* $\boldsymbol{\pi}'$, $\boldsymbol{\pi}''$, *and* $\boldsymbol{\pi}'''$ *be objects of* **Sched** *then*

1. *for any* $\langle c, n\rangle \in \mathcal{I}^{\pi'}$ *and* $\langle \boldsymbol{\pi}', k, \sigma, \boldsymbol{\pi}''\rangle \in \textbf{Sched}(\boldsymbol{\pi}', \boldsymbol{\pi}'')$, *the item* $\langle \sigma(c), k(n)\rangle$ *belongs to* $\mathcal{I}^{\pi''}$;
2. *for any morphisms* $\langle \boldsymbol{\pi}', k_1, \sigma_1, \boldsymbol{\pi}''\rangle$ *and* $\langle \boldsymbol{\pi}'', k_2, \sigma_2, \boldsymbol{\pi}'''\rangle$, *the following is true*

$$\boldsymbol{F}(\langle \boldsymbol{\pi}'', k_2, \sigma_2, \boldsymbol{\pi}'''\rangle \circ \langle \boldsymbol{\pi}', k_1, \sigma_1, \boldsymbol{\pi}''\rangle) = \boldsymbol{F}(\langle \boldsymbol{\pi}'', k_2, \sigma_2, \boldsymbol{\pi}'''\rangle \circ \boldsymbol{F}(\langle \boldsymbol{\pi}', k_1, \sigma_1, \boldsymbol{\pi}''\rangle));$$

3. $\boldsymbol{F}\langle \boldsymbol{\pi}', 1, 1, \boldsymbol{\pi}'\rangle$ *is the identity mapping from* $\mathcal{L}^{\pi'}$ *into itself.*

Proof. Item 1 follows immediately from the definition of a schedule morphism (see Definition 8).
Item 2 and Item 3 are checked by direct calculation. □

Corollary 3. *Formulae* (2a) *and* (2b) *determine a functor*

$$F\colon \mathbf{Sched} \to \mathbf{LinStruct}.$$

Below the following lemma is also needed for us.

Lemma 2. *For any linear C-structure $\mathcal{L}$ there exists C-schedule $\boldsymbol{\pi}$ such that $\mathcal{L}^{\pi}$ is isomorphic to $\mathcal{L}$.*

Proof. Let $\mathcal{L} = (C, \mathcal{I}, \gamma, \preccurlyeq)$ then we assign $\pi[n] = \gamma(\mathcal{I}[n]) \subset C$. Hence, the sequence $\boldsymbol{\pi} = (\pi[0], \pi[1], \dots, \pi[n], \dots)$ is a C-schedule.
Let us calculate $\mathcal{L}^{\pi} = (C, \mathcal{I}^{\pi}, \gamma^{\pi}, \preccurlyeq)$.
By construction $\mathcal{I}^{\pi} \subset C \times \mathbb{N}$ and $\langle c, n\rangle \in \mathcal{I}^{\pi}$ if $c \in \pi[n]$. In other words, $\langle c, n\rangle \in \mathcal{I}^{\pi}$ if $c = \gamma i$ for some $i \in \mathcal{I}[n]$. Corollary 2 guarantees that such i is uniquely determined. Thus, we can determine the mapping $f\colon \mathcal{I}^{\pi} \to \mathcal{I}$ by the following conditions $f\langle c, n\rangle = i \in \mathcal{I}[n]$ if and only if $\gamma i = c$. Due to Item 2 of Proposition 5, (f, id_C) is a morphism from $\mathcal{L}^{\pi}$ into $\mathcal{L}$.
Now consider the mapping $g\colon \mathcal{I} \to \mathcal{I}^{\pi}$ determined as follows $g\, i = \langle \gamma\, i, n\rangle$ where $i \in \mathcal{I}[n]$. The correctness of g is ensured by the construction of $\mathcal{L}^{\pi}$. Item 1 of Proposition 5 ensures that (g, id_C) is a morphism from $\mathcal{L}$ into $\mathcal{L}^{\pi}$.
It is evident that by construction $g\,(f\langle c, n\rangle) = \langle c, n\rangle$ for any $\langle c, n\rangle \in \mathcal{I}^{\pi}$ and $f\,(g\, i) = i$ for any $i \in \mathcal{I}$.
Thus, we have proven that $\mathcal{L}$ and $\mathcal{L}^{\pi}$ are isomorphic. □

Now we have all necessary tools to prove Main Theorem.

Proof (of Main Theorem). Our proof is based on applying Theorem 2 with the constructed above functor $\boldsymbol{F}\colon \mathbf{Sched} \to \mathbf{LinStruct}$. In accordance with the mentioned theorem, it is sufficient to prove that functor $\boldsymbol{F}$ holds two following properties

1. for any $\mathcal{L} \in \mathbf{LinStruct}$, there exists $\boldsymbol{\pi} \in \mathbf{Sched}$ such that $\boldsymbol{F}\,\boldsymbol{\pi}$ is isomorphic to $\mathcal{L}$;
2. for any C'-schedule $\boldsymbol{\pi}'$ and C''-schedule $\boldsymbol{\pi}''$ the mapping

$$\langle \boldsymbol{\pi}', k, \sigma, \boldsymbol{\pi}''\rangle \in \mathbf{Sched}(\boldsymbol{\pi}', \boldsymbol{\pi}'') \mapsto \boldsymbol{F}\langle \boldsymbol{\pi}', k, \sigma, \boldsymbol{\pi}''\rangle \in \mathbf{LinStruct}(\boldsymbol{F}\,\boldsymbol{\pi}', \boldsymbol{F}\,\boldsymbol{\pi}'')$$

   is bijective.

Lemma and the method of constructing the functor $\boldsymbol{F}$ guarantee the validity of the first property.
The mapping in the second property is injective. Indeed if $\boldsymbol{F}\langle \boldsymbol{\pi}', k_1, \sigma_1, \boldsymbol{\pi}''\rangle = \boldsymbol{F}\langle \boldsymbol{\pi}', k_2, \sigma_2, \boldsymbol{\pi}''\rangle$ then $\langle \sigma_1(c), k_1(n)\rangle = \langle \sigma_2(c), k_2(n)\rangle$ for all $\langle c, n\rangle \in \mathcal{I}^{\pi'}$. Taking into account that for each $n \in \mathbb{N}$, there exists $c \in C$ with $\langle c, n\rangle \in \mathcal{I}^{\pi'}$, and that $C' = \bigcup_{n \in \mathbb{N}} \pi'[n]$ one can conclude that $k_1 = k_2$ and $\sigma_1 = \sigma_2$.
The mapping in the second property is surjective. Really, if (f, σ) is a morphism from $\mathcal{L}^{\pi'}$ into $\mathcal{L}^{\pi''}$ then $f\langle c, n\rangle = \langle \sigma(c), k_f(n)\rangle$. Reasoning as above we obtain the function $k_f\colon \mathbb{N} \to \mathbb{N}$, such that $k_f(n) = m$ iff $f(\mathcal{I}^{\pi'}[n]) \subset \mathcal{I}^{\pi''}[m]$.
It is evident that

$$\mathcal{I}^{\pi'}[n] = \left\{c \in C \mid \langle c, n\rangle \in \mathcal{I}^{\pi'}\right\} \times \{n\}. \tag{3}$$

As it was noted above the morphisms of global clock correspond to monotonous mappings of $\mathbb{N}$. Further, Proposition 11 ensures that $k_f(n)$ is monotonous. Thus, $\langle \boldsymbol{\pi}', k_f, \sigma, \boldsymbol{\pi}'' \rangle$ is a morphism from $\boldsymbol{\pi}'$ into $\boldsymbol{\pi}''$.
Now taking into account equality (3) and definition (2b) one can easily derive that $\boldsymbol{F}\langle \boldsymbol{\pi}', k_f, \sigma, \boldsymbol{\pi}'' \rangle = f$. □

4 Conclusion

Summarising the presented results we can claim that refuse from fixing the set of clocks for all clock structures forming the corresponding category is very productive because it provides new possibilities, in particular, to construct finite co-products of clock structures and to describe global clocks and synchronisations in the terms of category-theoretic language.

In addition, it has been proven that all principal properties of the categories of clock structures and schedules introduced earlier are preserved in this new, more general situation.

Thus, we have reasons to hope that further development of the category-theoretic approach in the research area under consideration allows obtaining methods for specifying and analysing causality relationships in distributed systems, which would be applied for the expertise of real projects of complex systems.

References

1. UML profile for MARTE: Modeling and Analysis of Real-Time Embedded Systems. Specification Formal/2011-06-03, OMG (2011)
2. André, C., Mallet, F.: Clock constraints in UML MARTE CCSL. Research Report RR-6540, INRIA (2008)
3. Lamport, L.: Time, clocks, and the ordering of events in a distributed system. CACM **21**(7), 558–565 (1978)
4. Mac Lane, S.: Categories for the Working Mathematicians, 2nd edn. Springer, Heidelberg (1998)
5. Mallet, F.: Clock constraint specification language: specifying clock constraints with UML/MARTE. Innov. Syst. Softw. Eng. **4**(3), 309–314 (2008)
6. Mallet, F.: MARTE/CCSL for modeling cyber-physical systems. In: Drechsler, R., Kühne, U. (eds.) Formal Modeling and Verification of Cyber-Physical Systems, pp. 26–49. Springer, Wiesbaden (2015). https://doi.org/10.1007/978-3-658-09994-7_2
7. Novikov, B.: Lecture Notes on Category Theory. Luhansk National Pedagogical University (2004)
8. Zholtkevych, G., El Zein, H.K.: Logical time models to study cyber-physical systems. In: Ermolayev, V., et al. (eds.) Information and Communication Technologies in Education, Research, and Industrial Applications. Integration, Harmonization and Knowledge Transfer. Workshop Proceedings, vol. 1844. CEUR-WS, May 2017
9. Zholtkevych, G., El Zein, H.K.: Two approaches to modelling logical time in cyber-physical systems. In: Bassiliades, N., et al. (eds.) ICTERI 2017. CCIS, vol. 826, pp. 21–40. Springer, Cham (2018). https://doi.org/10.1007/978-3-319-76168-8_2
10. Zholtkevych, G., El Zein, H.K., Polyakova, L.: Category methods for analysis of two approaches to modelling logical time based on concept of clocks. In: Ermolayev, V., et al. (eds.) Information and Communication Technologies in Education, Research, and Industrial Applications, vol. 2104, no. 2, pp. 696–706. CEUR-WS (2018)

A Mixed Method of Parallel Software Auto-Tuning Using Statistical Modeling and Machine Learning

Anatoliy Doroshenko, Pavlo Ivanenko, Oleksandr Novak, and Olena Yatsenko(✉)

Institute of Software Systems of National Academy of Sciences of Ukraine, Glushkov prosp. 40, Kyiv 03187, Ukraine
doroshenkoanatoliy2@gmail.com, {paiv, oayat}@ukr.net

Abstract. A mixed method combining formal and auto-tuning approaches and aimed at maximizing efficiency of parallel programs (in terms of execution time) is proposed. The formal approach is based on algorithmic algebra and the usage of tools for automated design and synthesis of programs based on high-level algorithm specifications (schemes). Parallel software auto-tuning is the method of adjusting some structural parameters of a program to a target hardware platform to speed-up computation as much as possible. Previously, we have developed a framework intended to automate the generation of an auto-tuner from a program source code. However, auto-tuning for complex and nontrivial parallel systems is usually time-consuming due to empirical evaluation of huge amount of parameter values combinations of an initial parallel program in a target environment. In this paper, we extend our approach with statistical modeling and neural network algorithms that allow to reduce significantly the space of possible parameter combinations. The improvement consists in automatic training of a neural network model on results of "traditional" tuning cycles and the subsequent replacement of some auto-tuner calls with an evaluation from the statistical model. The method allows, particularly, transferring knowledge about the influence of parameters on program performance between "similar" (in terms of hardware architecture) computing environments for the same applications. The idea is to reuse a model trained on data from a similar environment. The use of the method is illustrated by an example of tuning a parallel sorting program which combines several sorting methods.

Keywords: Algorithmic algebra · Automated program design · Auto-tuning · Machine learning · Neural network · Parallel computation · Statistical modeling

1 Introduction

The problem of optimal use of computing resources has always been important in the process of development of any software—from mobile applications to complex client-server systems. The auto-tuning paradigm [1, 2], which has become a standard for solving the problem of software application optimization over the last decade, allows to fully automatize this process for any computing environment. Its popularity lies on the

V. Ermolayev et al. (Eds.): ICTERI 2018, CCIS 1007, pp. 102–123, 2019.
https://doi.org/10.1007/978-3-030-13929-2_6

simplicity of use and independence from qualitative characteristics of a computer and an operating system. Auto-tuning traditionally uses empirical data for obtaining a qualitative evaluation of code being optimized (the quality usually refers to program execution time and accuracy of output results). It automates the search for the optimal program version out of a set of provided possibilities by running each candidate and measuring its performance on a given parallel architecture. Its main benefit is a high level of abstraction—a program is optimized without explicit knowledge of hardware implementation details, such as a number of cores, cache size or memory access speed on various levels. Instead, it needs to use subject domain concepts such as a number and a size of independent tasks.

In [3–8] we have developed a theory, methodology and tools for automated program design, synthesis and auto-tuning, based on Glushkov's system of algorithmic algebra (SAA) [4, 5] and term rewriting technique. The model for parallel programs optimization and auto-tuning framework named TuningGenie [7], aimed at automating the adjustment of programs to a target platform, were proposed. The framework operates with a source code of parallel software and performs source-to-source transformations by using facilities of a rule-based rewriting system called TermWare [3].

The main drawback of the auto-tuning approach is in a significant one-time cost of the optimization process: if the number of program versions is large enough, the optimization process may run for many hours and even days. In our previous work [9], we have introduced a hybrid auto-tuning approach, enhanced by means of statistical modeling and machine learning. This approach allowed to significantly prune the tuner's search space by replacing a part of empirical performance estimates with an evaluation from the statistical model. In this paper, we consider knowledge transfer between similar (in the sense of hardware architecture) environments for the same applications. The idea is to reuse models, trained on data from a similar environment, when applicable. The paper describes the mixed method of software development combining formal algebra-algorithmic and rewriting rule facilities, and auto-tuning tools extended by statistical modeling and neural networks.

The rest of the paper is organized as follows. In Sect. 2, the Glushkov's SAA and the toolkit for automated design and synthesis of programs, which is based on the use of the algorithmic algebra, are described. Section 3 is devoted to the developed auto-tuning framework TuningGenie and the general approach to application of machine learning in software auto-tuning. In Sect. 4, the example of usage of the developed tools and machine-learning technique for developing and tuning a parallel sorting algorithm is presented. The paper ends with a discussion of related work (Sect. 5) and conclusions (Sect. 6).

2 Automated Software Design Based on Algorithmic Algebra Facilities

In this section we give a brief consideration of algebra-algorithmic facilities applied for the formal design of algorithms and programs. The approach to the formal design is based on the use of Glushkov's system of algorithmic algebra [4, 5] intended for the formal representation, analysis and transformation of sequential and parallel algorithms.

Glushkov's SAA (or Glushkov's algebra, GA) is the two-sorted algebra $GA = <\{Pr, Op\}; \Omega_{GA}>$, where Pr and Op are the sets of predicates (logic conditions) and operators both defined on an information set *IS*; *IS* is a set of all data (input, output and intermediate) being processed by algorithms; Ω_{GA} is the signature consisting of logic and operator operations, which will be considered further. In this paper we use the natural linguistic form of operation representation; the algebraic form of the operations is given, for example, in [5]. The predicates take the values of the three-valued logic $E_3 = \{0, 1, \mu\}$, where 0 is for false, 1 is for true and μ is for unknown. The value μ is used to indicate that an error has occurred during the computation of a condition. The operators are mappings of *IS* to itself.

SAA is the basis for a language called SAA/1 [4], which has the advantage to be human-friendly, uses a natural linguistic representation of algorithms and can be translated to target programming languages C++, Java.

Algorithms represented in SAA/1 are called SAA schemes. Predicates and operators in SAA/1 can be basic or compound. Basic elements are considered in SAA schemes as primary atomic abstractions. Compound predicates are constructed from basic ones by means of the following generalized Boolean operations:

- disjunction: '*condition* 1' OR '*condition* 2';
- conjunction: '*condition* 1' AND '*condition* 2';
- negation: NOT '*condition*'.

Compound operators are built from elementary ones by using the following operations:

- serial execution of operators: "*operator* 1"; "*operator* 2";
- branching: IF '*condition*' THEN "*operator* 1" ELSE "*operator* 2" END IF;
- loop: WHILE '*condition*' LOOP "*operator*" END OF LOOP;
- asynchronous execution of n operators: PARALLEL(i = 0, …, n − 1)("*operator* i");
- synchronizer, which delays the computation until the value of the specified condition is true: WAIT '*condition*'.

The operations for formalization of main concepts of object-oriented programming (such as classes, objects, etc.) were also added to algorithmic algebra [8]. The use of the above operations for designing a parallel algorithm is considered in Subsect. 4.1.

The integrated toolkit for design and synthesis of programs (IDS) [4–6] supports automated construction of SAA schemes and generation of corresponding code in C++ and Java programming languages. IDS integrates three forms of design-time representation of algorithms: algebraic, natural linguistic and flowgraphs. The combination of these forms gives a comprehensive understanding of algorithm schemes and facilitates the achievement of demanded program quality. The toolkit consists of the following basic components (see Fig. 1):

- the scheme constructor, intended for automated design of algorithm schemes represented in SAA and synthesis of programs in target languages (C++, Java);
- the flowgraph editor, which is applied for editing a graphical representation of an algorithm;

- the database of SAA constructs, containing the description of SAA operations, basic predicates and operators, and also their program implementations;
- the generator of SAA schemes which operates on the basis of higher-level schemes (hyperschemes). The hyperscheme [5] is a parameterized algorithm for solving a certain class of problems and setting specific values of parameters; the subsequent interpretation of a hyperscheme allows obtaining algorithms adapted to specific conditions of their use.

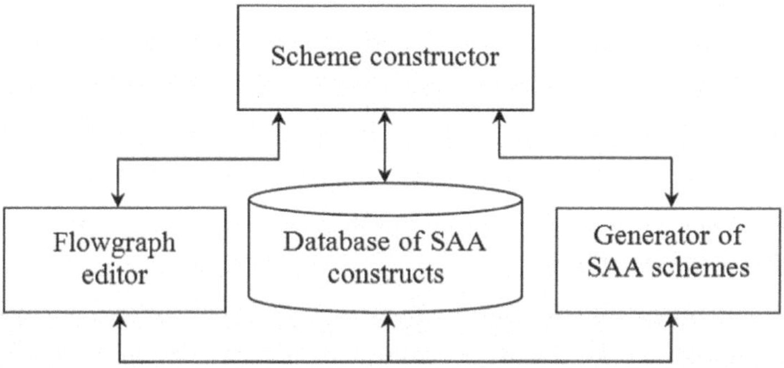

Fig. 1. The architecture of IDS toolkit.

The basic idea of the scheme constructor consists in descending design of algorithms by superposition of predefined SAA language constructs, which are considered as reusable components and are stored in the database. The design process is represented by a tree of an algorithm. The user selects the items from the list of SAA constructs and adds them to an algorithm tree. The list contains the logic and operator operations of SAA and also basic predicates and operators. On each step of the design process, the system allows a user to select only those operations, the insertion of which into a scheme does not break its syntactical correctness. The algorithm tree is then used for the automatic generation of SAA scheme text and a source code in a target programming language. The mapping of SAA operators to a text in a programming language is defined in a form of code templates in the database.

To automate the transformation (e.g., parallelization) of algorithms and programs, the IDS system is used together with the term rewriting system named Termware [6]. The facilities of SAA and developed software tools were applied for automated design and generation of parallel programs for multicore central processing units (CPUs) and graphics processor units (GPUs), particularly in the subject domain of weather forecasting [8].

IDS semi-automatically generates parallel source code based on SAA schemes. The application of formal methods in IDS and TermWare [6] enables the automation of manual work of programmers and a more advanced parallelization of algorithms.

However, a performance of programs being designed can be further increased by using the TuningGenie framework. Software auto-tuning is based on using the expert knowledge of the developer which is added to SAA schemes (during design in IDS) in the form of special comments (pragmas) describing tuning parameters. The TuningGenie pragmas are considered in the next section. The mixed use of IDS and TuningGenie for designing and tuning a parallel algorithm is illustrated in Subsect. 4.1.

3 Auto-Tuning Framework and Machine Learning

The TuningGenie framework has been developed in our previous work [7] and targets the automated generation of auto-tuning applications from a source code. The idea of an auto-tuner consists in the empirical evaluation of several versions of an input program and the selection of the best one with reduced execution time and higher result accuracy.

3.1 TuningGenie Facilities

TuningGenie framework works with program source code using the expert knowledge of the developer as well as certain automation facilities developed within it. The developer adds some metadata (parameter names and value ranges) to a source code in the form of special comments-pragmas, which will be considered further. By exploiting such expert knowledge, we can reduce the number of program versions to be evaluated and therefore increase auto-tuner performance.

The auto-tuning software implementation is based on the open-source rewriting rules system called TermWare [3]. It provides a language for describing rewriting rules that operate on data structures which are called terms, and a rule engine that interprets rules to transform terms. Informally, terms are tree-like structures of the form $f(t_1, t_2, ..., t_n)$, where subterms $t_1, t_2, ..., t_n$ are either tree nodes themselves or leaf nodes corresponding to constants or variables. TermWare works with arbitrary data encoded as terms. The Java source code is represented in a form of an abstract syntax tree (AST) as a term that is transformed with TermWare rules. The AST is derived from source code using the parser for Java language included into the TermWare. The general form of a TermWare rule is $source[condition] \rightarrow destination[action]$, where *source* is input sample, *destination* is a target sample, *condition* is a term defining the applicability of the rule, *action* is the operation executed when the rule triggers. Actions and conditions are optional components of a rule, which can call imperative code.

TuningGenie uses TermWare to extract expert knowledge from the program source code and generate a new version of a program on each tuning iteration while the knowledge base is used as an interlink between the pre-tuning and tuning phases. By manipulating terms representing an AST, TuningGenie can perform structural changes in a program using a declarative style. The current TermWare version contains components for interaction with C# and Java languages. To support other languages, both a parser to translate source code into the TermWare language as well as a pretty-printer should be implemented. The current TuningGenie version supports Java programs.

The common workflow of the TuningGenie framework is shown in Fig. 2. TuningGenie accepts as an input Java source code marked with pragmas describing configurations and transformations of a program that affect its performance. Pragmas are specified manually by program developers, using Java comments of a special form. During parsing of a source code into a term representing AST, the auto-tuner builds a set of configurations based on expert data extracted from pragmas. Then, these configurations are translated into rewriting rules. Also, on this preliminary stage some values of program parameters are calculated. The results of this stage include a program term, a set of parameterized rewriting rules and a set of rule configurations C that specify the specific parameter values. Each of these configurations specifies a unique version of the input program.

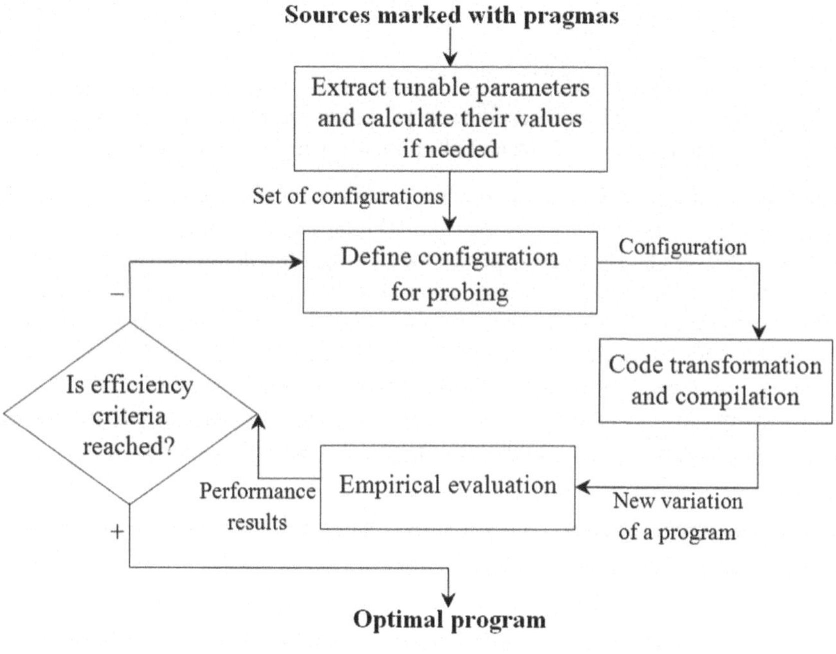

Fig. 2. Software tuning workflow in TuningGenie framework.

Then, TuningGenie searches for the most efficient configuration $C_{opt} \in C$ by iteratively performing the following steps.

1. Select a configuration to test $C^* \in C$. Configurations are selected sequentially from the set C. Size of C depends on values from pragmas. This means that developers are expected to use their expert knowledge to narrow the region of search (group pragmas only if they are really related, narrow boundary values for each pragma).

Additionally, a time limit can be used in cases where it is hard to estimate the overall optimization time (step 4).

2. Generate and compile the corresponding program variant. The parameters from a given configuration are substituted into the rules; then these rules are executed on the source term of the program. After the transformation is complete, the term is transformed into a Java source code using the TermWare facilities. The code is compiled using common JDK tools. The result of this step is a new version of the program ready for performance measurement.
3. Execute the program and evaluate its performance. A small launcher class is generated automatically. It runs the program and measures its execution time, therefore computing $f(C^*) = t$. Each configuration is executed several times. If the time consumed for each run differs a lot (more than 10%), then a warning message is logged.
4. If all configurations from set C have been evaluated, or if the optimization process has used up all the allotted time, then go to step 5, else go to step 1.
5. The optimal configuration is selected from a set of all configurations which have been evaluated: $C_{opt} = C^* : f(C^*) = f_{\min}$. For this configuration, an optimal program version is generated, as in step 2.

The program obtained as a result of steps 1–5 is saved and executed in a target environment. This program is considered optimal for a given architecture.

TuningGenie Pragmas. Currently, the TuningGenie framework supports three kinds of pragmas [7]: *bidirectionalCycle, calculatedValue* and *tunableParam*. The *bidirectionalCycle* pragma can be used to specify loops, where the iterations can be run in any order. The *calculatedValue* evaluates some function in a target environment and assigns the resulting value to a given variable. The *tunableParam* pragma specifies a search domain for an optimal value of a numeric variable. As we focus on tunable parameters in this paper, the *tunableParam* pragma is considered further.

The example of the *tunableParam* is given below. The pragma sets the possible values for a `threadCount` variable in a range [1...8] with a step 1:

```
//tunableParam name=threadCount start=1 stop=8 step=1
int threadCount = 1;
```

The *tunableParam* pragma is applicable to algorithms that use geometrical (data) parallelization: it allows to find the optimal decomposition of computation by estimating the size of a block which is executed on a single processor. It can also be applied when it is necessary to estimate the optimal number of some limited resources like a size of caches or a number of threads to be used in a program. Another use of *tunableParam* pragma is to find an optimal threshold value to switch to a different algorithm. Consider a sort algorithm which uses QuickSort or MergeSort for large arrays and applies the same sort recursively to subarrays. For some small size of a subarray, it is more efficient to switch to another sorting algorithm, such as InsertionSort. The optimal threshold value depends on processor architecture and can be determined experimentally. TuningGenie framework allows finding this value with minimal changes to source code:

```
//tunableParam name=threshold start=10 stop=100 step=5
int threshold = 10;
if (high - low < threshold) {
    insertionsort(array, low, high);
} else {
    addPartitionForQuickSort(array, low, high)
}
```

It should be noticed that by default all pragmas (and corresponding parameters) are treated as independent. It means that for each parameter's value from a single pragma, a separate version of a program is generated and measured; parameters from other pragmas are ignored (i.e., they take values specified in source code). So if there are n pragmas with a number of configurations contributed by each pragma denoted as N_i, then the total size of search space will be $\sum_{i=1}^{n} N_i$ and not $\prod_{i=1}^{n} N_i$. To mark parameters which need to be tested in correlation, optional *group* option must be specified in related pragmas.

Another important fact is that now all pragma's variables have a "local" scope and are tied to demarked instructions. That is why it is impossible to gather all pragmas in one place—they must be spread among the source code and duplicated for each section of duplicated code (if such is present).

Translating Pragmas to Rewriting Rules. For each pragma, a corresponding rewriting rule is generated. Then all generated rules are applied to the term representing the source code of the program. As an example, consider the following code fragment containing *tunableParam* pragma:

```
//tunableParam name=threshold start=1 stop=10 step=1
int threshold = 1;
```

The rule generated from this pragma (slightly simplified for readability) has the following form:

```
VariableDeclarator(Id("threshold"), Literal($value))
[$value != "newValue"]    ->
VariableDeclarator(Id("threshold"), Literal("newValue"))
```

Here `$value` is a TermWare variable. Notice the condition `[$value != "newValue"]` that was added to prevent infinite rule application. Rewriting rules are applied repeatedly until the moment when none of them is applicable. In this example, we need to prevent the infinite rule application in case when the substitution value `newValue` is equal to the initial `$value`. The actual value of `newValue` is taken from a range of values defined by the pragma. In this particular case, the `newValue` parameter is in the range [1...10] and each of its values would be automatically sequentially probed and estimated by the auto-tuner.

All rewriting rules, together with the strategy for their application and the knowledge base (facts base) [10], are combined into a single term system. The system performs a reduction of the initial term and obtains a new term. This term is translated back into a program source code using a target language (Java).

3.2 Application of Machine Learning

Auto-tuning for complex and nontrivial program systems usually takes a lot of time empirically estimating a large number of parameter combinations of an input program in a target environment. TuningGenie uses the expert knowledge of a developer of the program to be optimized to form the set C of parameters combinations. In this paper, we propose to optimize the auto-tuning method by using statistical modeling and machine learning. The improvement consists in reducing the number of auto-tuner launches by means of constructing an approximation model which allows dismissing the parameter combinations that are unlikely to lead to a fast program execution. The model approximation often results in a reduction of dimensionality of input parameters of the set C; this means that there can be a significant auto-tuning process speed-up.

Generally, machine learning methods are based on the concept of learning some knowledge from data [2, 11]. The learned knowledge can be of different nature, e.g., a classification, a model of a function. In the context of auto-tuning, the knowledge to be learned, for example, can be the program performance at different settings of program parameters. A machine learning method first evaluates several alternatives within the search space for n different input programs $P_1, \ldots, P_n$, defined by configurations $C_1, \ldots, C_n$. The set of evaluated alternatives is called *training data*. Ideally, a set of input applications should cover all possible program features. Some features of the input programs with the training data results are correlated once the training data has been evaluated. The process of generating and evaluating the training data and learning knowledge from this data is called *training*. Once the training is completed, and given a new version of program P' to be evaluated, execution of P' is replaced with an estimate, obtained from the trained model. The advantage of machine learning methods is that the training has to be done only once and tuning of a new program only requires querying the selected machine learning method with new program features [2].

Machine learning is closely linked to (and often overlaps with) computational statistics [12]. All statistical algorithms (including machine learning algorithms) require a significant number of statistical data for analysis and model construction. In the context of auto-tuning tasks, the collection of much statistical data can be a long process. Therefore, the problem of selecting the algorithms narrowing the search space at a minimal number of real launches of an auto-tuner is very acute. For a partial solution of the mentioned problem, in this work we use a neural network for data extrapolation (see Sect. 4). In this case, a relatively small number of real launches is required for the construction of an approximate model, after which the neural network model can be used by other algorithms according to the black box principle.

In general, the approach to solving machine learning tasks consists of the following steps [13].

1. Analysis of records and preparation of "raw" data for loading. Since the data may come from different sources and in different formats, everything must be converted to a single form.
2. Data preparation. At this stage, the issues, such as data incompleteness, noise, value conflicts, are solved. The main tasks of this stage are cleaning (filling of absent values, removal of distorted data), transformation (data normalization for reducing distortions), consolidation (creation of datasets for separate attributes or groups of attributes) and discretization (conversion of continuous attributes to categorical ones).
3. Data analysis and model construction.
4. Verification of the model on a test data set.

One of the methods for evaluation of the model accuracy is based on the confusion matrix [14]. The confusion matrix contains information about actual and predicted classifications done by a classification system. Performance of such systems is commonly evaluated using the data in the matrix. Table 1 shows the confusion matrix for a two-class classifier.

Table 1. Confusion matrix.

	Predicted positive	Predicted negative
Actual positive	True positive (*TP*)	False negative (*FN*), type II error
Actual negative	False positive (*FP*), type I error	True negative (*TN*)

The entries in the confusion matrix have the following meaning: *TP* is the number of correct predictions that an instance is positive; *FN* is the number of incorrect predictions that an instance is negative; *FP* is the number of incorrect predictions that an instance is positive; *TN* is the number of correct predictions that an instance is negative.

The accuracy Acc is the proportion of the total number of predictions that were correct. It is determined according to the following formula:

$$Acc = \frac{TP + TN}{TP + TN + FP + FN}.$$

4 A Case Study

In this section we illustrate the mixed method of software development combining the algebra-algorithmic and auto-tuning tools extended by statistical modeling and neural networks. In our mixed design process, we take a top-down formal transformational style provided by the IDS toolkit (see Sect. 2 and works [4–6]) with a subsequent auto-tuning stage. We begin with a high-level specification presented as a generalized scheme of an algorithm in SAA/1 [4]. The program generated in the toolkit is further optimized via the TuningGenie framework. The method is illustrated with an example of tuning a hybrid parallel sorting algorithm which applies a merge sort or an insertion sort depending on the block size of an input numerical array.

4.1 Sorting Algorithm Design

The initial SAA scheme of the hybrid sorting algorithm is given below and consists of a description of `ParallelMergeSort` class and a nested class `MergeSortTask`. The identifiers of basic operators and compound operators (class methods) in the SAA scheme are written with double quotes. The implementations of class methods begin with a string of "=" characters. For example, `"parallelMergeSort(array)"` is the method of the `ParallelMergeSort` class, where the `array` is the name of the formal parameter. The operator `"Declare a variable (parallelism) of type (int) = (1)"` is the example of a basic element with values of its parameters written in parentheses. The identifiers of basic predicates are written with single quotes, e.g., `'All threads completed work'`. The TuningGenie pragmas are written with the help of the basic operator `"Comment (...)"`. The `tuneAbleParam` pragma included in the method `parallelMergeSort(array)` of `ParallelMergeSort` class specifies the search domain for optimal values of variable `parallelism`, which defines the number of parallel threads. The `compute` method of `MergeSortTask` class contains `tuneAbleParam` pragmas specifying the domain for searching the optimal values of two variables: `insertionSortThreshold` (the size of a block at which the insertion sort is applied) and `mergeSortBucketSize` (the threshold for block size to be sorted sequentially within one thread). The algorithm scheme was translated to Java code using the IDS toolkit.

```
SCHEME PARALLEL HYBRID SORT

  CLASS ParallelMergeSort

    CLASS METHODS

    "parallelMergeSort(array)"
    ==== "Comment (tuneAbleParam name=parallelism start=1
                   stop=8 step=1)";
         "Declare a variable (parallelism) of
          type (int) = (1)";
         PARALLEL(i = 0, ..., parallelism - 1)
         (
            "MergeSortTask(array)"
         );
         WAIT 'All threads completed work';

    CLASS MergeSortTask EXTENDS RecursiveAction

      CLASS FIELDS
      "Declare an array (array) of type (int)";

      CLASS METHODS

      "MergeSortTask(array)"
      ==== (this.array := array);

      "compute"
      ==== "Comment (tuneAbleParam
                     name=insertionSortThreshold
                     start=10 stop=200 step=10)";
           "Declare a variable (insertionSortThreshold)
            of type (int) = (100)";

           "Comment (tuneAbleParam
                     name=mergeSortBucketSize
                     start=10000 stop=100000000
                     step=10000)";
           "Declare a variable (mergeSortBucketSize)
            of type (int) = (50000)";

           IF 'Length of array (array) is less or equal
               to (insertionSortThreshold)'
           THEN "insertionSort(array)"
           ELSE
```

```
              IF 'Length of the array (array) is less or
                  equal to (mergeSortBucketSize)'
              THEN
                 "sequentialMergeSort(array)";
              ELSE
                "Declare an array (left) of type (int)";
                "Declare an array (right) of type (int)";
                (left := "Copy from array (array) the
                          elements in the range from
                          (0) to (array.length / 2)");
                (right := "Copy from array (array) the
                           elements in the range from
                           (array.length / 2) to
                           (array.length)");
                ("MergeSortTask(left)"
                 PARALLEL
                "MergeSortTask(right)");
                WAIT 'All threads completed work';
                "merge(left, right, array)";
              END IF
           END IF;

    END OF CLASS MergeSortTask
  END OF CLASS ParallelMergeSort
END OF SCHEME PARALLEL HYBRID SORT
```

The auto-tuning of the above algorithm using the TuningGenie framework is considered in the following subsections.

4.2 The First Phase of Auto-Tuning

In the auto-tuning experiment, a set of 2×10^7 random integer numbers was used as an input of the hybrid sorting algorithm. The auto-tuner parameters are $C = \{T_{cn}, T_s, T_h\}$, where T_{cn} is a number of parallel threads (*parallelism* variable), T_s is a threshold for block size (*mergeSortBucketSize*) to be sorted sequentially within the current thread (blocks with $size > T_s$ are split into smaller blocks and assigned to different threads), T_h is a block size (*insertionSortThreshold*) at which insertion sort is used.

We consider knowledge transfer between two similar (in terms of hardware architecture) environments for the sorting application. The model trained on data from the first environment is reused in the second environment. The first environment is the following: 2.7 GHz Intel Core i7 processor (6820HQ) with 4 cores and 8 MB L3 cache; 16 GB 2133 MHz RAM; MacOS 10.12. The second computing environment is 2.3 GHz Intel Core i7 processor (4850HQ) with 4 cores and 6 MB L3 cache; 16 GB 1600 MHz RAM; MacOS 10.13.

The experiment consisted of two phases. In the first phase, the auto-tuner was executed without a statistical model to estimate how quick the tuned algorithm can be [9]. In the second phase, the statistical modeling was plugged in to understand how heavily the search space can be pruned while preserving the near-optimum performance of the tuned algorithm.

The results of the first phase are given in Table 2. Three configurations are listed: *slow* ("default" configuration that behaves almost as classical sequential merge sort); *optimal* (the quickest one that was automatically picked by the auto-tuner) and *intuitive* (values are filled in by intuition with respect to known hardware specifications and algorithms details). The *optimal* configuration is 4.93 times quicker than the *slow*. This result is quite good for 4-core processor and was achieved primarily by a combination of two factors: optimal usage of processor caches (by switching to in-place sorting for small data sets) and efficient parallelization schema (merge sort is easy to parallelize with "divide and conquer" method). The *intuitive* combination was 3.1 times faster than the *slow*, which is also a decent result, but it was easy to guess due to the relative simplicity of the test algorithm. Usually, optimal configurations are not so obvious for real-life parallel programs. The *optimal* configuration is still substantially quicker—by 58%, so we can say that it was worth the time spent on tuning.

Table 2. The results of the first auto-tuning phase.

Configuration	*Slow*	*Optimal*	*Intuitive*
Parallelism level T_{cn}	1 (one thread)	8	4
Insertion sort threshold T_h	0 (do not switch to insertion sort at all)	120	30 (common notion is to set couple dozen as a threshold for this trick)
Threshold for sequential sorting T_s	100 000 000 (it's bigger than the test data size, so no data decomposition is applied)	50 000	10 000
Test data size	20 000 000 integers		
Average sorting time	4432 ms	898 ms	1426 ms

4.3 The Second Phase of Auto-Tuning

Now, let's move to the second phase to see how the auto-tuner's search space can be reduced with the help of statistical analysis methods. The T_s parameter was excluded from the model during the primary analysis phase due to its minor impact on the overall performance: once the number of subtasks after the decomposition of input data is a couple of times bigger than the parallelism level, it makes almost no difference what value is used for it. This can be explained by the high effectiveness of the Java's *RecursiveAction* [15] mechanism that was used in the implementation. The *RecursiveAction* is a recursive *ForkJoinTask*, which is "a thread-like entity that is much lighter weight than a normal thread. Huge numbers of tasks and subtasks may be hosted by a small number of actual threads in a ForkJoinPool, at the price of some usage limitations" [16]. The experiment proved that the computational overhead on executing new *RecursiveAction* is negligible.

The primary analysis of data was performed in Python language with a help of the open-source machine learning library named Scikit-learn [17], which offers various classification, regression and clustering algorithms. Further analysis was implemented by means of R [18], which is a programming language for statistical computations, analysis and graphical representation of data. The experiment consists of several stages: preparation and loading of auto-tuner results to the R environment, data preparation (including normalization), building a neural network model on a training dataset and evaluating the model on a test dataset.

The data analysis process is shown in Fig. 3. At first, the auto-tuner performs N experiments and saves the result data to a separate file. The data is used by the neural network for training. After the training, the neural network extrapolates the data, generates the new dataset, which is written into a separate file. In the end, both datasets are analyzed and compared by a human. As a neural network, a multilayer perceptron with three input neurons, three hidden layers (20-10-5 neurons per layer) and one output neuron were applied. The rectified linear function $f(x) = \max(0, x)$ was used as an activation function. The backward propagation of errors was used as a machine learning method and the Broyden-Fletcher-Goldfarb-Shanno algorithm [19] was applied for the optimization of weighting factors.

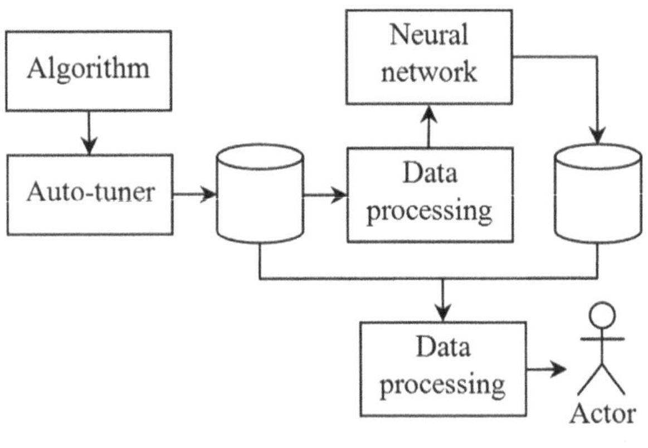

Fig. 3. The process of analysis.

The initial neural network was built based on the results of 3300 launches. Then it was used for further data generation. The use of the neural network for initial approximation allowed to reduce the search region by 58% (from 10^6 to 4.2×10^5). For estimating the quality of the obtained results, more than 30000 real launches (evenly distributed over the combinations set) of the auto-tuner was performed. Then, the results of neural network prediction (on the basis of 3 thousand launches) were compared to actual results. The evaluation of the accuracy *Acc* of the model is based on a confusion matrix (see Subsect. 3.2). Figure 4 shows the dependency of the model accuracy *Acc* from 10 neural networks on the ratio of sample data used for training. For the case of using 10% of all measurements (i.e., 3300 measurements), the confusion matrix is as shown in Table 3. The model accuracy is $ACC = 97\%$.

Table 3. The confusion matrix at using 3300 measurements.

	Predicted positive	Predicted negative
Actual positive	19978	570
Actual negative	488	9506

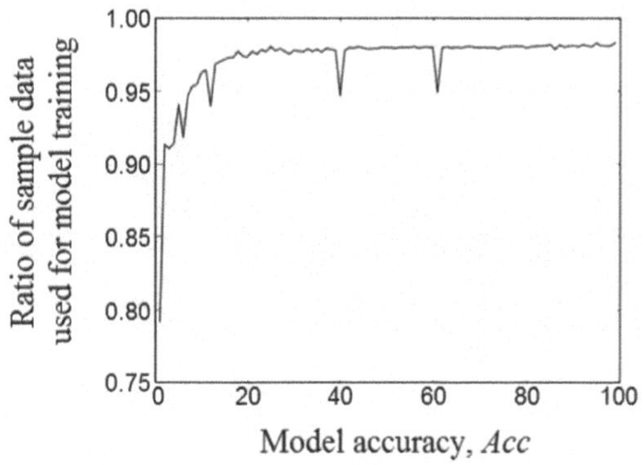

Fig. 4. The dependency of the model accuracy *Acc* on the ratio of sample data used for model training.

It could be concluded that at solving the task of decreasing the search space for the auto-tuner, the system needs a relatively small number of measurements for obtaining sufficiently accurate results.

4.4 Analysis of Parameters' Influence on Application Performance

To reuse the obtained model in a similar computing environment, it is necessary to estimate to what extent the parameters of the sorting algorithm influence the application performance. There are three tuning parameters defined in the program: *parallelism*, *mergeSortBucketSize* and *insertionSortThreshold*. Figures 5, 6 and 7 visualize how computation time (in milliseconds) depends on each pair of them in the first computing environment.

Figure 5 shows the dependency of execution time on *parallelism* and *insertionSortThreshold* parameters. The *mergeSortBucketSize* parameter is fixed to 500000 elements. The blue zone depicts the quickest configurations, the gray one represents the slowest. As can be seen, latency heavily depends on the used number of threads (*parallelism* parameter)—the best results are achieved at $parallelism \geq 8$. This is explained by the number of cores (four) and hyper-threading technology of Intel processors. At small values of *parallelism*, not all processor cores are utilized, and computation is almost four times slower than quick configurations of the blue zone. The *insertionSortThreshold* parameter has a less impact on performance. At large values of *insertionSortThreshold* (200–400) and *parallelism* (13–20), the results become unstable, which is shown as peaks in the right side of the chart. This can be explained by competition of threads for resources at large values of *parallelism*. The efficiency of L1 caches usage falls, since data blocks are larger than the cache size. Therefore, the insertion sort becomes slower.

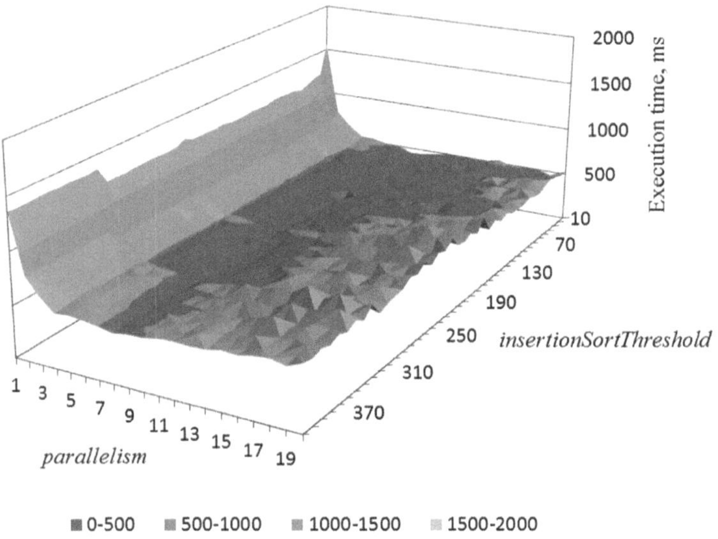

Fig. 5. The dependency of execution time on *parallelism* and *insertionSortThreshold* at a fixed value of *mergeSortBucketSize* = 500000.

Figure 6 shows the results of experiments at fixed *parallelism* = 12. As it can be seen, *mergeSortBucketSize* and *insertionSortThreshold* parameters impact the overall application's performance in a reduced way with respect to the number of used threads (the execution time lies approximately in the range 400–600 ms).

In Fig. 7, the results of experiments at fixed *insertionSortThreshold* = 100 are given. Again, as can be seen, the number of threads (*parallelism*) has a much more significant influence on performance and execution time varies between 400 and 1200 ms. The quickest configurations are in small extremum zones of orange color. The quickest configurations are in the gray zone.

The general conclusion to be drawn from the obtained results is that *parallelism* is the most influential parameter. The quickest configurations are at values of *parallelism* close to 8 (4 cores with using hyper-threading) when threads do not compete for resources. The model obtained for the first environment proved to be correct for the second (similar) environment.

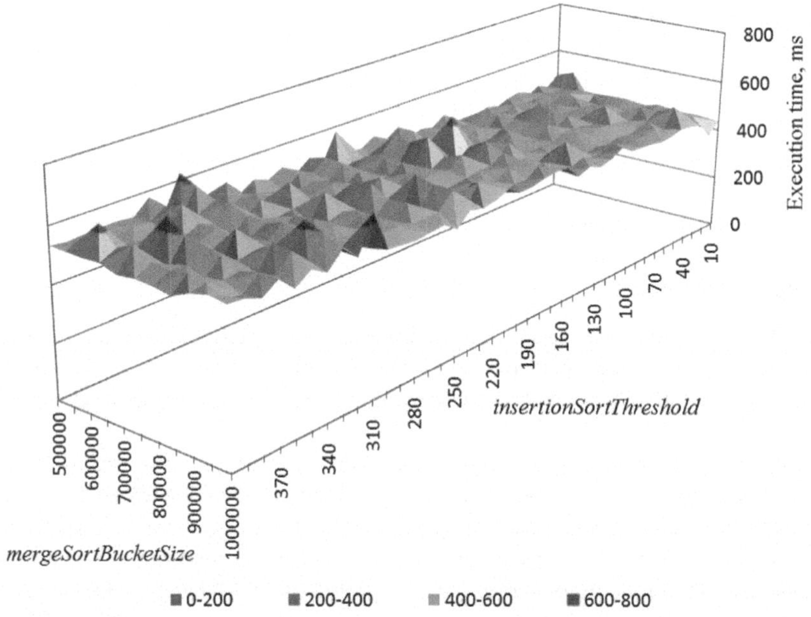

Fig. 6. The dependency of execution time on *mergeSortBucketSize* and *insertionSortThreshold* at a fixed value of *parallelism* = 12.

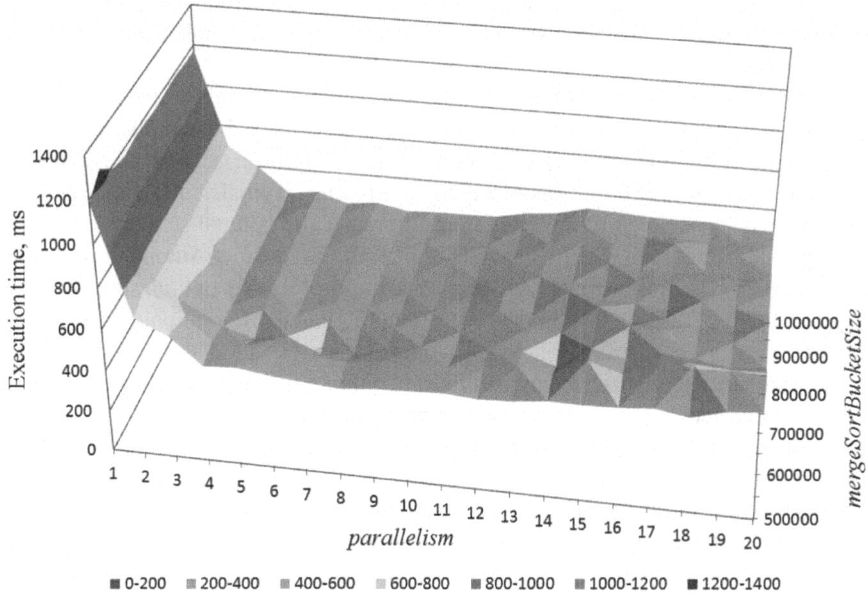

Fig. 7. The dependency of execution time on *parallelism* and *mergeSortBucketSize* at a fixed value of *insertionSortThreshold* = 100.

5 Related Work

Many software solutions have been proposed for the problem of auto-tuner development. Auto-tuners usually are distributed among the following categories [20] according to their implementation details:

- independent libraries;
- stand-alone applications;
- part of operating systems;
- compiler extensions.

Well-known examples of specialized libraries are ATLAS [21] and FFTW [22], which introduce the high-performance implementation of some specific functions. Unlike representatives of other categories, they are tied to a specific domain and language.

In the case of an auto-tuner as a stand-alone software application, the tuner is created separately from the software being optimized. The framework that we use for the generation of auto-tuners, i.e., TuningGenie, falls into this category. TuningGenie is quite similar to Atune-IL [23], a language extension for auto-tuning. It also uses pragmas and is not tied to some specific programming language. Other familiar examples of auto-tuners of this category are ActiveHarmony [24], POET [25] and FIBER [26].

Auto-tuner as an operating system extension [1, 20] is one of the most promising approaches, where the universal auto-tuner is a part of an operating system and is involved in task scheduling. Unfortunately, because of architectural complexity of such solution and variety of operating systems, an effective software implementation has not been developed yet.

Examples of auto-tuning compiler extensions are considered, for instance, in [27–29]. In [27], machine learning techniques are used to decrease the compile time for the static commercial compiler TPO (Toronto Portable Optimizer), while preserving the execution time. In [28], an open-source self-tuning compiler Milepost GCC is described, which exploits machine learning to predict the optimal setting of compilation flags for a program when using GCC. In [29], a Tool for Automatic Compiler Tuning (TACT) is presented, which performs multi-objective optimization of GCC compiler options using the improved Strength Pareto Evolutionary Algorithm (SPEA2).

Software auto-tuning uses many methods, beginning from simple random search to more advanced techniques like machine learning or evolutionary search [2]. In [30], sequential model-based optimization strategies were developed for algorithm configuration problems based on using both Gaussian stochastic process models and random forests. The work in [31] focuses on hyperparameter optimization using the sequential model-based Bayesian optimization framework (SMBO) [30]. In [32], neural networks are used for learning the knowledge about a given program transformation (parametric loop tiling) for different values of input parameter (tile size); the model is then used to search for optimal parameter values. In paper [33], a machine learning approach is applied for automatic optimization of task partitioning for OpenCL, a framework for writing programs that execute across heterogeneous platforms consisting of CPUs, GPUs and other processors or hardware accelerators. The optimization is considered for different input problem sizes and different heterogeneous architectures.

In our work, we use neural networks for learning on results of tuning cycles (program execution time at different values of internal program parameters) with subsequent replacement of some auto-tuner calls with an evaluation from the model. The main advantage of the TuningGenie used in our paper in comparison to related solutions consists in using the term rewriting engine for source code transformation. Representing program code as a term allows modifying the program structure in a declarative way, which significantly increases the efficiency of software auto-tuning. As well as the authors of the above-mentioned work [31], we use the machine learning model to reduce the time needed for searching the optimal parameter values. However, the mentioned work applies the Bayesian optimization methods, which use a probabilistic model to describe the relationship between parameter configuration and its performance, and therefore requires more costs to configure the model and reach a higher level of accuracy. At the same time, in contrast to our work, which focuses on optimizing numerical (i.e., integer-valued) parameters, paper [31] additionally targets categorical (i.e., discrete-valued and unordered) domains. Compared to work [32], which considers the parametric loop tiling, our approach is more general and can be applied for tuning arbitrary numerical parameters. Works [32, 33] use machine learning techniques to automatically build a model and optimize a program, while our method on some stage requires a developer to analyze the result statistical data. The full automatization of the method is one of the subjects of future work.

6 Conclusion

This paper proposes a mixed method for software development by combining formal algebra-algorithmic facilities and rewriting rule techniques, as well as auto-tuning tools extended by statistical modeling and neural networks. The method works at the design-time phase of development and allows:

- to generate a basic view of the parallel application code in a semi-automatic manner through program transformation reducing notably the manual labour of a developer;
- to get rid substantially of the main weakness of the auto-tuning methodology, namely, significantly accelerate the search for an optimal program version by the automatic training of a neural network model on the results of regular tuning cycles and subsequent replacement of some auto-tuner calls with an evaluation from the obtained model.

Furthermore, the use of a perceptron at the primary analysis stage helps to identify the most important input parameters (i.e., which have the largest influence on a result). The method allows to transfer knowledge about the influence of parameters on program performance between similar (in terms of hardware architecture) computing environments for the same applications. The idea consists in reusing the model trained on data obtained in a similar environment.

The use of the method is illustrated by the example of performance tuning of a hybrid parallel sorting program that exploits the developed earlier TuningGenie framework. The results of the experiment confirm the efficiency of the proposed methodology and are encouraging for its further development for more complex approximation functions and conducting experiments with more computationally and semantically complex programs.

References

1. Naono, K., Teranishi, K., Cavazos, J., Suda, R.: Software Automatic Tuning: From Concepts to State-of-the-Art Results. Springer, Berlin (2010). https://doi.org/10.1007/978-1-4419-6935-4
2. Durillo, J., Fahringer, T.: From single- to multi-objective auto-tuning of programs: advantages and implications. Sci. Program. **22**(4), 285–297 (2014)
3. Doroshenko, A., Shevchenko, R.: A rewriting framework for rule-based programming dynamic applications. Fundamenta Informaticae **72**(1–3), 95–108 (2006)
4. Andon, P.I., Doroshenko, A.Y., Tseytlin, G.O., Yatsenko, O.A.: Algebra-Algorithmic Models and Methods of Parallel Programming. Akademperiodyka, Kyiv (2007). (in Russian)
5. Yatsenko, O.: On parameter-driven generation of algorithm schemes. In: Popova-Zeugmann, L. (ed.) CS&P'2012, pp. 428–438. Humboldt University Press, Berlin (2012)
6. Doroshenko, A., Zhereb, K., Yatsenko, O.: Developing and optimizing parallel programs with algebra-algorithmic and term rewriting tools. In: Ermolayev, V., Mayr, H.C., Nikitchenko, M., Spivakovsky, A., Zholtkevych, G. (eds.) ICTERI 2013. CCIS, vol. 412, pp. 70–92. Springer, Cham (2013). https://doi.org/10.1007/978-3-319-03998-5_5
7. Ivanenko, P., Doroshenko, A., Zhereb, K.: TuningGenie: auto-tuning framework based on rewriting rules. In: Ermolayev, V., Mayr, H., Nikitchenko, M., Spivakovsky, A., Zholtkevych, G. (eds.) ICTERI 2014. CCIS, vol. 469, pp. 139–158. Springer, Cham (2014). https://doi.org/10.1007/978-3-319-13206-8_7
8. Doroshenko, A., Ivanenko, P., Ovdii, O., Yatsenko, O.: Automated program design – an example solving a weather forecasting problem. Open Phys. **14**(1), 410–419 (2016)
9. Doroshenko, A., Ivanenko, P., Novak, O., Yatsenko, O.: Optimization of parallel software tuning with statistical modeling and machine learning. In: Ermolayev, V., et al. (eds.) ICTERI 2018. Communications in Computer and Information Science, vol. 2105, pp. 219–226. Springer, Cham (2018)
10. TermWare Tutorial. http://www.gradsoft.ua/rus/Products/termware/docs/tutorial_eng.html. Accessed 30 Nov 2018
11. Mitchell, T.M.: Machine Learning, 1st edn. McGraw-Hill Education, New York (1997)
12. Givens, G.H., Hoeting, J.A.: Computational Statistics, 2nd edn. Wiley, Chichester (2012)
13. Han, J., Kamber, M., Pei, J.: Data Mining: Concepts and Techniques, 3rd edn. Morgan Kaufmann, Burlington (2011)
14. Fawcett, T.: An introduction to ROC analysis. Pattern Recogn. Lett. **27**(8), 861–874 (2006)
15. Class RecursiveAction (Java SE 9 & JDK 9) – Oracle Help Center. https://docs.oracle.com/javase/9/docs/api/java/util/concurrent/RecursiveAction.html. Accessed 30 Nov 2018
16. Class ForkJoinTask (Java SE 9 & JDK 9) – Oracle Help Center. https://docs.oracle.com/javase/9/docs/api/java/util/concurrent/ForkJoinTask.html. Accessed 30 Nov 2018
17. Pedregosa, F., et al.: Scikit-learn: machine learning in Python. J. Mach. Learn. Res. **12**, 2825–2830 (2011)
18. Crawley, M.J.: The R Book, 1st edn. Wiley, Chichester (2012)

19. Fletcher, R.: Practical Methods of Optimization, 2nd edn. Wiley, Chichester (2000)
20. Karcher, T., Schaefer, C., Pankratius, V.: Auto-tuning support for manycore applications: perspectives for operating systems and compilers. ACM SIGOPS Oper. Syst. Rev. **43**(2), 96–97 (2009)
21. Whaley, R., Petitet, A., Dongarra, J.J.: Automated empirical optimizations of software and the ATLAS Project. Parallel Comput. **27**(1–2), 3–35 (2001)
22. Frigo, M., Johnson, S.: FFTW: an adaptive software architecture for the FF. Acoust. Speech Sig. Process. **3**, 1381–1384 (1998)
23. Schaefer, C.A., Pankratius, V., Tichy, W.F.: Atune-IL: an instrumentation language for auto-tuning parallel applications. In: Sips, H., Epema, D., Lin, H.-X. (eds.) Euro-Par 2009. LNCS, vol. 5704, pp. 9–20. Springer, Heidelberg (2009). https://doi.org/10.1007/978-3-642-03869-3_5
24. Tapus, C., Chung, I.-H., Hollingsworth, J.K.: Active harmony: towards automated performance tuning. In: 2002 ACM/IEEE Conference on Supercomputing, SC 2002, pp. 1–11. IEEE Computer Society, Los Alamitos (2002)
25. Yi, Q., Seymour, K., You, H., Vuduc, R., Quinla, D.: POET: parameterized optimizations for empirical tuning. In: Parallel and Distributed Processing Symposium 2007, IPDPS 2007, p. 447. IEEE Computer Society, Piscataway (2007)
26. Katagiri, T., Kise, K., Honda, H., Yuba, T.: FIBER: a generalized framework for auto-tuning software. In: Veidenbaum, A., Joe, K., Amano, H., Aiso, H. (eds.) ISHPC 2003. LNCS, vol. 2858, pp. 146–159. Springer, Heidelberg (2003). https://doi.org/10.1007/978-3-540-39707-6_11
27. Pekhimenko, G., Brown, A.D.: Efficient program compilation through machine learning techniques. In: Naono, K., Teranishi, K., Cavazos, J., Suda, R. (eds.) Software Automatic Tuning, pp. 335–351. Springer, New York (2010). https://doi.org/10.1007/978-1-4419-6935-4_19
28. Fursin, G., et al.: Milepost GCC: machine learning enabled self-tuning compiler. Int. J. Parallel Program. **39**(3), 296–327 (2011)
29. Plotnikov, D., Melnik, D., Vardanyan, M., Buchatskiy, R., Zhuykov, R., Lee, J.-H.: Automatic tuning of compiler optimizations and analysis of their impact. In: 8th International Workshop on Automatic Performance Tuning (iWAPT 2013). Procedia Computer Science, vol. 18, pp. 1312–1321. Elsevier B.V., Amsterdam (2013)
30. Hutter, F., Hoos, H.H., Leyton-Brown, K.: Sequential model-based optimization for general algorithm configuration. In: Coello, C.A.C. (ed.) LION 2011. LNCS, vol. 6683, pp. 507–523. Springer, Heidelberg (2011). https://doi.org/10.1007/978-3-642-25566-3_40
31. Eggensperger, K., Hutter, F., Hoos, H.H., Leyton-Brown, K.: Efficient benchmarking of hyperparameter optimizers via surrogates. In: 29th AAAI Conference on Artificial Intelligence (AAAI 2015), pp. 1114–1120. AAAI Press, Palo Alto (2015)
32. Rahman, M., Pouchet, L.-N., Sadayappan, P.: Neural network assisted tile size selection. In: 5th International Workshop on Automatic Performance Tuning (iWAPT 2010), pp. 1–15. Springer, Berkeley (2010)
33. Kofler, K., Grasso, I., Cosenza, B., Fahringer, T.: An automatic input-sensitive approach for heterogeneous task partitioning. In: 27th ACM International Conference on Supercomputing (ICS 2013), pp. 149–160. ACM, New York (2013)

ICT in Education and Education Management

Evolutionary Revision Model for Improvement of Computer Science Curriculum

Hennadiy Kravtsov and Vitaliy Kobets(✉)

Kherson State University, 27 Universitetska st., Kherson 73000, Ukraine
kgm@ksu.ks.ua, vkobets@kse.org.ua

Abstract. The onrush of information technology raises the task of revising curricula of specialties in computer science. The *objectives of the study* are to define the requirements and develop a model of the curriculum revision system in computer science.

Subject of the research: curriculum revision system in computer science.

Research methods: review and analysis of scientific publications, modeling of complex systems, questionnaires, expert method of evaluation.

The model of the revision system of the curriculum was created in accordance with the modern requirements of employers to the competence of university graduates in computer sciences. The architecture of this system is based on the research on the functionality of such elements as the Top Competences (required in the labor market), Employers' Requirements, Learning outcomes (expected results of teachers) and the formation of the ICT curriculum. The expert method is used to assess the quality of the revision model of the curriculum on the example of a specialty in the field of computer science.

Results of the research: the requirements of the labor market to the competence of university graduates were investigated, systematized and formulated in the field of computer science. These requirements are the basis for the formation of a new or improved university curriculum.

Keywords: Curriculum in computer science · Curriculum revision model · Requirements for the competences · Learning outcomes · Expert method

1 Introduction

Universities play an important part in modeling, creating and developing innovative systems in the economy. To promote innovative systems successfully, universities must effectively manage the dissemination of knowledge [1]. Entrepreneurial skills and innovative efforts in the field of information technology increase the scale, complexity and connect an increasing number of key stakeholders who can offer key innovative changes in the curriculum of ICT. In an era of growing global competition, it is asserted that innovative and entrepreneurial competencies are key sources of the competitive advantage of the curriculum in the field of computer science [2]. The importance of crossing and revising the global, national, industrial and local needs of stakeholders that form the ultimate demand for university graduates should be taken into account in an innovative curriculum in computer science. This approach offers a way to develop

V. Ermolayev et al. (Eds.): ICTERI 2018, CCIS 1007, pp. 127–147, 2019.
https://doi.org/10.1007/978-3-030-13929-2_7

and improve the curriculum in the field of computer science as a link between education, technological innovation, production and development [3].

It is especially important to reach a high level of education in the field of computer science in Ukrainian universities because by 2020 the IT industry of Ukraine has every chance to take the second place in the country's export structure reaching 7.7 billion dollars. Within five years, the number of IT professionals in Ukraine should reach 200000 people. The need for highly qualified specialists also occurs in all developed and developing countries. Broad and intensive development of IT technologies, their use in the economy determines the need for specialists in new specialties in the field of computer science, and also requires more frequent updating of university curricula.

The purpose of this study is to develop a model for creating a new system and revising the existing curriculum in computer science for a master's program using the requirements of stakeholders at the international and local levels.

The paper is organized as follows: Sect. 2 describes modern requirements for the curriculum, Sect. 3 is devoted to development of model of the curriculum revision system in computer science, Sect. 4 includes implementation of the model into the educational process, Sect. 5 explains the expert method in the model for improvement of computer science curriculum, and the last part concludes.

2 System Analysis

2.1 Modern Requirements for the Curriculum

Law of Ukraine "On Education" [4] and law of Ukraine "On Higher Education" [5] provides National Qualifications Framework [6], which creates framework for curriculum but does not indicate requirements to all content of curriculum, competences and learning outcomes (LO). Traditional approach in former curricula criticizes LO as unethical and against tradition because the role of teacher is eliminated. At the same time LO are important as they are the basis for recognition process according to Lisbon Recognition Convention. LO approach is more relevant to the labor market and is certainly more flexible, taking into account issues of lifelong learning, non-traditional forms and informal learning. Standards of Higher Education in the field of Computer Science, which includes competences and LO, were developed by Ministry of Education and Science of Ukraine [7].

The model of the curriculum is designed to eliminate the difference between the traditional approach to the development of curricula with normative disciplines and the modern approach to the development of curricula with normative competences. According to modern MSIS 2016 approach [8] a curriculum is specified using graduate competencies as its foundational element instead of courses or knowledge areas, units, and topics (Fig. 1). We support advantages of competence-oriented approach, which change our learning and teaching paradigm from what student knows to what student is expected to be able to do. Traditionally, most ICT curriculums have been structured around a typically hierarchical Knowledge Area – Knowledge Unit – Topic structure that together forms a Body of Knowledge. Knowledge areas include several knowledge units. Each knowledge unit, in turn, is divided into topics [9].

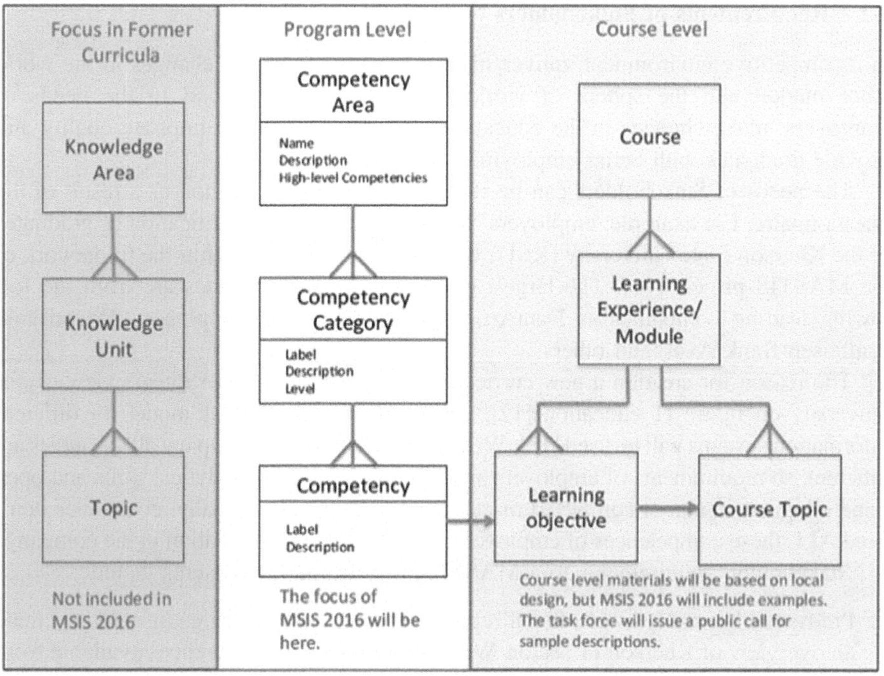

Fig. 1. Curriculum model for traditional and modern approach [8].

A potential problem with the traditional structure is that it focuses mostly on cognitive aspects of learning and leaves requirements of labor market out. A curriculum based on a knowledge area structure conveys relatively few skills and attitudes on what the graduates are able to do at the time of graduation. Instead of defining a knowledge area or a set of courses, a modern approach to the specification of the curriculum defines a set of graduate competencies, through which they can use their knowledge, skills and attitudes to successfully perform the assigned tasks.

The curriculum specifies high-level competency areas. The competency areas are divided into competency categories which consist of actual competencies. Competency areas and competency categories are much more stable and less technology-dependent than the competencies themselves which have relatively high likelihood of changes and local variations than at the higher levels. Each competency area has a name, a brief description and a few high-level dimensions. Each category and competency within a category is specified with a name and a brief description too. Competence area covers high-level competencies which enable the graduates to achieve their goals. Competence categories consist of low-level competences which are the tactical tasks to achieve strategic goal.

2.2 Requirements of Stakeholders

In a competitive environment, universities take into account the changes in the world labor market and the sphere of world education. They respond to the needs of employers; make changes in the educational process in order to improve quality and provide graduates with better employment opportunities.

The needs of Stakeholders can be studied by the survey method as a result of the questionnaire. For example, employers' requirements for the qualification of graduates of the Kherson State University (KSU) were investigated [10] within the framework of the MASTIS project [11]. The largest employers of our graduates are from the following leading companies as DataArt, Logicify, Wezom, Autoplaneta, PrivatBank, Raiffeisen Bank Aval, and others.

The reason for creation a new curriculum is new insights and strategic view of our university on future IT education [12], where simulation business model for different information systems will be used [13]. We conclude that in each company all vacancies are different, so requirements of employers are different too. Good analytical skills and open mind are great results of completed master degree education. Actually, companies don't need ALL these competences of employees; it depends on their position in the company.

Stakeholders requirements to a MASTER in Information Systems include:

1. Pre-research – analysis of official reports, research and strategies in order to create an overview of Kherson IT sector. We have used data and references available from official data of Ukrainian statistics, Ministry of Science and Education, Reports of IT Industry 2015. KSU has been engaged in collecting and maintaining lists of IT companies and IT departments, with CEO & Heads' names, phone numbers, e-mail and web addresses, etc.
2. Online questionnaire of IT companies (https://goo.gl/N7QbFp).
3. Interviews and meetings with stakeholders. CISEEC of KSU organized separate interviews with different stakeholders.
4. Collection and analysis of information obtained from pre-research, online questionnaire and meetings with stakeholders to assess the courses according to the requirements to specialists training in the field of information systems. CISEEC of KSU is engaged in the analysis of the results.

Among our stakeholders there were the representatives of small, medium and large companies [10].

After processing the online questionnaire of employers prepared by 6 Ukraine universities (Simon Kuznets Kharkiv National University of Economics, Ukraine National Technical University of Ukraine "KPI", Lviv Polytechnic National University, Vinnytsia National Technical University, Kherson State University, National Technical University "Kharkiv Polytechnic Institute") and 2 Montenegro universities (University of Donja Gorica, University "Mediterranean" Podgorica), a list of required competences for the Master of Science in Information Systems (MSIS) was generated (Table 1). Thus this table includes requirements of national labor market. Global labor market is presented by employers of EU universities (members of MASTIS project), among them are University Lumiere Lyon 2 (France), Guido Carli Free International University for Social Studies (Italy), University of Münster (Germany), Kaunas University of

Technology (Lithuania), University of Maribor (Slovenia), University of Agder (Norway), Lulea University of Technology (Sweden), University of Liechtenstein (Liechtenstein), Italian Association for Informatics and Automatic Calculation (Italy).

Relevance of each competency area for MSIS were calculated as average rank of all employers (from 69 = Most important, to 1 = Less important), after that Relevance of area for each item was estimated as ratio of average rank of this item to sum of average ranks of all items (expressed as a percentage).

Table 1. Stakeholders grades of MSIS 2016 competency areas [10].

Competency areas	Relevance of area, %
Business Continuity and Information Assurance	8.78
Systems Development and Deployment	10.61
Data, Information and Content Management	11.33
Ethics, Impacts and Sustainability	12.19
Enterprise Architecture	12.55
IS Strategy and Governance	12.84
Innovation, Organizational Change and Entrepreneurship	13.27
IS Management and Operations	9.47
IT Infrastructure	8.96

After an interview with main employers in Kherson (Ukraine) we obtained the following results using previous method of average ranks (Fig. 2).

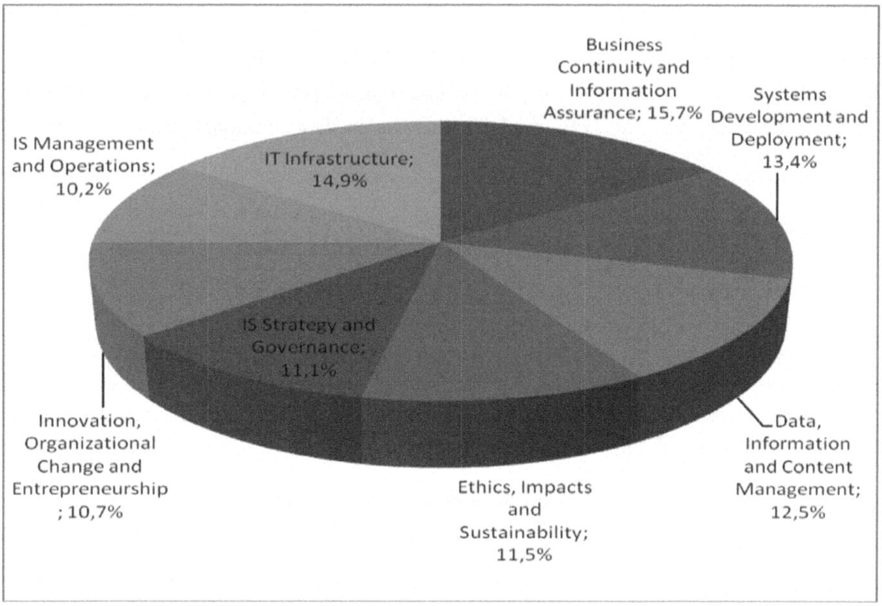

Fig. 2. Relevance for competency areas of Kherson employers, %.

After comparing the requirements of global (Table 1) and local Kherson labor markets (Fig. 2) we have got a significant overestimation for IS Management and Operations (+3.1%) and a significant underestimation for Business Continuity and Information Assurance (−6.92%) and IT Infrastructure (−5.4%).

Questionnaire of local Kherson employers according to National Qualifications Framework [6] and employers of global companies according to MSIS 2016 Global Competency Model [8] demonstrates different results using average rank method (from 5 = Most important, to 1 = Less important). Specific subject competencies are more important for local employers whereas international companies value both generic and specific subject competencies (Table 2).

Table 2. Analysis of local and global stakeholders' requirements to generic and specific subject competences of graduates for MSIS.

Online questionnaire competence specification grades	(1 - min; 5 - max)	Average result of stakeholders' grades for AIS MSIS 2016 competences	(1 - min; 5 - max)
Systematic competences	3.86	IS Strategy and Governance	3.58
		IS Management and Operations	2.64
Methodological competences	4.17	Data, Information and Content Management	3.16
Social/personal competences	3.37	Ethics, Impacts and Sustainability	3.40
		IT Infrastructure	2.50
Professional competences: analysis, design, and project management	4.17	Systems Development and Deployment	2.96
Professional competences: implementation and systems administration	4.10	Business Continuity and Information Assurance	2.45
		Enterprise Architecture	3.50
Research and academic/ analytical competences	4.13	Innovation, Organizational Change and Entrepreneurship	3.70

3 Model

When building a model of the curriculum revision system in computer science, we will use the E-Competence Framework 3.0 specification as an educational standard [8, 14]. The employers' requirements for the master program in Information Systems (MPIS) are the combination of the following realms (Fig. 3):

1. Computing/ICT subject area (IT industry);
2. IS Management (IT departments of small and medium enterprises and large companies);
3. Domain of practice (or Internship of master students);
4. Generic individual skills (soft skills).

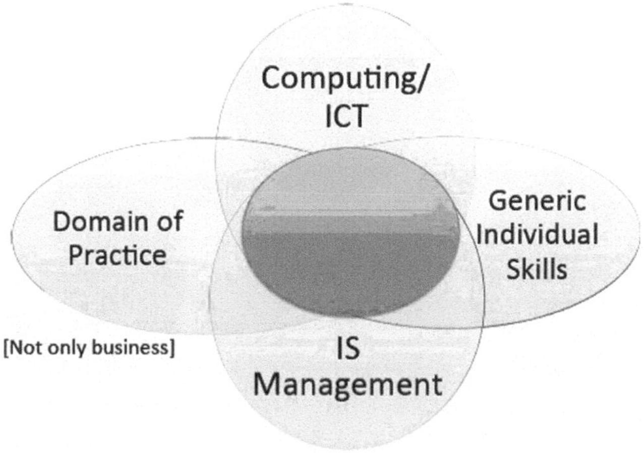

Fig. 3. Elements of MPIS.

After processing several hundred stakeholders' requirements (according to competences area of international Master of Science in Information Systems (MSIS) standards all project members revealed top competences. Stakeholders range competences from different competency area of MSIS 2016 standard according to their experience on labor market. To cover the key competences required by employers in labor market teachers prepare necessary learning outcomes (as indicators of forming competences and statements of what student is expected to be able to do as a result of a learning activity) which later will form the disciplines of ICT curriculum. Competence-oriented approach means that employers can form main requirements to graduates and curriculum because they are final consumers of labor force (Fig. 4).

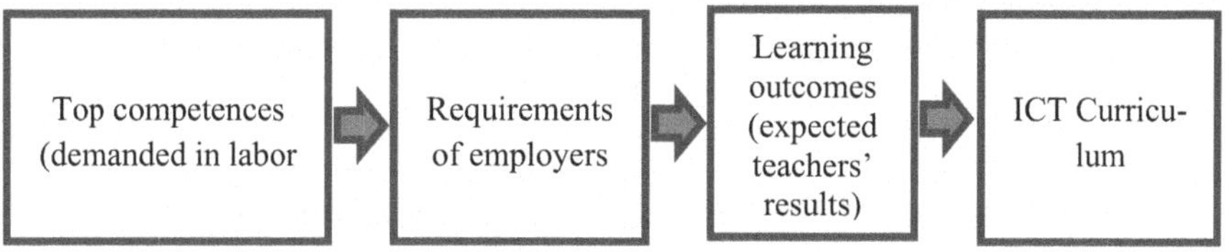

Fig. 4. Architecture of ICT curriculum as competence-oriented approach.

When revising curricula, universities should take into account international accreditation standards for providing training, such as the Association for Promoting Collegiate Schools of Business (AACSB). Learning provision (AoL) refers to a systematic process of collecting data on learning outcomes, reviewing and using it for the continuous development and improvement of degree programs (Fig. 5).

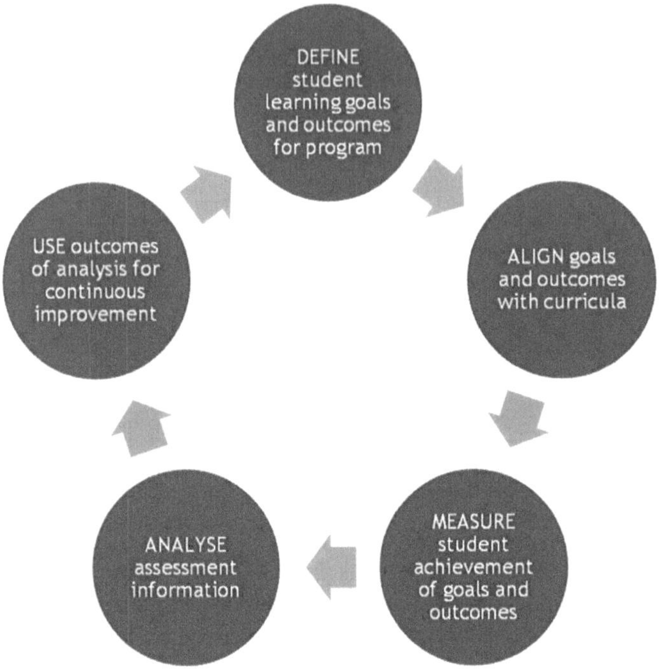

Fig. 5. AoL process modelling.

AoL ensures that university graduates achieve goals and outcomes if universities manage the quality of curricula. AoL supports the continuous improvement of curricula. Accreditation agencies suggest critical revision of module compendium from courses of ICT curriculum after 3–4 years, e.g. in the process of re-accreditation.

The expert method should be used to assess the quality of the curriculum revision model in the field of computer science.

Thus, the curriculum is built using competencies as the main element, rather than courses or knowledge areas, units and topics.

4 Implementation of the Model in the Educational Process

The model of the curriculum revision system in computer science was implemented into the educational process in the development of the curriculum of the new specialty "Information Systems and Technologies" in Kherson State University. The presented model was used to create a new master program in Information Systems and Technologies at the Department of Informatics, Software Engineering and Economic Cybernetics of KSU. The competences of the graduates were determined in accordance with the requirements of the stakeholders described above.

Each competences area (described in Table 1 and Fig. 2) includes defined competences according to MSIS standard [8]. We developed correspondence between competences and learning outcomes for general (Table 3) and specific subject competences (Table 4) for MPIS using requirements of global and local labor market.

Table 3. Generic competences for MPIS.

Learning Outcomes		Competences					
		1	2	3	4	5	6
		Developing a business plan	Know and apply widely used Project Management tools and techniques	Analyzing and documenting business activities	Maintaining an ethical culture	Engaging entrepreneurial thinking	Understanding enterprise architecture principles and the value it provides to business
1	to coordinate different needs and expectations of various project stakeholders, including team members, resource managers, senior management customers, and sponsors	X					
2	to plan, execute, and control tasks, phases, and deliverables of the project based on the identified project goals and objectives using Project Management		X				
3	to be able to reasonably choose modeling method to construct an adequate model of the system or process using modem computer tools to interpret and analyze the simulation results			X			
4	to use own methods and work techniques in CASE-tools, methods and techniques of work in the simulation system design and development of computer programs			X			
5	to use tools for detecting plagiarism and checking the originality of submitted reports, seminar works and theses; ability to comply with ethical requirements and to comply with copyright in professional activities				X		
6	counteraction to attempt to get honour of work that was done by somebody else				X		
7	to apply different concepts and skills in various business contexts using case studies and hands-on exercises with leading software applications					X	
8	to have an understanding of business intelligence and analytics and how businesses use them						X
9	to extract, manipulate and transform data from different sources						X

Table 4. Specific subject competences for MPIS.

	Learning Outcomes	Competences					
		7	8	9	10	11	12
		Selecting and using appropriate analytics methods	Integrating and preparing data captured from various sources for analytical use	Implementing and managing quality audit processes	Specifying and documenting systems requirements	Designing systems	Selecting between systems development approaches
10	to apply different concepts and skills in various business contexts using case studies and hands-on exercises with leading software applications	X				X	
11	to be able for metadata and data quality management to integrate data and convert data in any format for storage and delivery to any system		X	X			X
12	to use the requirements, specifications and application of international quality standards to apply methods for determining of indicators and quality criteria			X	X		X
13	to be able to simulate and manage the process of quality monitoring in the company			X	X		
14	to be able to reasonably choose the modeling method to construct an adequate model of the system or process using modern computer tools to interpret and analyze the simulation results				X		X
15	To use own methods and work techniques in CASE-tools, methods and techniques of work in the simulation system design and development of computer programs	X	X	X	X		
16	to be able to cooperate in a team in order to deal with a system case study	X				X	
17	Be able to cooperate in a team in order to deal with a system case study		X			X	
18	to use methods of structural and object-oriented analysis and design of modern software complexes to reveal the business processes requirements of enterprises				X		X

Considering competencies proposed by our employers we have created the following courses for Curriculum Master of Information System (Table 5) which take into account global and regional requirements to alumni of MPIS. These requirements are presented in the European e-Competence Framework (e-CF 3.0) which contains a reference to 40 competencies as applied at the Information and Communication Technology (ICT) workplace, using a common language for competencies, skills, knowledge and proficiency levels that can be understood throughout Europe [14].

Table 5. Developed curriculum for master in information system.

Number	Mandatory courses	Elective courses
1	IS Development and Deployment	Scientific research and intellectual property
2	MIS and Data Warehousing	Standardization and certification of information technologies
3	Enterprise Architecture Management	Business process modeling
4	Management of IS Projects	Advances in Programming and IS
5	IT Infrastructure	E-commerce and e-business systems
6	Innovations and Entrepreneurship	Formal methods of analysis and verification of IS
7	IS Strategy	Data Mining
8	IS Security	Elective courses from university pool

Each course of Computer Science Curriculum is aimed to cover demanded competencies of labor market. Each year at a meeting of employers and faculty members of the department, the list of generic and specific subject competencies should be revised. It creates the need to revise the relevant learning outcomes (LO) that are determined by university professors and are formed in students during their study of the course.

To revise the adequacy of competencies, learning outcomes, topics and disciplines to the needs of the labor market, a group of experts put points from 0 to 10. Experts are selected among the main employers for graduates of the department, the main faculty members, and representative students to review the adequacy of competencies, learning outcomes, topics and disciplines to meet the needs of the labor market. An example of an expert evaluation is presented in Table 6.

Table 6. Assessment form for expert x about topics of discipline.

Expert x	Competence t		
List of topics	$LO_1^{(t)}$	…	$LO_{N_t}^{(t)}$
Topic 1	$e[x]_{1,1}^{(t)}$	…	$e[x]_{1,N_t}^{(t)}$
Topic 2	$e[x]_{2,1}^{(t)}$	…	$e[x]_{2,N_t}^{(t)}$
…	…	…	…
Topic K	$e[x]_{K,1}^{(t)}$	…	$e[x]_{K,N_t}^{(t)}$

where $e[x]^{(t)}_{y,z}$ – expert's grade, x – expert's number, t – number of competence, y – topic's number, z – number of LO in the group corresponding to the competence t, N_t – number of LO in the group within the competence t, in accordance with Tables 3 and 4. For example, if $e[x]^{(t)}_{y,m} = 0$, $m = 1, \ldots, N_t$, then, according to expert x, topic y does not cover competence t. After experts' assessment each topic covers 2–4 competences which correspond to 3–5 learning outcomes according to correspondence matrix between competences and course learning outcomes. Expert evaluation of a one course will consist of 120 individual grades on average.

After questionnaire of all experts we will determine the average grade and the degree of variation obtained after each round. When the grades of the experts will no longer go out of the first and third quartiles then the poll will be stopped.

So, the curriculum is specified using the competencies as its foundational element, rather than courses or knowledge areas, units, and topics. Global labor market forms requirements to list of mandatory courses, whereas local labor markets – list of elective courses revealed by project members of Kherson State University.

5 Expert Method in the Model for Improvement of Computer Science Curriculum

To determine how topics of the discipline cover the formation of competences we propose to use the following methodology on the example of the discipline 'Enterprise Architecture Management' (EAM) in the specialty 'Information Systems'. To do this, we compose Table 7, where each competence corresponds to a set of learning outcomes (LO). The task of our research is to determine with the help of experts which competences are formed in each discipline and whether those topics, which do not cover any of the competencies, are demanded by the labor market? To establish this fact, an expert method is used when each expert determines the formation of student outcomes for each topic of discipline using the 10-point scale during final knowledge control (1 – LO is absent, 10 – fully formed LO).

The discipline is intended for the formation of 3 following competencies:

Competence 1. Understanding enterprise architecture (EA) principles and the value it provides to business.
Competence 2. Communicating and deploying an EA.
Competence 3. Engaging in IS strategic planning.

Each competence corresponds to 7 learning outcomes:

EAM1 – to understand and model of EAM opportunities for business and IS strategy alignment
EAM2 – to understand an enterprise architecture management frameworks and its standards
EAM3 – to be able to enhance organization's competitiveness by use of EAM
EAM4 – to be able to form purpose and tasks of IS and IT strategy alignment
EAM5 – to be able to effectively use architecture methodologies and tools
EAM6 – to be able to argue, justify and present their decision and plans
EAM7 – to be able to make decision and take responsibility for them.

Table 7. Competences, learning outcomes and topics of discipline 'Enterprise Architecture Management'

Competences	**Competence 1. Understanding enterprise architecture principles and the value it provides to business**							**Competence 2. Communicating and deploying an EA**							**Competence 3. Engaging in IS strategic planning**						
Learning outcomes	EAM1	EAM2	EAM3	EAM4	EAM5	EAM6	EAM7	EAM1	EAM2	EAM3	EAM4	EAM5	EAM6	EAM7	EAM1	EAM2	EAM3	EAM4	EAM5	EAM6	EAM7
List of topics	**Grades of expert #1**																				
Topic 1. Management IS	7	2	4	7	9	5	3	5	7	6	2	9	5	3	6	3	7	6	1	8	3
Topic 2. Architecture management IS	9	5	9	5	8	7	7	8	1	6	2	1	5	6	6	6	6	5	7	6	4
Topic 3. Objectives and principles of EAM	5	5	9	4	4	6	9	6	2	8	6	2	1	3	6	6	2	2	3	1	3
Topic 4. An EAM stakeholders and architecture impact	7	8	3	8	7	7	9	3	3	4	8	2	2	6	4	7	2	6	4	8	7
Topic 5. EAM governance and organization	8	8	7	3	7	4	7	4	6	1	4	8	8	4	5	9	8	2	2	1	3
Topic 6. Embedding EAM into strategic planning	9	3	5	9	4	4	4	9	5	6	4	7	5	2	5	1	6	4	4	7	1
Topic 7. Embedding EAM into the project life cycle	3	6	9	2	2	1	1	6	1	4	4	1	5	4	3	4	2	2	2	7	2
Topic 8. Embedding EAM into operation and monitoring	6	6	6	5	9	8	7	5	4	3	8	7	3	3	9	2	8	2	4	4	3
Topic 9. EA frameworks, modeling and tools	7	9	3	2	7	4	9	2	2	6	5	4	2	3	3	3	9	7	4	1	4

According to the results of the expert method, it is necessary to establish how topics of disciplines cover these competencies. Discipline ‘Enterprise Architecture Management’ consists of 9 topics. The example of results of one of the eight experts’ poll is demonstrated in Table 7. Described model can be used for other experts in a similar way.

Take, for example, topic 9, which is described by expert assessments of all 8 experts regarding the formation of Competence 1 using concordance coefficient. To check the consistency of expert assessments, we use the data ranking method [15], which is a procedure for streamlining expert assessments. After rankings of experts’ evaluations, the results obtained are shown in Table 8 and in Fig. 6.

Table 8. Ranking data for checking the consistency of expert assessments on the formation of Competence 1 on topic 9 “EA frameworks, modeling and tools” of discipline “Enterprise Architecture Management”.

LO	e1	e2	e3	e4	e5	e6	e7	e8
1	3.5	1.5	2.5	1.5	1	1	2.5	2.5
2	1.5	1.5	5	3	5	2.5	1	2.5
3	7	3	2.5	4	3	4	2.5	5
4	6	5	7	5	6.5	7	7	7
5	3.5	7	6	6.5	4	5	4	2.5
6	5	4	2.5	1.5	2	2.5	5	2.5
7	1.5	6	2.5	6.5	6.5	6	6	6

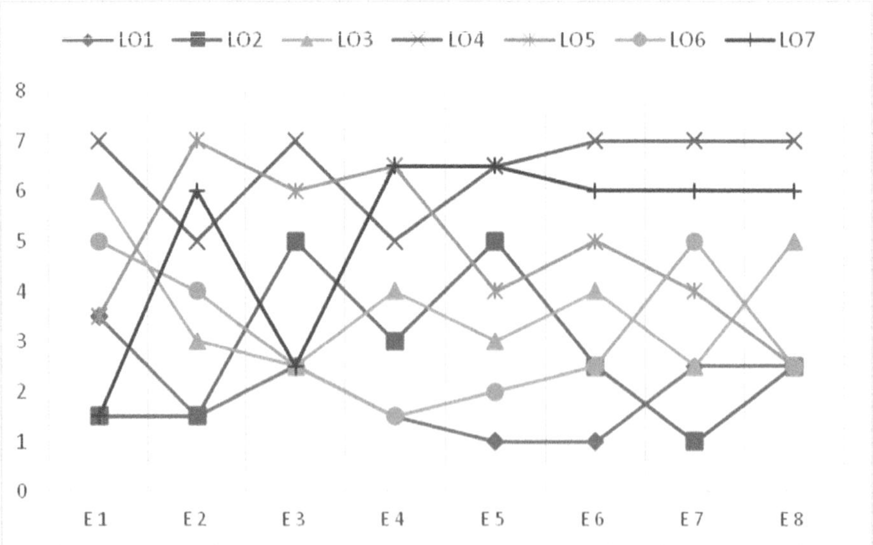

Fig. 6. Distribution of expert assessments about learning outcomes in topic 9 of EAM

Each column in Table 8 reflects the j-th expert's estimation. Each row in the table shows how each expert evaluates the availability of the i-th learning outcome in this topic.

Concordance coefficient W is calculated by the formula [16]:

$$W = \frac{12S}{k^2(n^3 - n) - kT}, \ S = \sum_{i=1}^{n}\left(\sum_{j=1}^{k} r_{ij} - \overline{S}\right)^2, \ \overline{S} = \frac{1}{n}\sum_{i=1}^{n}\sum_{j=1}^{k} r_{ij},$$
$$T = k\sum\nolimits_{j=1}^{k}\sum\nolimits_{t_j}\left(t_j^3 - t_j\right), \tag{1}$$

where n is the number of objects for evaluation (LO), k is the number of experts, r_{ij} is the rank of the i-th indicator (i = 1, …, 7), defined by the j-th expert (j = 1, …, 8), t_j – the number of identical ranks that assigns to different alternatives the j-th expert. If $W = 0$, it indicates a complete difference in opinion of the expert about the presence of LO in this topic of discipline, if $W = 1$, this confirms the complete consistency of experts' opinion about presence (absence) of LO in the discipline.

To assess the significance of the concordance coefficient, chi-square test is used:

$$\chi^2 = \frac{12S}{kn(n+1) - \frac{1}{n-1}\sum\nolimits_{j=1}^{k}\sum\nolimits_{t_j}\left(t_j^3 - t_j\right)}, \tag{2}$$

Based on the data in Table 8, using formula (1), we obtain:

$$W_9^{(1)} = \frac{12}{8^2(7^3 - 7) - 8 \cdot 1392} \cdot 911.93 = 0.54 \tag{3}$$

According to (3) expression $W_9^{(1)}$ indicates the average consistency of expert opinions regarding the fact that **Competition 1** is formed in topic 9 (we can use same procedure for the rest topics of the discipline). After similar calculations of concordance coefficient for **Competence 2** and **Competence 3** respectively, we obtain: $W_9^{(2)} = 0.34$, $W_9^{(3)} = 0.33$. It means that the opinions of the experts differ significantly in that competence 2 and 3 are formed in topic 9. Consequently, it means that topic 9 covers competence 1 and does not cover competences 2 and 3.

So far as number of experts is more than 7, the reliability of the results will be verified using the χ^2 criterion. We have a critical value of 12.59 for n – 1 freedom degrees and a significance level of a = 0.05. After substitutions we obtain (2): $\chi^2 = 50.03 > \chi^2_{cr} = 12.59$. Since the actual value is more than critical one, it confirms a significance of calculated concordance coefficient.

On the basis of the calculations we got the following result about the formation of LO concerning topics of discipline EAM (Table 9):

Table 9. Verification of the consistency of expert assessments about formation of Competence 1 for the topic 9 EA frameworks, modeling and tools of discipline 'Enterprise Architecture Management'

Topics of EAM	Competence 1	Competence 2	Competence 3
Topic 1. Management IS	+	+	+
Topic 2. Architecture management IS	+	–	+
Topic 3. Objectives and principles of EAM	+	+	+
Topic 4. An EAM stakeholders and architecture impact	+	–	+
Topic 5. EAM governance and organization	–	–	+
Topic 6. Embedding EAM into strategic planning	–	+	–
Topic 7. Embedding EAM into the project life cycle	–	–	–
Topic 8. Embedding EAM into operation and monitoring	+	+	+
Topic 9. EA frameworks, modeling and tools	+	–	–

From the results it follows that each topic of EAM covers one or more competences of the course, except of topic 7. It means the need to revise the curriculum to the relevance of this topic in this course or its transfer to another discipline, where it will cover other required competencies.

After excluding topics which do not cover the competences of the discipline, it is necessary to check the correspondence between topics and LO. For this purpose, the algorithm discussed in [17] is used. Let's demonstrate it on the example of EAM discipline for **Competence 1**.

1. For topics of discipline which cover Competence 1 we calculate the average expert estimates (Table 10).

Table 10. Verification of the consistency of expert assessments about formation of competence 1 for the topic 9 "EA frameworks, modeling and tools"

Competences	Competence 1. Understanding enterprise architecture principles and the value it provides to business						
Topics of EAM	EAM1	EAM2	EAM3	EAM4	EAM5	EAM6	EAM7
Topic 1. Management IS	5.13	5.50	4.25	5.50	6.38	5.50	4.38
Topic 2. Architecture management IS	5.63	6.00	5.75	4.50	4.75	7.50	4.50
Topic 3. Objectives and principles of EAM	6.50	3.00	5.38	5.38	6.63	6.38	6.00
Topic 4. An EAM stakeholders and architecture impact	5.00	5.38	3.63	5.25	6.25	7.00	5.63
Topic 8. Embedding EAM into operation and monitoring	3.75	7.13	5.25	4.13	4.25	5.13	4.38
Topic 9. EA frameworks, modeling and tools	5.63	6.63	6.00	4.38	4.00	6.38	5.13

2. Then we define variation measure for each expert using the data from previous tables, applying the formulas $ei_{j\_k}^{(s)} - \bar{e}_{j\_k}^{(s)}$ as the difference between the expert assessment and the average estimate of all experts. For expert 1 example of calculations is given in Table 11.

Table 11. Estimation of variability measures expert 1 about discipline 'Enterprise Architecture Management'

Competences	Competence 1. Understanding enterprise architecture principles and the value it provides to business						
Enterprise Architecture Management	EAM1	EAM2	EAM3	EAM4	EAM5	EAM6	EAM7
Topic 1. Management IS	3.52	12.25	0.06	2.25	6.89	0.25	1.89
Topic 2. Architecture management IS	11.39	1.00	10.56	0.25	10.56	0.25	6.25
Topic 3. Objectives and principles of EAM	2.25	4.00	13.14	1.89	6.89	0.14	9.00
Topic 4. An EAM stakeholders and architecture impact	4.00	6.89	0.39	7.56	0.56	0.00	11.39
Topic 8. Embedding EAM into operation and monitoring	5.06	1.27	0.56	0.77	22.56	8.27	6.89
Topic 9. EA frameworks, modeling and tools	1.89	5.64	9.00	5.64	9.00	5.64	15.02

3. After the survey of all experts, average estimates of variability for sample observation are determined (Table 12).

Table 12. Average measure of variation of all experts in discipline 'Enterprise Architecture Management'

Competences	Competence 1. Understanding enterprise architecture principles and the value it provides to business						
Enterprise Architecture Management	EAM1	EAM2	EAM3	EAM4	EAM5	EAM6	EAM7
Topic 1. Management IS	7.84	11.43	4.21	8.57	6.84	6.57	6.55
Topic 2. Architecture management IS	8.84	9.71	7.64	5.43	8.21	2.29	5.43
Topic 3. Objectives and principles of EAM	4.29	2.29	3.70	5.13	7.13	8.55	11.43
Topic 4. An EAM stakeholders and architecture impact	5.43	9.41	6.84	10.79	3.07	1.71	6.55
Topic 8. Embedding EAM into operation and monitoring	7.07	4.41	9.93	2.98	8.79	4.13	6.27
Topic 9. EA frameworks, modeling and tools	5.98	7.13	8.00	7.70	5.71	3.98	5.84

Then the coefficient of variation $V_{x\_y}^{(t)}$ is calculated as the ratio of the root of variation to the average grade. If $V_{x\_y}^{(t)} \leq \frac{1}{3}$, then the opinions of the experts are considered as compatible. Whereas, if $V_{x\_y}^{(t)} > \frac{1}{3}$, then the opinions of experts have low degree of consensus, and then they need further revisions of the disciplines of curriculum. If expert assessments are also low, then the corresponding LO and the topics (or their sections) are recommended for removal from the discipline for the next academic year. If expert assessments are high, then the appropriate LO and topics are recommended to be remained in the academic discipline for the next academic year.

Table 13. The average measure of variability of all experts in the discipline 'Enterprise Architecture Management'

Competences	Competence 1. Understanding enterprise architecture principles and the value it provides to business						
Enterprise Architecture Management	EAM1	EAM2	EAM3	EAM4	EAM5	EAM6	EAM7
Topic 1. Management IS	0.55	0.61	0.48	0.53	0.41	0.47	0.59
Topic 2. Architecture management IS	0.53	0.52	0.48	0.52	0.60	0.20*	0.52
Topic 3. Objectives and principles of EAM	0.32*	0.50	0.36	0.42	0.40	0.46	0.56
Topic 4. An EAM stakeholders and architecture impact	0.47	0.57	0.72	0.63	0.28*	0.19*	0.46
Topic 8. Embedding EAM into operation and monitoring	0.71	0.29*	0.60	0.42	0.70	0.40	0.57
Topic 9. EA frameworks, modeling and tools	0.43	0.40	0.47	0.63	0.60	0.31*	0.47

For cells selected by (*), there is a consensus among the experts that competency data is covered by these topics at the intersection of rows and columns (Table 13). For the first topic, all LOs have low expert consensus, and therefore Topic 1 does not cover **Competence 1**. For LO_3, LO_4 and LO_7 there is no expert agreement about impact of these LO in the formation of **Competence 1**. Based on the results of the expert method, the matrix of correspondence between LO, **Competence 1** and course topics we can represent as following Table 14. Based on the results of the application of two methods – the concordance coefficient and improvement of the discipline, the following refinements were made (Table 14):

(1) The number of topics covering competency 1 decreased from 9 to 5;
(2) The number of learning outcomes that form competencies 1 decreased from 7 to 4;
(3) The correspondence between 4 LO and 5 subjects of discipline is determined.

Our author method can be used to revise the matrix of compliance between **Competence** and LO, as well as between topics of discipline and Competence + LO, considering the requirements of stakeholders to the educational process.

Table 14. Revealed correspondence between competences and learning outcomes in the discipline 'Enterprise Architecture Management'

Competences	Competence 1. Understanding enterprise architecture principles and the value it provides to business			
Enterprise Architecture Management	EAM1_to understand and model of EAM opportunities for business and IS strategy alignment	EAM2_to understanding of enterprise architecture management frameworks and its standards	EAM5_to be able to effectively use architecture methodologies and tools	EAM6_to be able to argue, justify and present their decision and plans
Topic 2. Architecture management IS				❶
Topic 3. Objectives and principles of EAM	❷			
Topic 4. An EAM stakeholders and architecture impact			❸	❹
Topic 8. Embedding EAM into operation and monitoring		❺		
Topic 9. EA frameworks, modeling and tools				❻

6 Conclusions and Outlook

The model of the curriculum revision system in computer science was designed and developed, the curriculum of ICT was improved, which includes the following provisions:

1. Analysis of the results of the stakeholder survey in the local and global labor market determines the key ICT competencies of graduates in accordance with the requirements of employers.
2. Requirements for the competence of graduates determine the expected learning outcomes in accordance with the specifications of e-CF v.3.0.
3. To obtain the expected learning outcomes, a list of curriculum courses is formed, the compliance of which is determined by the European e-Competence Framework (e-CF 3.0).
4. Expert method for obtaining the correspondence between competences, learning outcomes and topics was used.

Evolving of the curriculum revision system in computer science is related to the optimization of the developed model by introducing quality management elements in accordance with ISO standards.

The experimental approbation of the presented model of revision of the curriculum was held at the Kherson State University in the development and introduction of the new specialty "Information Systems and Technologies" into the educational process.

References

1. Eggink, M.E.: The need for a change in roles of universities as participants in innovation systems. In: Soliman, K.S. (eds.). 28-th International Business-Information-Management-Association Conference, pp. 68–77 (2016)
2. Tomar, T.S.: Entrepreneurship development and innovation management: key connections. In: Grant, K.A., Wise, S. (eds.) 4th International Conference on Innovation and Entrepreneurship, pp. 263–271 (2016)
3. Kruss, G., et al.: Higher education and economic development: the importance of building technological capabilities. Int. J. Educ. Dev. **43**, 22–31 (2015)
4. Law of Ukraine "On Education". http://zakon.rada.gov.ua/laws/show/2145-19
5. Law of Ukraine "On Higher Education". http://zakon.rada.gov.ua/laws/show/1556-18
6. National Qualifications Framework. http://zakon4.rada.gov.ua/laws/show/1341-2011-п
7. Standards of Higher Education in the field of Computer Science (projects). https://mon.gov.ua/ua/osvita/visha-osvita/naukovo-metodichna-rada-ministerstva-osviti-i-nauki-ukrayini/proekti-standartiv-vishoyi-osviti
8. MSIS 2016 Global Competency Model for Graduate Degree Programs in Information Systems. https://www.acm.org/binaries/content/assets/education/msis2016.pdf
9. Revising the MSIS Curriculum: Specifying Graduate Competencies. http://cis.bentley.edu/htopi/MSIS2016_Draft_03-21-2016_Part1.pdf
10. Kravtsov, H., Kobets, V.: Implementation of stakeholders' requirements and innovations for ICT curriculum through relevant competences. In: Ermolayev, V., et al. (eds.) Proceedings of the 13-th International Conference ICTERI 2017, Kyiv, Ukraine, 15–18 May 2017, pp. 414–427. CEUR-WS.org/Vol-1844, ISSN 1613-0073 (2017). http://ceur-ws.org/Vol-1844/10000414.pdf
11. Establishing Modern Master-level Studies in Information Systems. https://mastis.pro
12. Moergestel, L., Keijzer, A., Stappen, E.: Tips and pitfalls for blended learning: redesigning a CS curriculum using IT. In: Ermolayev, V., et al. (eds.) Proceedings of the 12-th International Conference ICTERI 2016, Kyiv, 21–24 June 2016, pp. 273–283. CEUR-WS.org/Vol-1614, ISSN 1613-0073 (2016). CEUR-WS.org/Vol-1614/ICTERI-2016-CEUR-WS-Volume.pdf
13. Kobets, V., Weissblut, A.: Nonlinear dynamic model of a microeconomic system with different reciprocity and expectations types of firms: stability and bifurcations. In: CEUR Workshop Proceedings, vol. 1614, pp. 502–517 (2016). (Indexed by: Sci Verse Scopus, DBLP, Google Scholar). CEUR-WS.org/Vol-1614/ICTERI-2016-CEUR-WSVolume.pdf
14. e-Competence Framework 3.0: A common European Framework for ICT Professionals i all industry sectors (2014). http://www.ecompetences.eu
15. Brazdil, P.B., Soares, C.: A comparison of ranking methods for classification algorithm selection. In: López de Mántaras, R., Plaza, E. (eds.) ECML 2000. LNCS (LNAI), vol. 1810, pp. 63–75. Springer, Heidelberg (2000). https://doi.org/10.1007/3-540-45164-1_8

16. Legendre, P.: Coefficient of concordance (2010)
17. Kravtsov, H., Kobets, V.: Model of the curriculum revision system in computer science. In: Ermolayev, V., et al. (eds.): ICT in Education, Research and Industrial Applications. Proceedings of the 14th International Conference Workshops ICTERI 2018, Kyiv, Ukraine, 14–17 May 2018, vol. II, pp. 488–500. CEUR-WS.org/Vol-2104, ISSN 1613-0073 (2018). http://ceur-ws.org/Vol-2104/paper_253.pdf

Study of Digital Competence of the Students and Teachers in Ukraine

Olena Kuzminska[1](✉), Mariia Mazorchuk[2], Nataliia Morze[3], Vitaliy Pavlenko[2], and Aleksander Prokhorov[2]

[1] National University of Life and Environmental Sciences of Ukraine, Kiev, Ukraine
o.kuzminska@nubip.edu.ua

[2] National Aerospace University "KhaI", Kharkiv, Ukraine
mazorchuk.mary@gmail.com, pavlenko_vitalii@ukr.net, o.prokhorov@khai.edu

[3] Boris Grinchenko Kyiv University, Kiev, Ukraine
n.morze@kubg.edu.ua

Abstract. Professional fulfillment of the personality at the conditions of the digital economy requires the high level of digital competency. One of the ways to develop these competencies is education. However, to provide the implementation of digital education at a high level, the digital competency of the teachers and students is a must. This paper presents explanations on the level determination of the digital competencies for teachers and students in Ukraine according to the DigComp recommendations. We tried to identify the main factors that reflect the degree of readiness teachers and students for digital education based on their self-evaluation. We also attempted to estimate the level of digital competencies based on the analysis of Case-Studies execution results. The complex analysis let us assess the connection between respondents' self-evaluation and their real competencies. Here we provide a methodology and a model of level competencies determination by means of a survey, expert case rating and the results of the statistical analysis. On the basis of the obtained results, this paper suggests further research prospects and recommendations on the digital competency development in educational institutions in Ukraine.

Keywords: Digital competencies · Case studies · Survey · Principal component analysis · Education

1 Introduction

Modern digital technologies are the catalyst for world transformation [1]. Digital transformation has a huge impact on business and social life, providing ways to unlock economic and social benefits. The Digital Economy (DE) Theme is supporting research to rapidly realize the transformational impact of digital technologies on aspects of community life, cultural experiences, future society, and the economy [2]. DE brings together a unique community of researchers from diverse disciplines, including social science, engineering, computer science, the arts and medical research; and users;

V. Ermolayev et al. (Eds.): ICTERI 2018, CCIS 1007, pp. 148–169, 2019.
https://doi.org/10.1007/978-3-030-13929-2_8

including people, businesses and government; to study, understand and find solutions to real problems.

Most European countries approved development strategies until 2020. The Digital Agenda presented by the European Commission belongs to the seven main strategies and suggests wide usage of the Information and Communication Technologies (ICTs) potential in order to foster innovation, economic growth, and progress [3]. Likewise, the Digital Agenda 2020 was approved in Ukraine [4]. The Digital Agenda must help to make maximum use of digital technologies [5] since the qualified professional availability is crucial for creating a digital society and providing competitiveness of individual countries and their citizens [6]. However, as of 2017, according to the "digital skills" index of the European digital economy and society index (DESI), almost half (44%) of the EU population lacks skills in using digital technologies [7]. This, undoubtedly, is a large-scale problem that must be solved.

A number of researches [8] are devoted to the problem of reducing the gaps in digital competencies understanding by different categories of people. The EU recommendations on monitoring the Digital Economy & Society 2016–2021, suggest indicators for measuring digital skills [9]. Implementation of digital technologies influences many spheres and aspects of society's activities, thus, for example, the possibility of employment, education, leisure, attraction and participation in society are transformed. The digital competence, as a confident use of information and communication technology (ICT) tools, is vital for a person to participate in today's socio-economic life. That is why digital literacy (or digital competence) is recognized by the EU as one of the eight key competencies for a full life and activity. In this regard, the problem of improving (transforming) the education system as a social institution for human development for the training of competent specialists, considering the needs of the market and the current trends in the development of digital technologies, is being actualized.

The purpose of this study is to define how well education participants in Ukraine are prepared to use their educational environment as a space for using and improving their digital competencies. The research is concentrated on the study of both the use of information technologies by participants of the learning process and the level of teachers and students' digital competencies during their occupational problem-solving sessions. The two above mentioned are the factors having the effect on education quality in the whole.

2 System of the Digital Competency DigComp: Theoretical Facet

2.1 Structure and Methods of Estimation of the Digital Competency

There exist a few frameworks those allow to define the level of digital competencies. Among them, there are European e-Competence Framework for ICT Professionals [10], European Computer Driving License [11], ICT Literacy Competencies, Global Media and Information Literacy Assessment Framework [12]. In our research, we based on the European system of the digital competency, known also as DigComp, that provides a general approach to defining and describing the main spheres of the digital competency of people and is the general mark in the European level [13]. DigComp agrees with other frameworks and has experience of implementation in European countries, for example, integration into Europe's CV system, which allows applicants to evaluate their own digital competence and to present the results of this assessment in CV [14].

The DigComp has three main directions: (1) policies formation and support; (2) training and employment programs planning; (3) evaluation and certification. In this paper, the second direction is considered, in particular, readiness to implement open education [15]. In addition, digital competence DigComp refers to the necessary conditions for digital education implementation in The Digital Agenda 2020 Ukraine.

In 2017 EU suggested a new framework Digital Competence (DigComp 2.1) that has 5 dimensions [16]:

Dimension 1: Competence areas identified to be part of digital competence. There were defined areas: (1) information and data literacy (IL); (2) communication and collaboration (C); (3) digital content creation (DC); (4) safety (S); (5) problem solving (PS).

Dimension 2: Competence descriptors and titles that are pertinent to each area. There were defined 21 competencies [16, p. 11].

Dimension 3: Proficiency levels for each competency. There are 4 main levels (foundation, intermediate, advanced and highly specialized) and their decompositions. Each level represents a step up in citizens' acquisition of the competence according to its cognitive challenge (Cd), the complexity of the tasks (Ct) they can handle and their autonomy (A) in completing the task [16, p. 13].

Dimension 4: Knowledge, skills, and attitudes applicable to each competence [16, p. 19].

Dimension 5: Examples of use, on the applicability of the competence to different purposes. There were provided scenarios for two areas of use: employment and learning [16, pp. 19–20].

To evaluate the digital competencies on the base of the DigComp framework, there were developed special methodologies and online tools [17].

Traditional research approach includes three types of analytical evaluation methods for estimating the level of digital competencies of learning process participants [18]:

1. Sample survey;
2. Case Study (practical example);
3. Comparative study.

Use, advantages, and disadvantages of three types of analytical evaluation methods are presented in brief in Table 1.

Table 1. Comparison of design solutions for analytical assessment

Method	Use	Advantages	Disadvantages
Survey	Determine the scale of implementation of the approach or estimate the current state of the problem, and find out the respondents' opinion	- Comparatively low cost; - Comparative easiness of management; - You can analyze data comparatively easy (statistical analysis); - It is possible to approximate the results to a wide audience	Respondents often make mistakes and allow inaccuracies
Case Study (practical example)	Identify ways to implement the program, best practices and necessary changes	- A number of data and examples; - An effective method for sharing best practices; - It is possible to check the current real state of the problem	- High cost; - High time spent; - You can not reach conclusions about the great sample of respondents
Comparative Study	Compare the impact of two or more approaches or methods of activity	- It is possible to check the relative effectiveness of different variants; - It is helping to check the effectiveness of the approach before its large-scale usage; - It is possible to determine the cause-effect relationship; - You can easily analyze data; - Conclusions can be approximated to a wider audience	- High cost; - Comparative complexity of management

2.2 Analytical Evaluation of Digital Knowledge: Research Methodology

We used all three methods in the given research (see Table 1). The research consisted of two stages.

On the first stage methods of survey and comparative analysis were used for estimating the level of the teachers and students' digital competencies within the educational institutions of Ukraine:

– to describe the level of digital competencies by fields of occupation;
– to estimate the numbers on usage of available digital applications, a comparison of the level values for different groups of respondents, the strength of the connection between the various characteristic evaluation;

– to study of the cause-effect dependencies of the competence level and the properties of the respondents.

The authors developed a questionnaire containing 7 main sections according to recommendations of DigComp 2.1 (https://goo.gl/forms/h90Co24yF6vmU0JF2). They raised the questions according to the recommendation [16, p. 13]:

– Cognitive domain (Cd): appraising the competency level from Remembering to Creating according to Bloom's taxonomy;
– Complexity of tasks (Ct): from performing small tasks to solving the real issues;
– Autonomy level (A): from the necessity to integrate to contribute to professional practice and to guide others.

Sections 1–5 contain 21 questions that evaluate the level of digital competencies according to 5 areas of DigComp and consider the competencies usage in the field of education. The respondents have suggested a case: "You have to prepare a short report on the given subject and to provide it in the digital format". There was a suggestion: "Use different tools and methods on every stage of the process and communicate to different people (the examples below illustrate only some steps of work, as it doesn't refer to the subject). For each example write down how easy it was for you to do the task".

We suggest the next grading scale:

1. I am not sure I can perform this task on my own, I need some help (Foundation);
2. I can perform the task on my own, and I can solve the problems that appear during the work (Intermediate);
3. I can help others when performing the task, I can give some advice or help somebody to solve a problem (Advanced);
4. I can create a digital resource (a blog, a page in social networks, wiki, etc.) containing useful references, recommendations, instructions, and to provide help (lead a webinar, moderate the forum, etc.) (Highly specialized).

Section 6 contains 18 questions that must define online tools and information technologies that the respondents use to solve the tasks in Sections 1–5. This section contains closed questions of multiple choice. The questions' formulations consider Cognitive domain (Cd) and Autonomy (A). Based on the given questions we found the validity of the respondents' answers and the frequency of usage of specific tools in the process of preparation of the report.

We leveraged the Spearman's rank correlation coefficient, that reveals the correlation between the respondents' estimates and the number of tools, selected to solve one task. If the Spearman's rank correlation coefficient is less than 0.4, the final estimate is corrected by multiplying by the adjustment factor. However, all of the respondents got the Spearman's rank correlation higher than 0.5, and we did not correct the esteems.

The last section contains the questions that we need to fill in the personal profile of the respondent (considering the age, field of occupation, the access level of IT, etc.).

We determine variables (est. conformity with areas of DigComp 2.1), the scale of evaluation and interval for the questions in our questionnaire (Table 2).

Table 2. The questionnaire specification

Groups of questions (DigComp 2.1)	Variables	Indication of the variable of points	Scale of evaluation	Intervals of evaluation
Processing data (IL)	V11–V13	V1	Ordinal	1…4
Communication (C)	V21–V26	V2	Ordinal	1…4
Creation of digital content (DC)	V31–V33	V3	Ordinal	1…4
Information security (S)	V41–V43	V4	Ordinal	1…4
Solving technical problems (PS)	V5	V5	Ordinal	1…4
Studying and data analysis (PS)	V61–V62	V6	Ordinal	1…4
Personal data	P1–P6, P8	–	Nominal	Categories depending on a variable (Table 1)
Questions to evaluate the usage of digital competency tools	I11–I13	–	Nominal, Multiple Response	1…8
	I21–I26			
	I31–I33			
	I41–I43			
	I5			
	I61–I62			

At the second stage, our task was to change the level of the respondents' digital competencies. The aim was to see how the respondents use their digital competencies to solve the real educational tasks and challenges (for teachers) and in the process of academic studying and research process (for students), and also to analyze which competencies affect the process of solving the tasks.

We used the case-study method and comparative analysis. The authors of this research are experienced in using case-study to evaluate the competency of masters at universities [19]. We used the results received after the first stage of our research (the questionnaire) both to plan the experiment based on the case-study and for further contacts with respondents, personal interviewing and collecting information.

This stage was difficult to organize, so we didn't manage to get information from all of the respondents, who participated in the first stage of our research. However, this fact didn't affect the final results much.

Experiment participants were offered a case: you need to prepare a short report on the topic of scholarly communication in high school and to provide it in the digital format. For that:

- analyze how deeply the problem is reviewed (analysis of scientific publications);
- present the result in digital format and share it with the community.

The expected results:

- preparing a brief digital report (presentation, thesis, etc.);
- passing the result for checking (sending by e-mail or handing into university representative);
- completing the table of task execution table (see Table 3).

These results help experts to estimate not only the resulting document but also the participants' level of their competencies in the following areas: (1) information and data literacy (IL); (2) communication and collaboration (C); (3) digital content creation (DC); (4) safety (S); (5) problem solving (PS). The completed table (Table 3) is to be sent to the stated e-mail address.

Table 3. Task execution table

Code	Task	Answer
IL1	What keywords and URLs of the resources do you use to search for information?	
IL2	What articles titles did you select for analysis? (provide their URLs)	
IL3	What method of storing and structuring data did you use?	
PS.1	What type of document did you use to present your analytics? What tool did you use to create it?	
DC.1	Describe the structure of the document you received as a result	
PS.2	What way of data presentation did you use?	
DC.2	How did you observe the copyrights?	
DC.3	What format did you choose to save your document? Where did you save it?	
S.1	How did you provide access to the documents you created? (viewing, commenting, etc.)	
C.2	How do you control your digital identity when presenting the results of your work?	
C.3	Specify the field contents when sending a letter	
PS.3	What difficulties did you have in this task? How did you solve them?	
PS.4	What skills or knowledge did you miss to do the task? How can you get the missing skills?	

The performance was estimated by experts with two estimates in the 100-point scale (we chose two estimates proceeding from the research results received after the first stage). The first estimate shows the results (areas of DigComp 2.1: IL, C; DC) that belong to the first aim of our research, which was to find the digital competences extent of use for solution of real tasks. The second estimate demonstrates a solution of a specific objective with use of information technologies and tools (includes estimation of S and PS areas of DigComp).

The experts estimated results within a case (Table 3) on a scale from 1 to 5 points, and the presented results and the final document – from 0 to 35 points. At this stage, we considered the Complexity of tasks (Ct) and Cognitive domain (Cd) levels [16, p. 13].

The levels of digital competency formation were defined according to points distribution of national rules according to ECTS grade scale [20]: 0–59 (F-FX, 1-low), 60–74 (E-D 2-sufficient), 75–89 (C-B, 3-good), 90–100 (A, 4-excellent). These levels correspond to the grading scale of DigComp: foundation, intermediate, advanced, highly specialized.

To process the interview and case-study results, and to provide comparative analysis of the received results we used a number of methods of the statistical analysis, namely: descriptive statistic methods used to identify the common patterns in answers and the level of competences estimation [21]; statistical inference methods used to prove the hypotheses of communications between various signs to the studied selection of educational process participants [21], the main component analysis for data reduction and search of hidden regularities [28]; planning and experiment analysis with use of expert estimation methods [22]; regression analysis methods used to estimate the contribution of separate characteristics to the resulting estimates variation, and to predict the competences level based on received estimates [23].

The received results reliability was confirmed on the basis of statistical criteria and with the logic of statistical researches [24, 25] with an application of modern information means and tools for collecting and data processing (Google-forms, SPSS).

3 Research Design

3.1 The Description of the Experiment Sample: Characteristics of Respondents and Research Hypothesis

On the first stage (a survey) to study the problem of readiness of teachers and students for digital education and living in the digital world we chose the cross-section single research scheme.

The survey created by the research authors (https://goo.gl/forms/h90Co24yF6vmU0JF2) was distributed with mailout and specific-purpose contacts with the educational institutions.

The sample of the population was formed of employees, teachers, and students of higher education in various fields: mathematics and informatics, humanitarian specialties, right and law, medicine and veterinary science, etc. The full list of the estimated features that reflect personal data of respondents is provided in Table 3. Since the aim of our research wasn't an exact assessment of competencies level in each field, but defining the communications between groups of the respondents those differ in age, gender sign, status (the student, the teacher), and field of occupation (technical or nontechnical), the error of representativeness didn't exceed 8% at total of the interviewed respondents (193 persons). Most of the respondents are teachers and students of higher educational institutions as the National University of Life and Environmental Sciences of Ukraine, National Aerospace University "KHAI" and Boris Grinchenko Kyiv University. The questionnaire was widespread in two ways: on the Universities' web pages and through the social networks. Every feature has calculated beforehand descriptive statistics and constructed frequency distributions. The main features (characteristics of respondents) are provided in Table 4.

One of the tasks was to evaluate the validity and reliability of the assessment tool, i.e. developed the questionnaire. We also needed to highlight the main components of digital competencies, which had significant differences for different groups of respondents. On the first stage of research these hypotheses were formulated:

1. The level of digital competencies among the majority of respondents is above the average for the entire sample.
2. The levels of competence in the competence of digital data processing, online communications, and protection, transmission, and storage of information depend on the gender, status, training directions, accessibility of technical and mobile means and the way knowledge and skills are acquired.

The second stage participants were the same as for the first stage, though some of them didn't submit the cases so the experts couldn't analyze the final results. The participants of the first stage (survey) received personal invitations to the second stage together with instructions and explanations for the task. Collecting and processing of results were carried out within 4 months after the end of the first analysis stage. In total, 178 people participated in the second stage.

For the second investigation stage the following hypotheses were formulated:

1. The respondents who master basic digital competencies can simply solve other problems related to the use of digital tools.
2. The respondents who received high scores from experts for the case tasks estimated their level as high; those respondents, who estimated their skills as average or low, received low expert scores for the case tasks.

The experts were the professors and lecturers from three universities: the National University of Life and Environmental Sciences of Ukraine, National Aerospace University "KHAI" and Boris Grinchenko Kyiv University. For each expert we collected data considering age, sex, scientific degree, position, publications, and recognition (Scopus and google Academy). The data on the experts' publications and recognition was necessary to estimate the quality of scientific publications analysis and of presented results. As a result, we selected 10 experts. There were 4 man and 6 women, with research degrees in pedagogics and technology. Every expert has a Scopus profile and is a member of the information technology department, so they can estimate the digital competencies level of the experiment participants. The estimation was conducted according to the rules described above. After that, each case estimation point result (Table 3) was analyzed from the inter-scorer agreement, and average estimates on each category of tasks were calculated.

Table 4. The main characteristics of the respondents

Feature	Category of a feature	Meaning	Percent / Descriptive statistics
Gender	1	Male	25,10%
	2	Female	74,90%
Status	1	Teacher (Professor)	45,50%
	2	Student (Magister)	33,00%
	3	Student (Bachelor)	21,50%
Occupation	1	Education	26,20%
	2	Humanities and Arts	6,30%
	3	Business and Economy	0%
	4	Natural sciences (chemistry, biology, geography, etc.)	9,40%
	5	Mathematics, computer programming, IT	27%
	6	Health or veterinary medicine	1,60%
	7	Construction and architecture	0%
	8	Engineering (purely technical areas, including geodesy and transport)	6,30%
	9	Agriculture and agricultural machinery	4,20%
	10	Sphere of service, public administration, social security	2,10%
	11	Social sciences, law and jurisprudence	14,70%
	12	Others	3%
Availability of mobile and technical devices	1	Always	84,30%
	2	Not always	15%
	3	The availability is restricted, I can hardly use devices	0,50%
Availability of the websites on the educational books and article	1	Always	23,00%
	2	Not always	55,00%
	3	The availability is restricted as the full access requires money	22,00%
How did you improve your digital competency?	1	I improved my skills on my own	43,10%
	2	I got the basic skills at school	14,90%
	3	I improved my skills in university.	14,90%
	4	I participate online courses, webinars, communicate with my friends on the topic of IT	21,20%
	5	Other	5,90%
Age		Age of respondents	Mean=31,01
			Median=23,5
			Mode=22,0

3.2 The Methods and Models of Data Processing

When analyzing we used a complex of methods and models that allow calculating all the descriptive statistics. The choice of certain indicators is influenced by the data type, the scale of assessment and the limitations of methods application. For calculations, we used the software tool for statistical processing data SPSS [24, 25].

On the first stage of the analysis, most of the features chosen to assess the level of digital competencies in the survey process were estimated in an ordinal 4-point scale. Therefore, in order to test the hypotheses, the method of analyzing two-dimensional frequency tables (contingency table) and the chi-square test was used at the first stage [24]. Also, the Cramer's V, contingency coefficient and the coefficient Phi, which are called measures of association, were calculated. These coefficients vary from 0 to 1 and allow us to conclude about the strength of the relationship between the features.

One of the analysis purposes is to estimate the reliability of the questionnaire [26]. To estimate of internal consistency of single questions of the questionnaire the coefficient Cronbach's alpha was used. Besides, for respondents' questions which purpose was to confirm the level of proficiency in these or those competencies have been offered. Such questions, as a rule, contains answers concerning the tools used for the solution of the tasks within digital competences. For a research of the communications between the main points of the questionnaire and questions concerning tools methods of the analysis of two-dimensional frequency, tables have also been used.

A number of features did not allow us to draw single-digit conclusions on the general tendencies of different groups of respondents' digital competences possession. Therefore, when data processing methods of data reduction were used. The first approach was based on the estimation of the total (aggregated) ball score on the groups displaying the main directions of digital competences. In Table 2 you can see the main groups on which score was calculated. For the analysis of distinctions of average summary points, the method of one-factor dispersion analysis (ANOVA) was used further [27]. The second approach was based on a method of the principal components [28] that allows transforming without loss of data to such variables which values cause the maximum value of the variance of the initial features. The further analysis of communication of factor values with groups of respondents was carried out on the basis of the frequency tables using methods of graphic visualization of data.

On the second stage of analysis, we used the Kendal concordance coefficient to estimate the experts' agreement. This coefficient varies from 0 to 1 and allows estimating if the experts are coherent in their actions. Further, we used average value as the resulting general assessment of the level of the respondent's competences.

To check a hypothesis concerning the connection of respondents estimates of their lever and expert estimates, we used the methods of the regression analysis [23]. We found it reasonable, to use linear regression to estimate the contribution of separate competences level indicators to the general estimates with step-by-step removal of insignificant signs from the model.

When testing statistical hypotheses at all analysis stages the decision is made on the basis of the size p-value which actually displays the probability of a mistake at a deviation of a zero hypothesis (an error of the first type). The p-value for a deviation of a zero hypothesis was accepted equal to 0.05.

4 Results of Research

4.1 Determination of Students and Teacher's Digital Competence Level

The main distributions of estimates of the digital competencies level obtained in the first stage and their descriptive statistics are presented in [29].

Estimation results show that for most of the competencies, respondents rated their abilities above average. The significance of the differences was confirmed by the value of the Student's t-test at the level $p < 0.05$. Thus, we can accept the hypothesis that the level of digital media and communications usage among teachers and students is quite high and above the average.

The expert estimate results of the second stage cases lead us to similar conclusions (Table 5).

Statistically, Ukrainian students and teachers' level of digital competences is above average ($p < 0.05$). Consider that the students and teacher levels differ insignificantly, that can be explained with assignment complexity. Scientific report preparation causes difficulties for some students.

Table 5. Distributions of digital competencies levels

Respondents (number)	The number of respondents according to the digital competencies' levels									
	1 (low)			2 (average)			3 (good)			4 (excellent)
	FX	X	Sum	E	D	Sum	C	B	Sum	A
Teachers (77)	0	4	4	3	25	28	16	19	35	10
Students (101)	0	14	14	5	31	36	26	18	44	7
Total	0	18	18	8	56	64	42	37	79	17

Thus, we can accept the hypothesis that the level of digital media and communications usage among teachers and students is quite high and above the average.

The analysis of two-dimensional frequency tables (cross-tabulations), and the criteria on the basis of which it is possible to assess whether there is a connection between such characteristics as the assessment of the level of one's own competencies and status, gender, and occupation proved that for most of the features of communication it is not observed for $p > 0.05$.

The coefficients of Cramer's V and contingency ranged from 0.086 to 0.366, that indicates a weak connection between the traits. Therefore, the study focused on the analysis of total scores by groups of competencies. Table 6 provides the values of the significance criteria for the differences in the total ball-point estimates for the main areas of digital competencies among the groups of respondents. The table shows the F statistics and p-value calculated using the ANOVA method.

Table 6. Criteria of value of scores on different fields of digital competencies among the groups of respondents

Measuring digital competences directions	Gender		Status		Occupation		Availability of mobile and technical devices		Availability of the websites on the educational books and article	
	F	p-value	F	p-value	F	p-value	F	p-value	F	p-value
Processing data	2.11	0.15	9.59	**0.00**	2.14	**0.03**	0.44	0.65	6.23	**0.00**
Communication	0.06	0.81	1.25	0.29	1.38	0.20	0.46	0.63	6.61	**0.00**
Digital content creation	3.24	0.07	1.46	0.23	3.31	**0.001**	1.9	0.15	5.96	**0.00**
Information security	2.82	0.10	0.43	0.65	2.44	**0.01**	0.72	0.49	10.29	**0.00**
Solutions of technical problems	5.51	**0.02**	0.29	0.75	1.47	0.16	1.39	0.25	8.25	**0.00**
Studying and analysis of data	1.92	0.17	4.04	0.02	1.54	0.14	1.74	0.18	7.20	**0.00**

We can see significant differences in competencies evaluation among teachers and students, among the respondents of different occupations, and among those who have limited access to websites with scientific books and articles (significance level was considered for $p < 0.05$). The difference between the groups was also tested by the Tukey criterion. We revealed the greatest differences between students and teachers. The teachers' scores are significantly higher. The level of competence among those whose occupations are related to mathematics, computer science, and information technology differs from the rest of the groups. The respondents with limited access or no access to websites with special literature have the levels of digital competencies significantly lower than those who have permanent access.

We analyzed the relation between the question "How to obtain digital competency?" and the final scores in the fields of digital competencies estimating. Since the question was presented on a scale with compatible alternatives, we perform the analysis on the basis of a two-dimensional frequency table. The analysis proved the level of competencies does not depend on which way knowledge and skills were obtained.

To analyze the relationship between age and total scores we used a linear regression model. The results showed a lack of connection between the features. The coefficient of determination (R squared), which shows the tightness of the connection, was 0.042, and the coefficient of linear correlation (Pearson's r) was 0.206, which indicates the absence of a linear relationship between the signs.

Thus, the hypothesis that the level of competences depends on gender, status, activities, and access to digital media, the way of teaching was partially confirmed.

In the framework of the questionnaire analysis reliability, we prepared the contingency table between the features, those reflect the respondents' assessment of their digital competencies and the tools used. Analysis of these tables proved that the higher is the respondent's self-esteem the more tools he owns and uses in his daily practice. The indicators reflecting the internal consistency of the questionnaire were also evaluated, namely, the Cronbach alpha was 0.944, Lambda Guttmann 0.89, the Spearman-Brown coefficient 0.889, and the intra-group correlation coefficient 0.49. These numbers indicate the questionnaire high reliability.

4.2 Description of Principal Components of Digital Competence

To reduce the data, we used the principal component analysis (PCA), which was based on 18 features with orthogonal rotation (varimax). The Kaiser-Meyer-Olkin measure confirmed the adequacy of the sample for analysis, KMO = 0.939 ("excellent" in [24]),

Table 7. Summary of exploratory factor analysis results for the digital competence questionnaire (N = 193)

Rotated Component Matrix		
	Component	
	PC 1 - digital competencies as mean of communication	PC 2 - competencies of professional usage digital resources
Preparation of the report	0.72	0.24
Search for sources of information	0.77	0.07
Information storing	0.76	0.35
Choice of communication tools	0.45	0.58
Use of mail and cloud services	0.59	0.43
Informing the public	0.66	0.49
Tools for joint activities	0.61	0.42
Netiquette rules following	0.55	0.34
Account management, creating accounts	0.64	0.44
Creation of animated presentations	0.64	0.37
Copyrights	0.70	0.24
Developing simple applications for websites or smartphones	0.24	0.67
Identification of risks when accessing dialers or digital platforms	0.16	0.87
The choice of the optimal protection means	0.32	0.80
Awareness risks	0.31	0.73
Technical tasks solutions	0.32	0.73
Ability to visualize data	0.55	0.54
Online Learning usage	0.55	0.59
Eigenvalues	9.499	1.304
% of variance	52.77	7.242
Rotated loadings	0.727	0.686

and all KMO values for individual traits were greater than 0.914, well exceeding the permissible limit of 0.5 [24]. Bartlett's test of sphericity $\chi^2(153) = 2251,953$, with $p < 0.0001$, proved that the correlations between the points were quite large for PCA. The initial analysis was performed to obtain the eigenvalues for each component in the data. Two components had similar values according to the Kaiser's criteria of 1 and higher, and in combination they explained 60.01% of the variance. Given not large sample, and the convergence of the scree plot and Kaiser's criterion on two components, this is the number of components that were retained in the final analysis.

Table 7 shows the load factors after rotation. The attributes are added to the main components by the absolute values of the coefficients of the rotated matrix (the cells are highlighted in color). Some characteristics can be attributed to both components (they are reflected at the bottom of the table), but they were assigned to the second component. The elements that are grouped on the same components assume that principal component 1 (PC1) is a digital competency, as a means of use and communication, component 2 (PC2) - the competence of the professional use of information tools.

In Fig. 1, you can see the graph of the analysis result of the main components method with the eigenvectors selected. We can say from the graph, that the initial correlation of characteristics separates the initial data no more than in two directions, which led to the selection of the two main components. At the same time, one can find it difficult to single out separate groups of attributes for some components. This suggests that the various digital competencies are closely related.

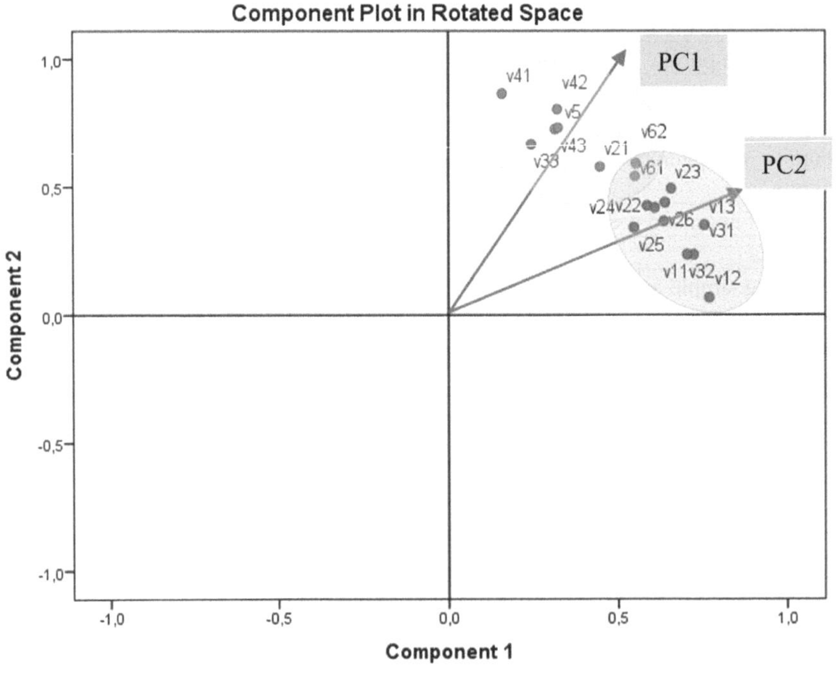

Fig. 1. The graph of the contribution of characteristic values to the main components

The further analysis of the obtained factor values on the basis of the method of principal components in the context of the groups of respondents (by sex, status, activities, availability of digital means) did not show significant differences in gender and availability of technical means. Teachers have a significantly higher level of factor values for the first component, while students have better competencies in the second component ($p < 0.05$). There are also significant differences between groups of respondents working or studying in different areas of activity. Significantly higher average factor values of the first component in the groups of humanitarian and healthcare respondents, while the second component identifies respondents whose activities are related to mathematics, information technology, and information technology, as well as engineering direction ($p < 0.05$). Those who have access to resources with scientific literature have higher averages for both components compared to groups of respondents whose access is limited.

4.3 Expert Analysis of Case-Study

This stage considered expert estimation of the case tasks that were suggested for our respondents. Each of 10 experts estimated the results in a scale from 0 to 5 (Table 3). The Kendell concordance coefficient has made 0.708 at significance value of $p < 0.05$ that proves rather a high level of inter-scorer agreement.

Table 8 contains descriptive statistics, and Fig. 2 shows the results of average estimates distribution on tasks of the first and second group of a case are shown.

The estimates distributions are close to normal, the asymmetry parameters variation and the excess are within 1. Apparently, from results, average assessment 2 is less, than assessment 1. The experts have estimated the level of the competences connected with the process of the task's solution using digital tools lower than the results representation competences. Distinctions of averages are significant at the level $p < 0.005$. However, the Pearson coefficient of correlation between Score1 and Score 2 has made

Table 8. Expert estimates descriptive statistics

Descriptive statistics		Score1	Score2
Number (N = 193)	Valid	178	178
	Missing	15	15
Mean		76.99	70.21
Std. error of mean		0.88	1.22
Median		78	71,15
Mode		72.1	60.1
Std. deviation		11.71	16.28
Skewness		−0.48	−0.266
Std. error of skewness		0.18	0.18
Kurtosis		0.67	−0.71
Std. error of kurtosis		0.36	0.36
Minimum		34.1	34.1
Maximum		99.1	98.4

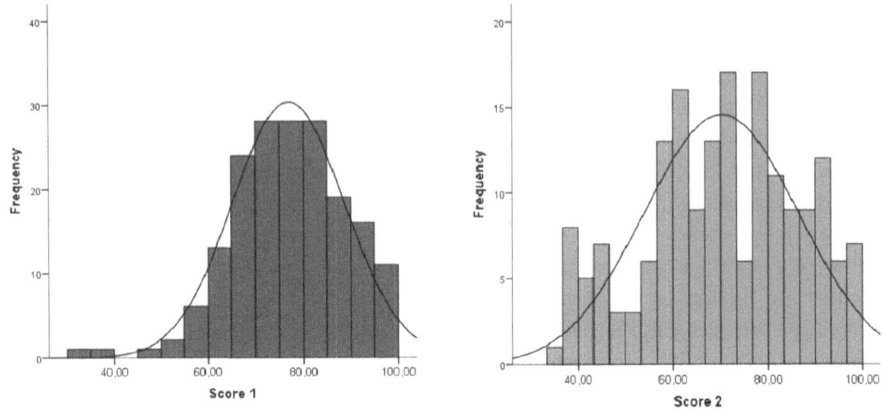

Fig. 2. Frequency Distribution Of Score1 and Score2 (Source: Own work)

Table 9. Correlation between experts' estimates and respondents' estimates

	Processing data V1	Communication V2	Creation of digital content V3	Information security V4	Solving technical problems V5	Studying and data analysis V6
Score1	0.586	0.638	0.61	0.532	0.438	0.511
Score2	0.394	0.614	0.548	0.739	0.639	0.637

0.468. That confirms the assumption, that the respondents having digital competences in use of information technologies can have also rather a high level of technical competences.

At the next stage, we estimated the correlations between the estimates received by experts and estimates received as a survey result at the first investigation stage. Linear Pearson correlation coefficient was calculated between the general estimates of Score1 and Score2 and V1–V6. From the results given in Table 9, it's clear that all of the estimates have strong correlation connection which is significant at the level $p < 0.005$.

There also was estimated correlation between the main component values (PC1 and PC2), displaying the levels of respondents' competences, and the expert estimates. From the results given in Table 10, we can conclude that the first estimate correlates from the first main component.

The distribution of the Score1 and Score2 general estimates, and the linear correlation dependence allow us to create a regression model, to estimate the contribution of

Table 10. Correlation between the experts' estimates and the main components

	Factor score of PC1	Factor score of PC2
Score1	0.515	0.385
Score2	0.286	0.714

separate competence level estimates V1–V6 received from the survey and the expert estimates results, and also to create models based on the main components. The regression analysis gave the following results. As we used the stepwise regression, the model of the first estimate Score1 included the following factorial variables: V1 - Processing data, V2 – Communication and V3 – Creation of digital content. In model of the second estimate Score 2 included the variables: V4 – Information security, V5 – Solution of technical problems and V6 – Studying and data analysis. The variables selection in models was carried out on the basis of F-criterion (F1 = 51.00 and F2 = 90.028 at $p < 0.05$). The model we received as a result corresponds to empirical data.

The R2 determination coefficient for Score1 has made 0.468 and for Score2 – 0.608, i.e. 46.8% of estimate of Score1 and 60.8% of estimate of Score 2 are caused by competences levels V1–V6. The autocorrelation of the regression models remains wasn't observed (coefficients of Durbin-Watson have made 1.969 for the first model and 2.178 for the second model that gets to limits from 1.5 to 2.5). Diagnostics of collinearity was performed on the basis of the analysis of VIF (variance inflation factor) coefficient. The average VIF value for the first model factorial variables was 2.485, and for the second – 2.033 that speaks about lack of a multicollinearity (values less than 10).

In multiple regression the model takes the form of equation and in that equation, there are several unknown quantities (the b-values). These b-values indicate the individual contribution of each predictor to the model. We defined the models for Score1 and Score 2 as follows:

$$Score1 = 38.758 + 0.935 \cdot V1 + 1.33 \cdot V2 + 1.06 \cdot V3 \quad (1)$$

$$Score2 = 26.319 + 2.943 \cdot V4 + 1.886 \cdot V5 + 3.48 \cdot V6 \quad (2)$$

The b-values tell us about the relationship between Scores1 - Score2 and each predictor V1–V6. All values are positive; therefore, we can tell that there is a positive relationship between the predictors and scores. So, as level digital competences increases, the quality level of case studies increase. The b-values tell us more than this, though. They tell us to what degree each predictor affects the scores if the effects of all other predictors are held constant.

V2 predictor – Communication influences the Scores1 the most, though this influence isn't significantly higher comparing to other factorial variables, as b-values at other predictors are also close to 1. The Score2 is influenced the most by V6 variable - Studying and data analysis. Respondents who estimated their problem-solving skills high received higher scores from the experts for tasks, which displayed the case-study process.

Consider the constructed models have been checked also for a homoscedasticity. Figure 3 displays the remains for Scores1 - Score2. The diagrams show the uniform variability of the values of observations, expressed in relative stability, the homogeneity of the random error variance. We can see, that assumption met.

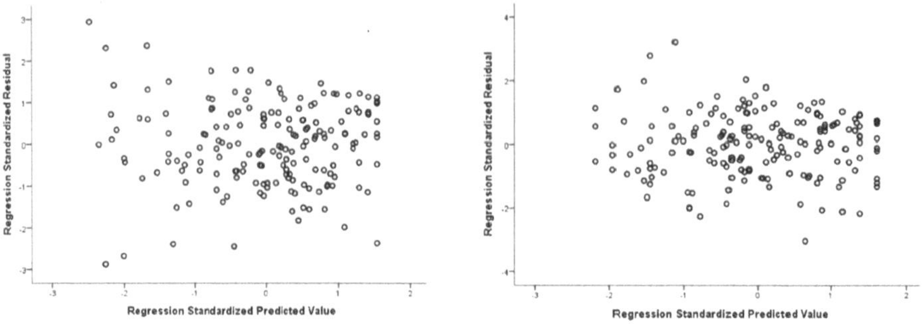

Fig. 3. Residuals of Score 1 and Score2 (Source: Own work)

Analogous, regression models for Scores1 - Score2 where the main components received after reduction of data on the basis of a method main a component acted as predictors were created. The models for Score1 and Score 2 was defined as follows:

$$Score1 = 75.837 + 6.474 \cdot PC1 + 4.916 \cdot PC2 \tag{3}$$

$$Score2 = 69.731 + 5.563 \cdot PC1 + 12.4 \cdot PC2 \tag{4}$$

The received from equations b-values mean that the contribution of the first component is higher for Score1, and the contribution of the second component is much higher for Score2.

The quality of predicted expert estimates based on the estimates received from the survey is also rather high. Figure 4 presents charts of dispersion that represent compliance of observed estimates to predicted estimates (models (1–2)). The graphic highlights the groups corresponding to the respondent's status: student-bachelor, student-master, and teacher. As we can see, prediction errors aren't big. Differences between the predicted average values and observed estimates, both for the Score1 model (t-Student = 1.131) and for the Score2 model (t-Student = 0.168) – aren't significant at the level p > 0.05.

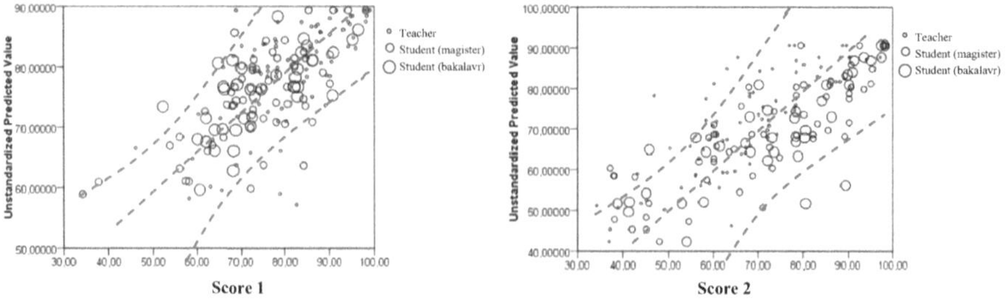

Fig. 4. Unstandardized predicted value as a function of Score1 and Score2 (Source: Own work)

The assumption that the respondents, who received high expert estimates for the case tasks estimated their skills high themselves and those respondents who estimated their skills as average or below average received low estimates were proved with results of regression analysis. We didn't only build models for Score1 - Score2 on the basis of the survey results predictors but also estimated contribution of separate competencies.

We confirmed that the respondents who master basic digital competencies can simply solve other "digital tools usage" related problems. The models received for Score1 - Score2 on the basis of the main components proved the significant influence of both main components on the case study performance. Thus, we conclude that respondents who know the basic digital competencies solve equally other problems related to the use of digital tools. However, there are some differences in the level of digital competencies between users of information resources solely for solving the problems of searching, presenting, storing and transmitting information, and respondents able to solve technical problems, providing reliable protection and processing of data by means of special means. Most people learn skills independently, regardless of the direction of activity, status, and access to technical and digital tools.

5 Conclusions

The digital competencies are essential for people to achieve success in the condition of the digital economy. The results of a survey in which participated 193 teachers and students of Ukrainian educational institutions aiming to define the readiness to implement digital education for obtaining the digital competencies, and the estimation results of 178 cases allow us to conclude:

1. The teachers and students have the above average level of usage of digital tools and communications. However, the level of competencies does not depend on the way that the skills were obtained.
2. The level of competency of professional usage of IT is much higher for students than for teachers. The teachers have a higher level of IT usage for performing educational tasks. The level of competencies in exact sciences differs from the others. The level of competencies of the respondents who have restricted access (or no access at all) to the resources with the literature is far lower than the level of those respondents who has full access to such resources.
3. There were defined no difference in gender, age and availability of technical means.
4. The respondents who received high expert estimates for the case tasks estimated their level as high; those respondents, who estimated their skills as average or low received low estimates from the experts.
5. The factors affecting the performance depend on digital competencies, and extent of competences influence varies.

Since the analysis of the obtained data confirms high reliability of the questionnaire developed by authors, we can formulate the further researches perspectives. It seems to be perspective to measure the digital competencies in each field of DigComp and to develop the training modules for formal or informal training.

The sufficient level of digital competencies of both students and teachers proves their readiness for digital training implementations. The difference of levels of students (as the developers of e-content), and teachers (as the competent users) can be used effectively to provide collaborative training online.

Consider that digital competencies influence the training programs structure, professional development of teachers and services and resources intended for students at the university. That is why there must be created a uniform environment of digital competencies management at the university. That allows providing within the university: common information space for control, development and a transfer of digital competencies; optimized communication between students, teachers, and administration of the university; individual planning, monitoring, and management of educational trajectory personally for every student.

References

1. Digital Transformation Initiative Telecommunications Industry World Economic Forum (2017). http://reports.weforum.org/digital-transformation/wp-content/blogs.dir/94/mp/files/pages/files/dti-telecommunications-industry-white-paper.pdf. Accessed 25 Sept 2018
2. Digital Economy. https://www.epsrc.ac.uk/research/ourportfolio/themes/digitaleconomy/. Accessed 25 Sept 2018
3. Europe 2020 strategy. https://ec.europa.eu/digital-single-market/en/europe-2020-strategy. Accessed 25 Sept 2018
4. Digital agenda for Ukraine. http://www.e-ukraine.org.ua/media/Lviv_Minich_2.pdf. Accessed 25 Sept 2018
5. Digital agenda for Europe. http://eige.europa.eu/resources/digital_agenda_en.pdf. Accessed 25 Sept 2018
6. New Skills for the Digital Economy. http://dx.doi.org/10.1787/5jlwnkm2fc9x-en. Accessed 25 Sept 2018
7. The Digital Economy and Society Index (DESI). https://digital-agenda-data.eu/datasets/desi/indicators. Accessed 25 Sept 2018
8. Akca, H., Sayili, M., Esengun, K.: Challenge of rural people to reduce digital divide in the globalized world: theory and practice. Gov. Inf. Q. **24**(2), 404–413 (2004)
9. Monitoring the Digital Economy & Society 2016–2021. http://ec.europa.eu/eurostat/documents/341889/725524/Monitoring+the+Digital+Economy+%26+Society+2016-2021/7df02d85-698a-4a87-a6b1-7994df7fbeb7. Accessed 25 Sept 2018
10. A common European framework for ICT Professionals in all industry sectors. http://www.ecompetences.eu/. Accessed 25 Sept 2018
11. ECDL Foundation: Computing and Digital Literacy: Call for a Holistic Approach ECDL Foundation. http://www.ecdl.org/media/PositionPaper-ComputingandDigitalLiteracy1.pdf. Accessed 25 Sept 2018
12. Global Media and Information Literacy Assessment Framework: Country Readiness and Competencies. UNESCO, France. http://www.unesco.org/new/en/communication-and-information/resources/publications-and-communication-materials/publications/full-list/global-media-and-information-literacy-assessment-framework/. Accessed 25 Sept 2018
13. The Digital Competence Framework for citizens. https://ec.europa.eu/jrc/en/digcomp/digital-competence-framework. Accessed 25 Sept 2018

14. Digital competence. http://europass.cedefop.europa.eu/resources/digital-competences. Accessed 25 Sept 2018
15. Learning and Skills for the Digital Era. https://ec.europa.eu/jrc/en/research-topic/learning-and-skills. Accessed 25 Sept 2018
16. The Digital Competence Framework for Citizens. http://publications.jrc.ec.europa.eu/repository/bitstream/JRC106281/web-digcomp2.1pdf_(online).pdf. Accessed 25 Sept 2018
17. Conoce el diagnóstico de tu nivel de competencias digitales. http://www.digcomp.andaluciaesdigital.es/#pregunta. Accessed 25 Sept 2018
18. James, T., Miller, J.: Developing a monitoring and evaluation plan. In: Wagner, D., Day, B., James, T., Kozma, R., Miller, J., Unwin, T. (eds.) Monitoring and Evaluation of ICT in Education Projects, pp. 57–76. infoDev, World Bank, Washington, DC (2005)
19. Morze, N., Glazunova, O., Kuzminska, O.: Training of E-learning managers at universities. In: Bassiliades, N., et al. (eds.) ICTERI 2017. CCIS, vol. 826, pp. 89–111. Springer, Cham (2018). https://doi.org/10.1007/978-3-319-76168-8_5
20. ECTS Users' Guide 2015. https://doi.org/10.2766/87592
21. Online Statistics Education: A Multimedia Course of Study. (http://onlinestatbook.com/). Project Leader: David M. Lane, Rice University. Moore, D., McCabe, G.: Introduction to the Practice of Statistics, 3th Ed. Freeman (1998)
22. Kendall, M.G., Smith, B.B.: The problem of m rankings". Ann. Math. Stat. **10**(3), 275–287 (1939). https://doi.org/10.1214/aoms/1177732186
23. Harrell, F.E.: Regression Modeling Strategies. Springer, Cham (2015). https://doi.org/10.1007/978-3-319-19425-7
24. Field, A.: Discovering Statistics Using SPSS (and Sex and Drugs and Rock 'n' Roll), Los Angeles [i.e. Thousand Oaks, Calif.].SAGE Publications (2009)
25. Levesque, R.: SPSS Programming and Data Management. A Guide for SPSS and SAS Users, 2nd edn. SPSS Inc., Chicago (2005)
26. Carmines, E., Zeller, R.: Reliability and validity assessment. Quantitative applications in the social sciences series no. 17. Sage Publications, London (1979)
27. Kutner, M., Nachtsheim, C., Neter, J., Li, W.: Applied Linear Statistical Models. McGraw-Hill/Irwin, Homewood (2004)
28. Jolliffe, I.: Principal Component Analysis, 2nd edn. Springer, New York (2002)
29. Kuzminska, O., Mazorchuk, M., Morze, N., Pavlenko, V., Prokhorov, A.: Digital Competency of the students and teachers in Ukraine: measurement, analysis, development prospects. In: Information and Communication Technologies in Education, Research, and Industrial Applications. Communications in Computer and Information Science, vol. 2104, pp. 366–379 (2018). http://ceur-ws.org/Vol-2104/paper_169.pdf

University-Enterprises Cooperation in Ukrainian Game Industry

Maryna Zharikova and Volodymyr Sherstjuk(✉)

Kherson National Technical University, Kherson, Ukraine
marina.jarikova@gmail.com, vgsherstyuk@gmail.com

Abstract. This paper describes the current state, the process, and the attitudes of cooperation between Kherson National Technical University (KNTU) and game industry (GI) companies in the frame of Erasmus+ project "University-Enterprises Cooperation in Game Industry in Ukraine" (GameHub). The work on the project in KNTU is reported in the paper. It covers monitoring of the competence profiles needed on the GI market, creating necessary infrastructure and developing the relevant curricula, study programs, education resources, and organization of the feedback loop with GI enterprises. One of the priorities is to equip students with soft competencies needed for GI such as creativity and academic integrity. The paper also presents the actual cooperation results, which could be helpful for Ukrainian higher education sector to respond to current and future education needs because mutually beneficial and sustainable university-enterprises cooperation can achieve a better match between competence profile of university graduates and those required by GI.

Keywords: Cooperation · Higher education · Gaming industry · Infrastructure · Curricula · Competencies

1 Introduction

For today, the state-of-the-art in the game industry (GI) in Ukraine is dynamically developing and becoming attractive for many international game application development companies. Developing Ukrainian GI is stimulated by the wide spreading of computer games among children and young people, active using of computer games in education, mature computer games marketing with well-established methods of monetization. On top of it, Ukrainian information technologies (IT) sector is proved its ability to develop and promote software of high quality.

Even with all things considered, preparing specialists for GI in Ukraine is complicated for a score of reasons. On the one hand, there is a deficiency of narrowly focused specialists such as graphic designers, content managers, storytellers, scriptwriters, sound designers, sound programmers, web-client programmers, sketchers etc. On the other hand, a majority of Ukrainian universities do not have neither necessary software and hardware to equip laboratories for training such specialists nor appropriate curricula and educational programs related to the development of competencies needed for the work in game development companies. Besides, the content of Ukrainian educational programs does not meet requirements of international standards

V. Ermolayev et al. (Eds.): ICTERI 2018, CCIS 1007, pp. 170–194, 2019.
https://doi.org/10.1007/978-3-030-13929-2_9

and employer requirements at international labor markets. At the same time, openness and attractiveness of international labor markets, along with the availability of remote work for international companies and the vigorous growth of outsourcing, are important incentives for the progress of Ukrainian IT education.

Thus, we have a significant contradiction. To solve it, we need to create a concept of extensible, creative IT education sector and understand its importance.

Fortunately, Ukrainian higher education system has a long-standing experience of combining work with education in the form of part-time study, and correspondence training, as well as in the form of full-time practical training at the real workplaces.

However, during the period of the transformation of the Ukrainian economy and higher education system, there was a devaluation of the traditions of the practical training in education, dissemination of imitation and falsification of practice, so many organizations generally refused to work with students, referring to the preservation of commercial secrets or complex competitive circumstances.

Because of corporatization and privatization, the government lost most of its administrative influence on companies, which allowed considering the practical training of students. At the same time, in condition of a lack of trained domestic specialists with contemporary and creative thinking, a significant part of the positive examples of the last decade are inextricably linked with the training of specialists in IT sphere, including specialist subjects in the development of computer games against a background of individual projects.

The nature of the GI sector requires from Ukrainian universities developing a new conception of Computer Game Development education: providing students with the solid technical knowledge and skills, enabling them to focus on the transversal components of computer game design at the same time. The Computer Game Development requires lectures that include computer science, physics, engineering, visual technology, music techniques, humanities, cognitive psychology, art study, and game design. Moreover, since international employers want graduates to be better prepared for the workplace, future specialists in the field of GI require not only theoretical knowledge, but also practical experience. Therefore, universities need to be thinking about the skills the graduates will need to do well in a job.

However, the truth is rather different. Ukrainian universities still produce graduates with the low level of hands-on experience, which blunts their quality and gives rise to a gap between educational supply and the demand of GI market. Mismatches between employers′ needs and what universities offer lead to skill gaps and economic underperformance. Therefore, the number of engineers-graduates with the competencies that fit the game industry employee profiles is almost zero.

Needs of the labor market force the enterprises and the universities to cooperate for overcoming the above-mentioned complexities on mutually beneficial conditions.

University-enterprise cooperation has been a subject of debates and focus of attention for many years and is commonly defined as all forms of interaction between universities and enterprises for the reciprocal and mutual benefit.

In the context of significant influence of globalization and internationalization [1–3] on the current stage of educational system in Ukraine, the university-enterprise cooperation is getting international. Globalization is defined as a force pushing higher education toward greater international involvement. The results of globalization include the integration of studying, the growing international labor market for scholars, and the use of international technologies, which facilitates communications, permits efficient storage, selection, and dissemination of knowledge [2]. The globalization results in internationalization, which involves many choices such as student mobility, double degree programs, European projects with the participation of Ukraine, providing scholarship exchange programs for Ukrainians to study in the European Union and for European Union citizens to study in the Ukraine etc. The goal of internationalization is a formation of a single educational environment with increased student mobility and cooperation between universities and enterprises in different countries.

In recent years, the international university-enterprise cooperation is being developed rapidly and markedly through implementing Erasmus+ projects in the Europe [4–7] such as "Integrating Entrepreneurship and Work Experience into Higher Education" (IE-WEXHE), "Embedding Entrepreneurship Education" (Triple-E), "European University-business cooperation" (UBC) and others. The European project "Integrating Entrepreneurship and Work Experience into Higher Education" (IE-WEXHE) is aimed at the integration of higher education and enterprises. The project assumes generating case studies of work-based learning involving four types of disciplinary sectors (hard-pure, e.g. natural sciences; soft-pure, e.g. humanities and social sciences; hard-applied, e.g. medicine and soft-applied, e.g. social work) covering work placements, traineeships, and entrepreneurship. A unique feature is an attention for Humanities for which the transition to the labor market is less transparent and mapped than for other sectors. The project "Embedding Entrepreneurship Education" (Triple-E) has been designed with the objective to increase the proportion of University students acquiring an entrepreneurial mindset and engaging in early-stage entrepreneurial activity [7]. The project "European University-business cooperation" (UBC) is aimed at conducting a Europe-wide study on cooperation between universities and business. This study is the largest study ever undertaken on the topic of UBC in Europe [8].

The internationalization requires all institutions to commit to its underlying values, and principles, including but not limited to: academic culture, mutual benefit, mutual respect, fair partnership. Nevertheless, universities and higher education institutions differ in how they address plagiarism, who is responsible for enforcing policies, and which learning practices are considered academically dishonest [9]. Significant differences in social and academic norms of Ukrainian and European universities pose additional challenges related to the interpretation and practice of academic culture, especially academic integrity.

Tolerance to dishonesty and corruption within Ukrainian academic community has led to the fact that academic dishonesty of some persons does not result in any negative effects for their careers. To be sure, internationalization itself significantly expands the possibilities for how different forms of fraud and corruption can be exchanged within and between universities and society. Obviously, the propensity for corrupt practices in academia greatly increases with internationalization. Therefore, there is a gap between requirements of internationalization and poor reputation of Ukrainian universities, which requires a wider effort to upgrade educational services to international standards [10]. Although the issues of academic misconduct is well discussed and studied in European Union and North America, there are many specific things related to the Ukrainian universities.

Thus, we have yet another contradiction. On the one hand, Ukraine is involved in internationalization of higher education. On the other hand, Ukrainian universities have seen their reputation diminished among enterprises on the international labor market that usually refuse to recognize Ukrainian diplomas. Ukraine has necessary legislation and strategies for university-business cooperation; however, in many cases they are not implemented yet, and thus need additional study and improvement.

One of the possible ways to overcome both of the above contradictions is to establish deep partnerships between universities and enterprises of IT industry. University-enterprises cooperation has not been sufficiently studied and practically implemented in Ukraine. Thus, over the past decades, we have seen a disconnection between Ukrainian education system and the labor market.

The aim of the paper is researching of the impact of extensible symbiotic University-Enterprises cooperation on the improvement of studying processes and employability of students against standards and requirements established on the international labor market. Therefore, the paper covers the following questions:

1. University-enterprises cooperation based on using such modern and mutually beneficial approaches as dual learning, flexible learning etc.
2. Internationalization of Ukrainian higher education, which requires transforming the learning process and the academic culture to the international standards, including academic integrity issues.

All above-mentioned problems have considered on the results of International Erasmus KA2 project "University-Enterprises Cooperation in Game Industry in Ukraine" (GameHub), which initialized in 2015 and aimed at maintaining University-Enterprises cooperation in Ukraine [11].

2 The Basic Problem of University-Enterprise Cooperation

The main problem that needs to be solved in the framework of cooperation between employers and universities is the inadequate level of readiness of many universities' graduates for their independent professional activities.

This problem has become a consequence of the long-term effects of the following factors:

- a formal education system is not motivated by the development of skills demanded by employers;
- a possibility of obtaining education outside universities remains limited;
- a lack of financing necessary for the creation of a modern university material and technical base and the formation of practical skills;
- a lack of a significant part of the scientific and pedagogical staff with the competencies necessary for the formation of actual practical skills of the applicants;
- a lack of reliable information on the current and future needs of employers in the competencies;
- a lack or complete absence of modern equipment and corresponding technologies in the universities;

- a non-inclusion of most higher education institutions (which are mostly in the status of budgetary institutions) in the current market relations and the uncertainty of the mechanism in the field of public-private partnership, which affects the training of education applicants for independent professional activity in a market environment.

Manifestations of the problem are:

- the readiness of graduates to work in a specialty;
- labor market dissatisfaction with the quality of education, which leads to the corresponding need for additional training in the workplace, an increase in the education system in the companies/corporations;
- ineffective work of universities, including low-cost use of budget funds;
- inefficient use of labor resources by employers, which overestimate the formal qualification requirements for universities' graduates, rather than trying to compensate for their reluctance to cooperate with educational institutions on the training of high-quality specialists;
- ineffective use of the best time to study of applicants for professional competences;
- requirements for having an experience of independent professional activity (or at least an experience of any work) from graduates of universities, which are arranged for work.

The solution of the problem is envisaged by implementing a set of measures for the development of models of mutually beneficial relations of universities and employers aimed at the practical training of applicants for independent professional activity and their social adaptation in production teams, conducting approbation, research, and updating of models as well as its recommendations for widespread use.

The proposed model of extensible and mutually beneficial relations of universities and employers is based on the modern dual learning approach.

The dual form of education was proposed for the training of graduates and specialists envisaged to establish an equal partnership of HEIs, employers and education providers with the aim of acquiring the latest experience in practical application of competencies and their adaptation in the context of real professional activities.

It is important to distinguish dual learning from the traditional practice of students and trainees. A dual form of education involves training in the workplace with the performance of official duties in accordance with the employment contract. In addition, dual education is directed solely at the adaptation of the education provider to the first workplace, which corresponds to a certain professional qualification corresponding to his educational specialty and qualifications.

At the same time, the dual form of education should not be absolute, so it is necessary to understand the limits in which the dual form of education is effective.

It is obvious that the overwhelming majority of modern IT educators will repeatedly change jobs, professions and activities during a long working life. At the same time, the objective of the university is to acquire a competency IT education provider that will allow them to adapt to various technological changes.

Thus, International Erasmus KA2 project GameHub was proposed to stipulate University-Enterprises Cooperation in Game Industry in Ukraine [11].

3 GameHub Project

GameHub project was aimed at the building of the infrastructure, which allows students to improve their skills and competencies needed to work in GI and intends cooperation between universities and enterprises.

The paper dwells on such kind of cooperation organized in Kherson National Technical University (KNTU), which is the partner of the international consortium of GameHub project.

The international project GameHub started in October 2015. The project is aimed at developing the infrastructure and resources for learning the target groups such as students, veterans of anti-terrorist operation (ATO veterans) and unemployed engineers the competencies and skills needed to create computer games.

The main tasks of the project was the following:

1. Developing the map of the competencies, which can determine a professional level in GI, as well as the instruments for monitoring the competence profiles [12–14];
2. Preparing and training the university staff;
3. Developing 18 bilingual learning modules.

It is clear that cooperation between universities, IT companies, unemployment center, and associations of ATO veterans will support GI market in Ukraine.

GameHub consists of four main components:

1. Pedagogical component, which includes methodology for learning the students, ATO veterans and unemployed people how to create game applications according to the developed modules;
2. Technological component, which includes creating and maintaining the work of game laboratory;
3. Methodological component;
4. Informational component, which involves providing information communication between the main branches of GameHub members:

- teachers and trainees/students, unemployed people, and ATO veterans;
- university management/GI representatives and employers;
- scientists/scientific society.

The work on the GameHub project in KNTU started with monitoring of the competence profiles needed on the GI market in Ukraine, creating necessary infrastructure and developing the educational resources.

The main features that differ the GameHub project in KNTU from other types of University and Industry collaborations from around the world are:

- Using the dual learning approach. The developed cooperation model uses the German experience of dual learning;
- Gamification of the learning process;
- Focusing on the graduates in software engineering at the both bachelor and master level.

4 Gamification Principles and Gamification of the Learning Process

Gamification is the use of game-play mechanics and game-design thinking for non-game applications. It is increasing in application to the learning sphere.

The objective of gamification in learning is to encourage both enjoyment and engagement through the learning experience by capturing the attention of learners and motivating them to continue learning [15]. Gamification in learning supports and motivates students, and can lead to enhanced learning processes and outcomes.

There are some principles of gamified learning [16]. They include the following:

- **Rewards.** Small increments of positive reinforcements build up the confidence of player during the learning process.
- **Interaction.** While the books are passive in the sense that the student cannot get them to talk back to in a real dialogue, the game reacts back, giving the player feedback and new problems. In a good game, words and deeds are all placed in the context of an interactive relationship between the player and the world.
- **Levels of progress.** Each game has a progress bar, and the learning process is also based on student progress. Therefore, the learning progress can be transformed in game levels. One thing to keep in mind when designing these levels of progress is to start easy and make things harder gradually. If students fail the first level of a learning game they might lose interest in it. But if they win, they'll get more ambitious to get over it and level-up.

Game based learning (GBL) is a type of game play that has defined learning outcomes. GBL describes an approach to teaching, where students explore relevant aspect of games in a learning context designed by teachers. Teachers and students collaborate in order to add depth and perspective to the experience of playing the game [17].

Game-based courses can draw students into virtual environments that look and feel familiar and relevant. Within an effective GBL environment, they work toward a goal, choosing actions and experiencing the consequences of those actions along the way. They make mistakes in a risk-free setting, and through experimentation, they actively learn and practice the right way to do things. This keeps them highly engaged in practicing behaviors and thought processes that they can easily transfer from the simulated environment to real life [16].

While similar, gamification is a different breed of learning experience. Gamification takes game elements (such as points, badges, leaderboards, competition, achievements) and applies them to a non-game setting. It has the potential to turn routine, mundane tasks into refreshing, motivating experiences [15].

Both game-based learning and gamification are the basic principles of the organization of the educational process in KNTU and the development of educational resources for modern education programs.

5 Goals and Objectives of GameHub Functioning in KNTU

GameHub in KNTU is a necessary and essential tool for the creation and implementation of new educational programs on Computer Games Development. GameHub is also a tool to overcome a gap between the insufficient technological infrastructure of the educational process and the demand of GI market with the help of developing the game learning laboratory that provides students with the entire scope of the necessary technical knowledge and skills. As a result, implementation of the project results helps the students to overcome some discrepancy between the knowledge gained at the university and the actual demand in the labor market.

GameHub in KNTU should unify connections between the university, game industry and society in general. The goal of GameHub is the stimulation of the students and trainees to acquire of knowledge and practical skills required for successful work in the computer games development sector.

The objectives of functioning GameHub in KNTU are:

- Providing conditions for creation and implementation of innovative education program on Computer Games Development as a specialization for the students of software engineering specialty;
- Adapting of education program on Computer Games Development to the GI market and employer's requirements;
- Increasing the level of quality of software engineering specialists through learning state-of-the-art technologies of Computer Games Development at the level of labor market requirements;
- Improving university and employers cooperation using dual learning techniques;
- Improving education process using game-based learning approaches;
- Implanting of academic culture, providing academic integrity through ensuring the creativity of learning;
- Creating the necessary conditions for implementation of ongoing, voluntary, and self-motivated long life learning education;
- Looking for the ways to improve organization as well as scientific and methodical ware of educational process;
- Providing consulting services to the base of university and game industry connection;
- Reciprocal exchange of experience, knowledge, educational materials and innovation practice of engineering education between partner universities.

The target audience of KNTU GameHub consists of students of Faculty of Cybernetics and System Engineering (bachelors, master's degree students, PhD students), as well as retraining students of the Institute of Postgraduate Education. The main emphasis is put on studying ATO veterans and unemployed people.

The principles of the GameHub organization in the University:

- Close collaboration with enterprises in the area of game industry on the regional, national and international levels;
- Implementation of modern innovative educational methods and methodologies;
- Adaptation of educational program to the labor market requirements;

- Combination of studying the students with practical activity;
- Gamification of the learning process;
- Intensive collaboration with partner universities.

6 Priorities of KNTU GameHub

The priorities of KNTU GameHub are illustrated in Fig. 1.

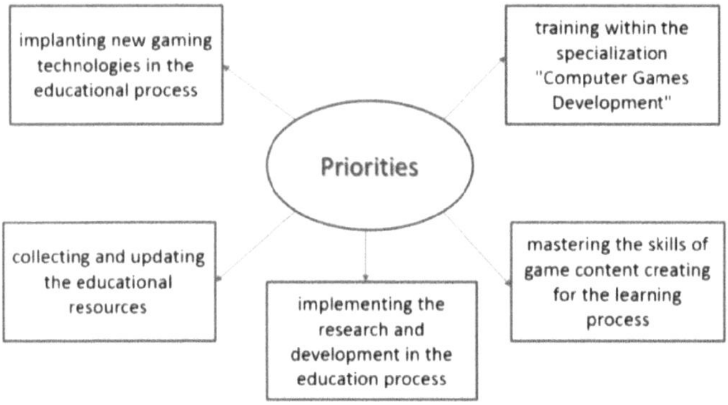

Fig. 1. Priorities of KNTU GameHub

All activities in KNTU within the project can be put into pedagogical, methodological, technological, informational, and academic components.

The ***pedagogical component*** (Fig. 2) of the GameHub project in KNTU consists of monitoring the competencies needed for employment in game industry and arrangement of conditions for obtaining these competencies by the students.

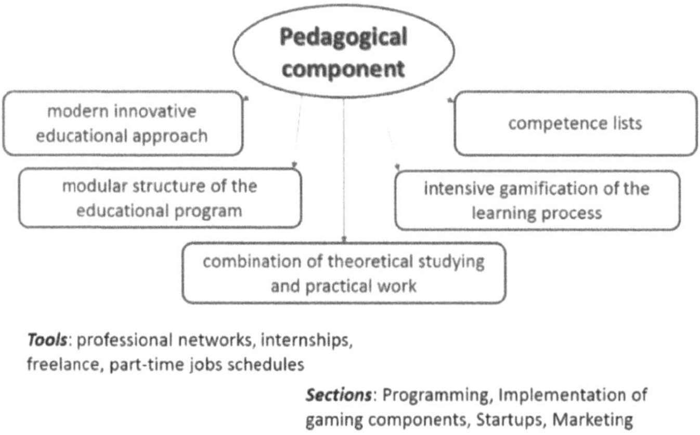

Fig. 2. The pedagogical component of KNTU GameHub

Thus, the pedagogical component includes the following stages:

1. monitoring the competencies needed for the GI market;
2. developing the competencies.

Monitoring the competencies was conducted by survey of specialists in game industry area using special questionnaires.

In consequence of monitoring the competencies, we have determined a set of competencies needed for the students of specialties related to the GI to meet GI market demands. Game developers tend to work in diverse teams that require more creativity and interpersonal communication skills than traditional software developers. The monitoring has shown that the main competencies for the game developer are creativity, academic integrity, problem-solving, teamwork.

Creativity appeared to be an especially important non-technical skill that could be enhanced in students headed for game development careers [18, 19]. It is what makes the games unique and competitive in a market. Creativity allows developing new ideas and coming up with the ways to hold consumers' attention.

Problem-solving is also important competence needed for game developers. All software, in general, is designed to solve some user problem and within that general solution is a wide array of smaller problems that make it up. Programmers are problem-solvers by occupation, which it is one of the most vital soft skills for success in the industry. After writing codes and creating programs, programmers also find and fix any issues that may appear. This is not often an easy task since even the tiniest of errors has the ability to wreak havoc on a program.

Ability to work in a team in many cases can be the first or most important skill for game developers. Games companies involve groups of people, working together, to achieve the same final goal. Game developers must be good communicators, who can cooperate with people working closely with programmers and receive feedback from testers ensuring that the functionality of the game is practical and balanced. Unless a game developer can effectively deal with other developers, managers, and even customers, he will constantly face trouble despite how good your ideas are or how valuable his skills are.

Another competence equally important for GI is *academic integrity* (AI) [20, 21]. Academic integrity means academic honesty and implies that students and teachers abide by a code of honesty, trust, fairness, respect, and responsibility related to the production, publication, assessment, and exchange of knowledge in learning, teaching, and research. Maintaining academic integrity is an issue of concern to all the students due to high and rising levels of plagiarism and other forms of cheating such as receiving unauthorized assistance. Courses related to computer programming require special consideration because they are connected with the intellectual property, and use of the computer permits easy copying and modification of programs. The accusation of AI by students has serious consequences in their future workplaces. In the workplace, since the profit of game developers and their employers depends upon the uniqueness and originality of the code, a plagiarism or stealing the code can potentially harm the career.

Any violation of academic integrity is a serious offense and is therefore a subject to an appropriate sanction or penalty. We propose to classify violations (Fig. 3) and choose a sanction according to a class of violation. Faculty members, college administrators, librarians, lab personnel, counselors, or other personnel noticing infractions of the standards of academic honesty and integrity may be responsible for instituting disciplinary procedures in response to the defined violations.

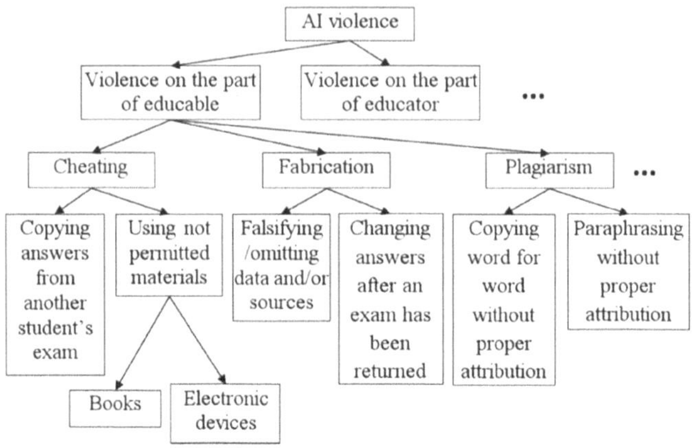

Fig. 3. Fragment of academic integrity violation classification

The relevant curricula and programs should be created to develop or/and improve the detected competencies specialized modular courses. Curricula and programs should be based on the alliance of theoretical learning providing a certain system of knowledge with the practical work providing a system of necessary skills through the active implementation of laboratory and practical tasks, training and work with potential employees, realization of creative ideas in the course works and graduation works, scientific work, taking part in startups etc.

The ***methodological component*** of GameHub (Fig. 4) in KNTU is based on the experimental realization of the innovative educational program through the development of learning modules (Table 1) for university courses and providing them with necessary educational recourses. Educational process with the use of the modules is based on problem-oriented learning, which develops creative thinking and cognitive abilities of students through solving problem-oriented tasks in the area of computer games development.

The educational program has module structure that covers all levels of necessary knowledge and skills, such as programming (Python, JavaScript, Java programming languages), startups, marketing.

Teachers, who passed appropriate training, have developed learning modules in the form of open educational recourses focused on using methods and forms of blended learning for students (trainees) on the base of learning game laboratory GameHub. The system of knowledge and skills provided by learning modules meet the developed profile of specialist's competencies.

The content of learning modules is provided through the verbal (lectures and consultations), scientific (presentations) and practical methods (laboratory practicals),

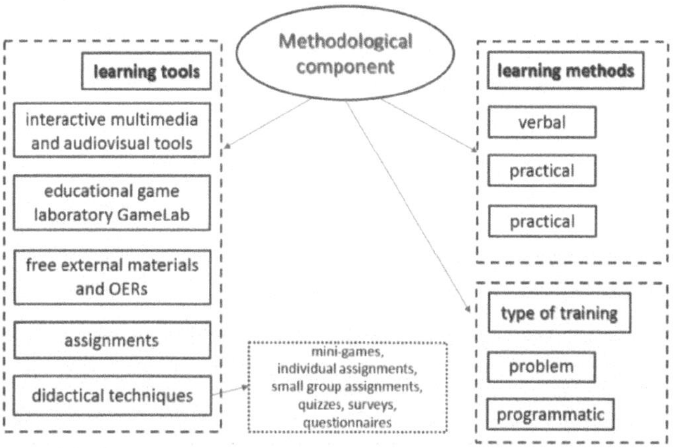

Fig. 4. The methodological component of KNTU GameHub

Table 1. Learning modules

	Module title	Master's degree students	Bachelors	ATO veterans and unemployed people
1	Computer game development using Unity3D	*	*	*
2	Developing Game Web-applications	*	*	
3	Developing network applications using Java programming language		*	
4	Visual game development			*

as well as through the intensive use of the method of project execution (performance of individual and group (in small groups) problem tasks).

The base of an educational process aimed at developing the necessary competencies is problem-based learning through a certain system of methods and tools, which form creative thinking and cognitive ability of students (trainees) through the solving of problem-based tasks in the area of computer games development and game content creation. Cognitive situations, according to which a student has a lack of available knowledge for practical situational tasks solving, are designed for this purpose.

Using programmed learning elements allows splitting learning material into certain portions, which fit in with specific elements of understanding, and by virtue of a problem task in each portion provides for individualization of learning with appropriate feedback and self-control in task performance.

Programmed and problem-based learning have a theoretical form of lectures and consultations and practical form of individual tasks and work in small groups.

The ***technological component*** (Fig. 5) includes creating and maintaining Game Learning Laboratory (GameLab).

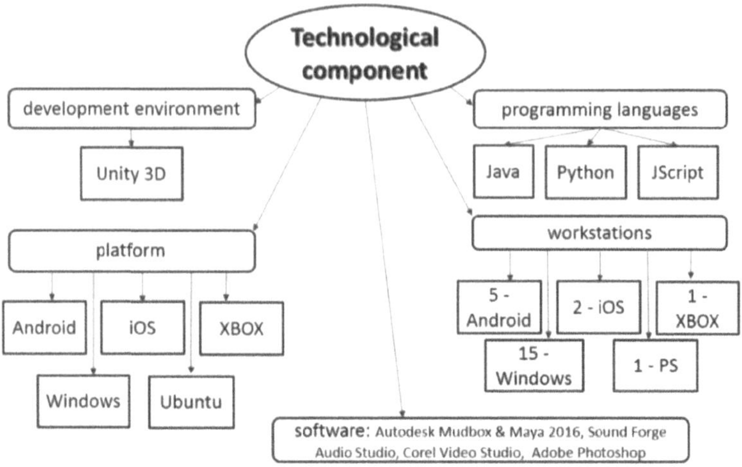

Fig. 5. The technological component of KNTU GameHub

The equipment and software of game learning laboratory GameLab meet the requirements of developed learning modules and will be used for learning and training the students, the trainees and teachers according to the educational program "Computer games development".

The ***informational component*** of the GameHub project in KNTU is aimed at wide dissemination of project results and involves providing information days, roundtables and job fairs.

The members of KNTU workgroup have organized information days for students that have been aimed at sharing the information about the goals of the project, as well as the opportunities for students in consequence of project realization in the university. The roundtables with potential employers and the students of Information Technologies department have been organized upon an initiative of the KNTU workgroup. Such roundtables are aimed at the development of partner relationships in University-IT-Enterprise format. One of such employers is Wezom web development agency that is a trustworthy website development and design company. Representatives of Wezom company have made aware the students of open vacancies in GI market, requirements to potential game developers. KNTU graduates who work for Wezom company have shared their working experience.

KNTU workgroup have also organized job fairs in Kherson Employment Centre, aimed at attracting certain target groups such as demobilized military, ATO veterans, resettlers from the Crimea and the eastern part of the Ukraine, as well as unemployed people to achieve professional competencies in the area of computer game design and development based on KNTU postgraduate centre.

The ***academic component*** of GameHub in KNTU consists of:

1. an educational program on "Program engineering" specialty, "Computer games development" area of study, which is being renovated and improved permanently through the feedback from potential employers;

2. three learning modules, which are being developed by KNTU workgroup for implementation into the educational program on "Computer games development" area of study:
   - Computer games development using Unity3d engine (Master's degree, Bachelor degree),
   - Developing Game Web-applications (Master's degree, Bachelor degree),
   - Network computer games development with Java (Bachelor degree);
3. two learning modules which are being developed by KNTU workgroup for trainees of Institute of Postqualifying Education (including unemployment people, veterans of anti-terrorist operation):
   - Computer games development using Unity3d engine,
   - Visual games programming;
4. learning modules which are being developed by KNTU partners, including 15 learning modules for Bachelor or Master's degree students and 5 learning modules for retraining of trainees, and which can be adapted for use in an educational program on "Computer Games Development" area of study in KNTU.

7 Concept of Game Learning Laboratory "GameLab"

Game learning laboratory can be subdivided into four work areas. They are as following (Fig. 6):

1. The area for developing and testing the games in Android, which contains work seats with the following hardware:
   - 15 work places for students developing games, each of which is provided with notebook, game keyboard, and manipulator.
   - 5 special work places, each of which is provided with the tablet, equipment of game virtual reality (OSVR), virtual reality helmets EMOTIV, and virtual reality headset.
2. The area for developing and testing the games in Apple iOS environment, which contains 2 specific work places, each of which is provided with the computer Apple iMac and the tablet Apple iPad Air 2.
3. The area of recording, processing and listening the audio content of the game, which is provided with the sound USB-interface with the microphones, cables, microphone stands and headphones for high-quality recording audio game content.
4. The area for developing photo- and video-content of the game, which is provided with mirror photocamera, scanner, laser printer for printing through WiFi and mobile devices.

Server HP ProLiant provides uninterrupted functioning the network Wi-Fi in the learning laboratory «GameLab».

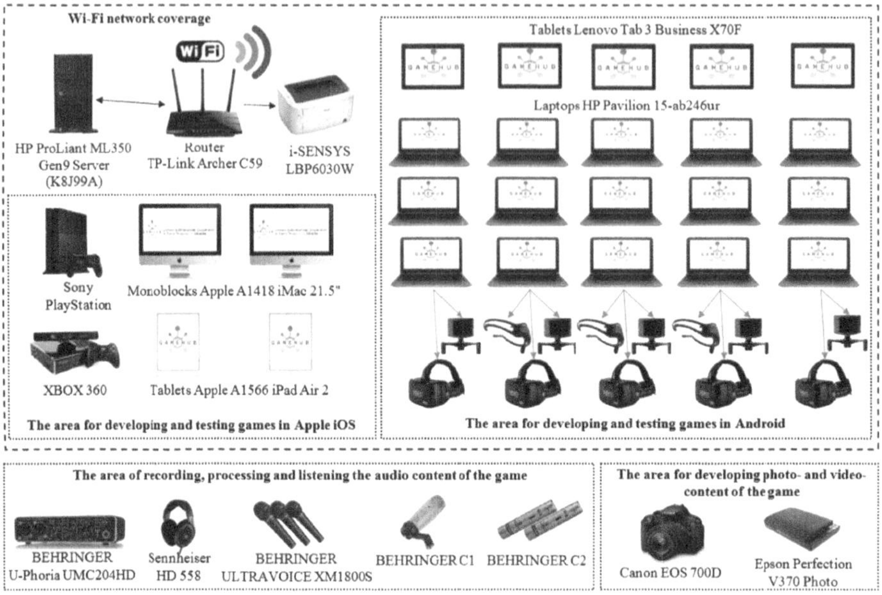

Fig. 6. The concept of GameLab

On top of it, game laboratory «GameLab» is provided with the game consoles Sony PlayStation 4 and XBOX 360 Kinect for game testing.

Open source software is primarily used in GameLab laboratory, such as cross-platform programming environments and development tools for operating systems Android, iOS, Windows, programming languages Java, JavaScript, Python, C++. This software is delivered to the students and teachers for free use.

However, commercial software is also used for development of game content, for which the open source analogous are absent. Commercial software contains such the following programs:

- cross-platform professional graphical programs for processing raster and vector images, photos, graphic design (Adobe Photoshop CC ALL Multiple Platforms);
- professional graphical programs for modelling high polygonal digital sculptures and texture painting of 3D models for creating digital 3D objects and 2D sketches with the wide spectrum of textures and colors (Autodesk Mudbox);
- specialized software for 3D-animation, modeling and visualization with applying video effects (Autodesk Maya);
- specialized software for preparing, editing and recording high quality audio-content (Sound Forge Audio Studio);
- specialized software for preparing, editing and recording video-content of different resolution and quality (Corel VideoStudio Pro).

Thus, the equipment and the software of game learning laboratory «GameLab» meets the modern requirements as to preparation of the specialists in development of cross-platform computer games and game content.

8 Dual Learning Model in KNTU

According to the model developed in the framework of the project, KNTU decides on the implementation of a dual learning, examines the potential of the labor market, determines a list of specializations for which dual educational programs will be developed, approves their list, adopts relevant internal documents, appoints those responsible for introducing the dual education, and conducts a preliminary selection of job-seekers for the establishment of cooperation.

Further, KNTU initiates and concludes contracts with enterprises, organizes discussion of the dual syllabus (study programs) in the specialty with employers or professional associations for compliance with professional standards and requirements of employers to the competences of future specialists. Upon learning of such a study program, KNTU provides continuous communication between all parties to eliminate possible disadvantages in organizing training and solving current problems that may arise.

To do this, they hold regular meetings with employers (1–4 times a year) and provide feedback from the education curriculum on the correspondence between the theoretical and practical parts of the study program and the quality of training in the production in the setting or organization with which KNTU has concluded contracts on the implementation of training in the dual education.

Employers can apply to KNTU with the initiative to introduce a dual education in the specialties in which they are interested. After the conclusion of a bilateral agreement, employers can take part in the formation of a relevant study program and an individual curriculum that involves the use of a dual education, the selection of applicants for the practice of their enterprise, and must ensure the performance of their duties in relation to the organization of the training on their own base, determined by the concluded interviews.

The dual form of education can be selected by those who study on a day-to-day basis and have expressed personal desire. The education provider concludes a trilateral agreement with the KNTU and the employer regarding the dual education and must fulfill its obligations under the contract.

The hours between the theoretical and practical components can be distributed in different ways, depending on the specifics of studying in a specialty.

There are three different methods for distributing hours and agreeing on the content of the training:

- **Integrated method:** a method of a divided week (several days during the week at KNTU, the other part of the week – at the enterprise);
- **Block method:** hours are distributed between KNTU and the enterprise by blocks (2 weeks, month, semester);

- **Partial method:** part of the hours of practice at the enterprise is covered by the hours of study at KNTU.

The university and the employer evaluate together the knowledge, skills, competence of education representatives.

The criteria for achieving results for educational establishments, education providers and job-seekers should be as follows.

1. For the university:
    - increase of competitiveness in the market of educational services;
    - access to up-to-date information on the current state of development of occupations and business activities for which the University prepares specialists;
    - improving the quality of education through the adaptation of study programs to the requirements of employers;
    - expansion of opportunities for applied research;
    - expansion of opportunities for the improvement of the qualifications of the teaching staff.
2. For the employer:
    - influence on the process for training a specialist with the necessary knowledge, skills and competencies;
    - obtaining skilled personnel who are ready to work qualitatively without additional expenses for initial familiarization with work processes at the enterprise or for retraining;
    - taking away (even while studying) the most talented graduates for employment after graduation.
3. For the applicant:
    - a combination of theoretical knowledge with practical experience in one or more enterprises;
    - increase the chances of getting the first job right after graduation;
    - availability before the completion of the training of work experience necessary for further professional growth, as well as a realistic vision of their own career path;
    - obtaining practical experience during studying and earning opportunities (receiving money rewards) in the learning process.

Training in the corresponding specialty in the dual form of education may end with the awarding of a professional qualification to a relevant qualification center, including a professional association, an enterprise, etc.

9 The Experiment Based on the Dual Learning Model

Now, as a result of the project implementation, an experimental study was carried out on the practical using of the developed model of dual education.

Thus, over the last year, three major employers have participated in the experiment, namely the regional companies Wezom, Metasoft, and Logicify. They took an active part in the development of study programs on the specialty "Software Engineering", with specializing in "Computer Games Development", "Web Application Development", and "Mobile Application Development", both at the bachelor level and master level.

Their participation was aimed at correcting the educational content as a list of the necessary competences of the graduates, as well as the corresponding list of educational courses and content modules, and the scope of their study.

For each of the specializations, a certain method was proposed for distributing the student's preparation between the university lecturers and the industrial environment of the employers' enterprises.

The curriculum includes both disciplines and modules that were developed by the university's working group within the framework of the project, and modules proposed by employers outside of the project.

The first results of the experiment are as follows. In cooperation with Wezom company prepared and released 16 people who received the relevant competencies and received a job in this company. According to the company's feedback, its participation in the pilot experiment on dual education has reduced the need for retraining and retraining of graduates by about half. The enterprise expects a significant increase in the readiness of KNTU graduates and their adaptability to their production conditions in the second year of the experiment.

Metasoft Company has trained 12 people who received the first jobs in this company after graduate. Reviews of the company's management are extremely positive.

Logicify company began to participate in the experiment a lot later, so the reliable results of the experiment have not yet been received. However, this employer intends to work actively in the field of dual education.

For the university, as a result of the experiment, students first got access to the most up-to-date technical environment, some kind of unloading of computer classrooms was achieved, leading specialists of employers' enterprises were involved in conducting training sessions, including in the form of guest lectures and master classes. In addition, almost all practical training of the student has been transferred to the employer's working environment.

Training in the corresponding specialty in the dual form of education may end with the awarding of a professional qualification to a relevant qualification center, including a professional association, an enterprise, etc.

Now, in the second step of the experiment, blocks of selective disciplines will be introduced, which will enable the student to graduate after completing their teaching and, subject to the successful completion of the certification exams, obtain certain qualifications, such as "Java Developer", "Python Developer", "Front-End Developer", "Computer Games Developer", etc.

In the third step, the university, together with employers, will create a training and certification center, in which students will be able to get not only the desired qualifications, but also immediately the first jobs from employers. Moreover, it is planned within this framework to carry out a significant amount of work for the development of the skills of programmers, that is, together solve the tasks that each company has been working on its own. This will significantly improve the quality and effectiveness of training specialists.

10 Providing Project Sustainability

For providing GameHub sustainability [22, 23] some conditions have been created in KNTU. They are as following (Fig. 7):

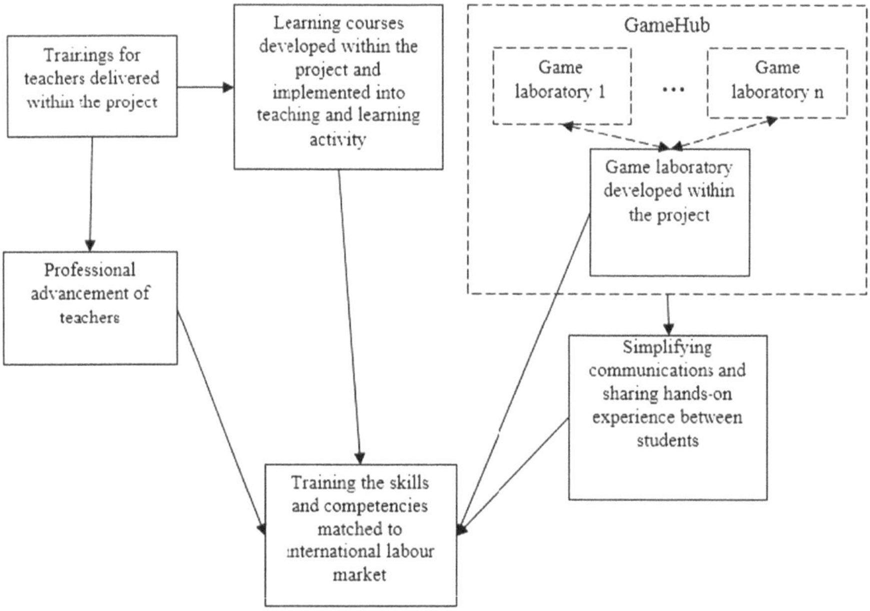

Fig. 7. Sustainability vision of GameHub project at KNTU

- GameHub learning laboratory provided with all necessary equipment and the staff for continued functioning is created;
- educational program on specialty "Program engineering", specialization "Computer games development" is developed;
- the agreements for collaboration with potential employers on the regional and national levels in terms of regular updates of competence profiles of specialists in the area of computer games development are signed;
- the agreement for collaboration with regional Employment Service is signed. The agreement provides for retraining and advanced training of unemployment people, ATO veterans and engineers, who are interested in the game development work in the area of game development, according to the Educational Program "Computer games development" in the Institute of Postqualifying Education of KNTU on the base of game learning laboratory GameLab;

- the learning modules, which constitute the Educational Program on specialization "Computer Games Development" are transformed and delivered on the base of the game learning laboratory GameLab into the open educational resource with the open public access;
- the game learning laboratory GameLab is used for delivery of course and graduate works by students and trainees, individual and scientific research tasks, for providing experiments by teachers and scientists while performing scientific and thesis research in the area of computer games development;
- the training, providing of consulting services in the area of computer games development on the base of the game learning laboratory GameLab for a wide variety of interested physical and juridic persons on the commercial basis are delivered;
- the GameHub infrastructure is used for the further implementation of innovative educational and scientific projects, related to computer games, their development technologies and gamification of education.

Within the GameHub project KNTU collaborates with eight IT companies specialized in computer game development. The closest collaboration is established with companies Wezom, Logicify and MetaSoft. We collaborated with them to create profiles of competencies needed by game studios using the questionnaires specially made for this purpose. 35 GI representatives were interviewed and questioned. The respondents evaluated common (core) and specific (professionally-oriented) knowledge and skills, which are necessary for the digital game design employees.

Based on the requirements of the employees, the work plan for the preparation of computer game developers and a set of necessary courses in KNTU has been developed. A set of learning modules within a frame of GameHub project was developed in each Ukrainian partner-university. In KNTU they are as following: Developing Game Web-applications, Computer Games Development with Unity 3D, Network Computer Games Development with Java, Visual Game Programming.

All modules are connected into Hub and will be accessible not only for the Partner Universities but for all interested people (Fig. 8).

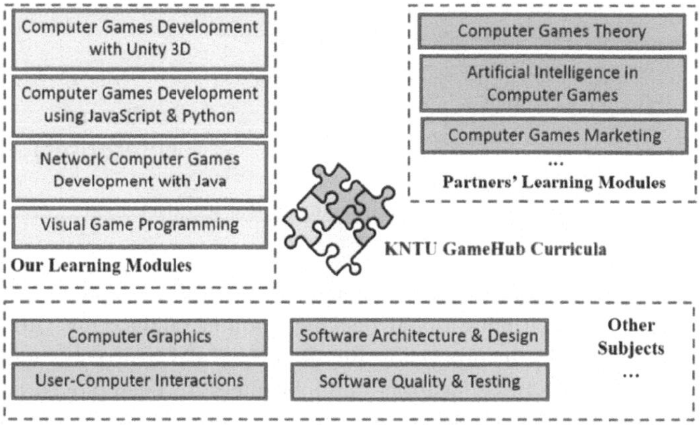

Fig. 8. KNTU pilot implementation: GameHub curricula

To control and monitor the development of learning modules the pilot has been delivered (Fig. 9). Approximately 180 University teachers, 500 students and 150 unemployed including ATO veterans will be trained during project pilot in Ukrainian universities-partners.

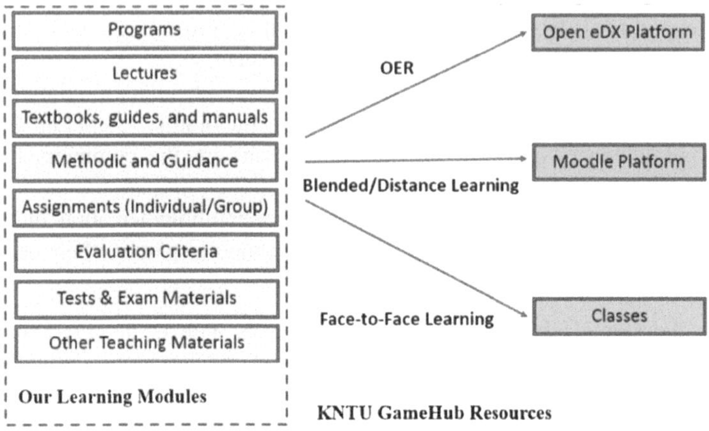

Fig. 9. KNTU pilot implementation: preparing stage

The implementation of the project allows preparing specialists in alignment with the demands of employees, which allow employees to save time for refresher courses and advances professional training of the graduates.

All partner universities are supposed to correct content of the modules based on the evaluation of using the modules in the educational process (Fig. 10).

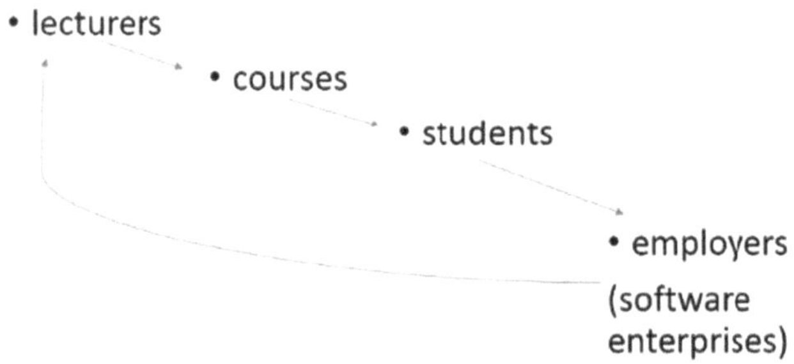

Fig. 10. KNTU GameHub educational feedback loop

Project sustainability is ensured by sustainable use of the GameLab and GameHub learning materials as open educational resources within engineering curricula in the universities, as well as by incorporation the project results and outcomes in the professional training provided in education/training centers for unemployed people, ATO veterans, and other interested individuals.

The KNTU GameHub RoadMap is presented in Fig. 11. It shows the main stages of planned feedback from GI enterprises within the framework of university-enterprise cooperation.

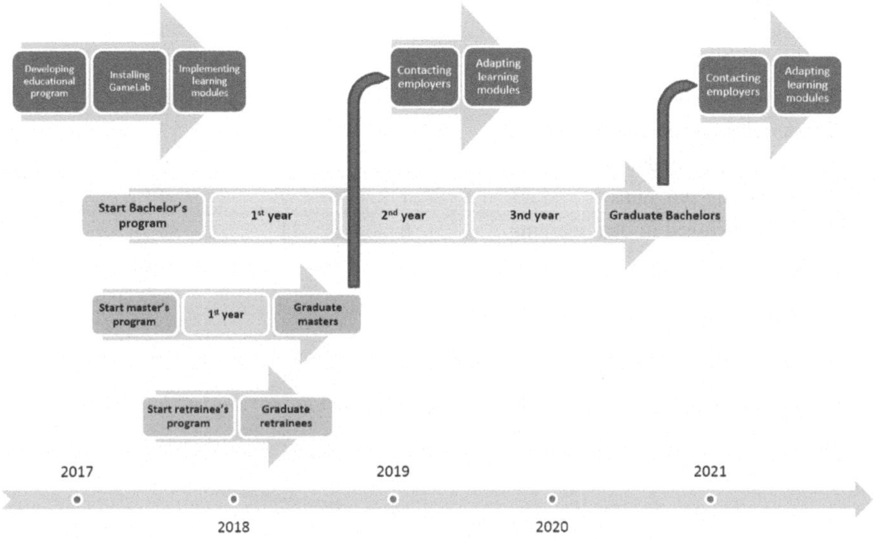

Fig. 11. KNTU GameHub RoadMap

In order to establish exploitation of the project results, the GameHub continuous to deliver and update the learning materials carried out by all trained teachers from Ukrainian Partner Universities during and after the project.

Table 2 provides data on some key indicators, which illustrate covering of KNTU teachers and students as well as employers' involvement in cooperation in the frame of the GameHub project. This data shows the interest to mutually beneficial and sustainable university-enterprises cooperation from IT companies on the regional and national level. The growing involvement of students in dual education programs shows their strong interest in obtaining high-quality IT education in cooperation with employers.

Thus, the mutually beneficial cooperation between KNTU and IT enterprises established during the project will ensure GameHub further growth and exploitation fostering in accordance with KNTU GameHub RoadMap as it is illustrated in Fig. 12.

Table 2. Key indicators

	Indicator	2016 year	2017 year	2018 year (expected)
1	Students trained			
	- bachelors	12	28	55
	- masters	3	9	24
2	Teachers trained	8	15	23
3	Employers in cooperation	2	3	8
4	Master classes conducted	-	3	7
5	Guest lectures conducted	1	2	6
6	Dual learning programs	1	2	4
7	Students involved in dual learning programs	6	11	38

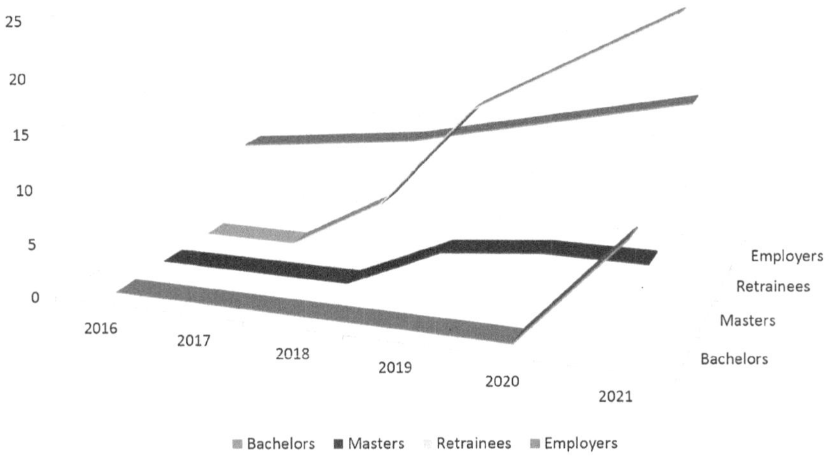

Fig. 12. Key indicators of the GameHub project implementation

11 Conclusions

The presented results show that dual education is a good basis for building extensible symbiotic cooperation between the university and employers in IT industry. This is particularly well illustrated by the results of the GameHub project, which will be helpful for Ukrainian higher education sector to respond to current and future education needs because mutually beneficial and sustainable university-enterprises cooperation can achieve a better match between competence profile of university graduates and those required by GI. This will provide graduates with high-level, employable skills, as well as the transferable skills that equip graduates for a fast-changing labor market.

Ukrainian graduates will meet the requirements of the international labor market and can enhance the integration of Ukraine into European IT sector.

Acknowledgement. This work was partially funded by the European Union in the context of the GameHub project (Project Number: 561728-EPP-1-2015-1-ES-EPPKA2-CBHE-JP) under the ERASMUS+ programme. This document does not represent the opinion of the European Union, and the European Union is not responsible for any use that might be made of its content.

We want to thank all GameHub partners for their invaluable contribution to our common work to develop the close cooperation between universities and enterprises of the game industry sector in Ukraine.

References

1. Rodchenko, V.B., Rekun, G.R., Prus, Y.I.: External academic mobility: case for Ukraine. Handel Wewnętrzny **1**(266), 56–68 (2017)
2. Jibeen, T., Khan, M.A.: The internationalization of higher education: potential benefits and costs. Int. J. Eval. Res. Educ. **4**(4), 196–199 (2015)
3. Pryschepa, N.P., Ostapenco, T.G., Hrashchenko, I.S., Teplinsky, G.V., Onoprienco, O.D.: Some issues of Ukrainian education. Sci. Rev. **7**(7), 60–61 (2017). Vol. 2
4. Darisi, T., Watson, L.: Strengthening youth entrepreneurship education. An Evaluation & Best Practices Report (2017)
5. Lu, Y.: Science and Technology in China: A Roadmap to 2050. Strategic General Report of the Chinese Academy of Sciences. Science Press/Springer, Beijing/Heidelberg (2010)
6. Fernandez-Nogueira, D., Arruti, A., Markuerkiaga, L., Saenz, N.: The entrepreneurial university: a selection of good practices. J. Enterp. Educ. **21**(3), 1–17 (2018)
7. Mahdi, R.: Myth and reality of entrepreneurial universities in Iran. In: ADVED 2016 2nd International Conference on Advances in Education and Social Sciences, Istanbul, Turkey, pp. 632–640 (2016)
8. Baaken, T., Davey, T.: University-Business cooperation in HEI across Europe with a focus on universities of applied sciences. Zeitschrift für Hochschulentwicklung **7**(2), 44–63 (2012)
9. Thomson, L.W., Bagby, J.H., Sulak, T.N., Sheets, J., Trepinski, T.M.: The cultural elements of academic honesty. J. Int. Stud. **7**(1), 136–153 (2017)
10. Sikorskaya, I.: Higher education internationalization in Ukraine: concepts and hope. Int. High. Educ. **91**, 10–12 (2017)
11. Dziabenko, O., Yakubiv, V., Zinyuk, L.: How game design can enhance engineering higher education: focused IT study. In: Auer, M., Zutin, D. (eds.) Online Engineering & Internet of Things. LNNS, vol. 22, pp. 619–627. Springer, Cham (2018). https://doi.org/10.1007/978-3-319-64352-6_58
12. Task Analysis; Development of Competence Profiles. GameHub project report. http://gamehub-cbhe.eu/wp-content/uploads/2016/10/GameHub_Deliverable_1.3.pdf
13. Report on analysis of existing programs and curricula at programme countries' universities. http://gamehub-cbhe.eu/wp-content/uploads/2016/10/D1.1_GameHub_final.pdf
14. Report on ICT and Digital Game Industry Technology Enhanced Learning and Service at Partner Country. http://gamehub-cbhe.eu/wp-content/uploads/2016/10/GameHub_D1.2.pdf
15. Joy, M.M.: Gamification: impact on learning and development with special reference to Deloitte leadership Academy. In: Proceedings of 2nd International Conference on HRD, ICHRD 2018 (2018)

16. Souza, M., Veado, L., Moreira, R., Costa, H., Figueiredo, E.: A systematic mapping study on game-related methods for software engineering education. Inf. Softw. Technol. **95**, 201–218 (2018)
17. Osborne O'Hagan, A., Coleman, G., O'Connor, R.V.: Software development processes for games: a systematic literature review. In: Barafort, B., O'Connor, R.V., Poth, A., Messnarz, R. (eds.) EuroSPI 2014. CCIS, vol. 425, pp. 182–193. Springer, Heidelberg (2014). https://doi.org/10.1007/978-3-662-43896-1_16
18. Kosa, M., Yilmaz, M., O'Connor, R.V., Clarke, P.M.: Software engineering education and games: a systematic literature review. J. Univ. Comput. Sci. **22**(12), 1558–1574 (2016)
19. Colton, S., Wiggins, G.A.: Computational creativity: the final frontier. In: De Raedt, L., et al. (eds.) 20th European Conference on Artificial Intelligence, ECAI, Montpellier, France, pp. 21–26 (2012)
20. Zharikova, M., Sherstjuk, V.: Academic integrity support system for educational institution. In: IEEE First Ukraine Conference on Electrical and Computer Engineering, UKRCON, Kiev, Ukraine, pp. 1212–1215 (2017)
21. David, L.T.: Academic cheating in college students: relations among personal values, self-esteem and mastery. Proc. Soc. Behav. Sci. **187**, 88–92 (2015)
22. Labuschagne, C., Brent, A.C.: Social indicators for sustainable project and technology lifecycle management in the process industry. Int. J. Life Cycle Assess. **11**(1), 3–15 (2006)
23. Silvius, G., Schipper, R.P.J.: Sustainability in project management: a literature review and impact analysis. Soc. Bus. **4**(1), 63–96 (2014)

ICT Solutions for Industrial Applications

Decentralized Autonomous Unmanned Aerial Vehicle Swarm Formation and Flight Control

Ihor Skyrda(✉)

National Aviation University, Kyiv, Ukraine
skyrda2@gmail.com

Abstract. Unmanned Aerial Vehicles (UAV) have become more popular for usage due to the low cost of deployment and maintenance. Single UAV employment allows remote area monitoring and transferring different payloads to inaccessible or dangerous zones for human. In order to deal with flight tasks that are more complex, UAV swarms are applied. The main challenge of UAV swarm formation and flight control is to avoid vehicle collisions. In this case, artificial intelligence is responsible for flight performance in the airspace in such way that collision is avoided. The main requirements to the method, which will provide conflict-free maneuvers, are safety (collision avoidance), liveness (decentralized control, destination area reachability) and flyability (UAV flight performance constraints are satisfied). Artificial force field method fulfills all of these demands. It allows to detect a potential conflict between multiple UAVs in a swarm and other static or moving obstacles found in airspace, to provide collision resolution by changing UAVs flight parameters through maintaining minimum separation distance, including cases when manned vehicles are found in the same airspace. There can be distinguished by a wide range of obstacles: static (buildings, restricted areas and bed weather conditions) and dynamic ones (other UAVs, manned aircraft). Method allows keeping UAV swarm shape on the flight path, taking into account ground speed and turn bank angle values restrictions according to UAV's flight performance characteristics.

Keywords: Autonomous unmanned aerial vehicle · Potential field · Vortex field · Swarm formation · Fixed wing · Three-dimensional space

1 Introduction

Unmanned Aerial Vehicles (UAV), also known as 'drones', are vehicles that fly without a pilot on-board with remote control or in an autonomous way. It is a part of Unmanned Aircraft System (UAS), which includes UAV, ground station (where an operator is located) and communication infrastructure.

A diverse range of systems and UAVs lies within the broad definition of UAS. Some differences between these UAS are immediately apparent features, such as size, weight or type of aerial platform (multi-rotor, fixed wing, single rotor) of UAVs. These systems have varying degrees of automation and autonomy, but usually include human remote operator controlling the vehicle from meters, kilometers or continents away.

V. Ermolayev et al. (Eds.): ICTERI 2018, CCIS 1007, pp. 197–219, 2019.
https://doi.org/10.1007/978-3-030-13929-2_10

UAVs are mostly applied for domestic functions such as environmental monitoring, security, emergency response, surveillance and recreation.

The main technical peculiarity of UAV is defined by the extent of autonomy and automation delegated from the operator to the system. Automation levels vary from those that are fully piloted from a remote location to fully automated. There are also several points in-between, with some maneuvers triggered automatically through autonomous conditions monitoring. Depending upon system priorities, autonomous maneuvers may have the priority over, or to be overridden by, the commands of a remote operator. The International Civil Aviation Organization (ICAO) and current European Commission (EC) plans will only permit the autonomous maneuvers to override operator command in extraordinary circumstances such as communication failure or imminent collision risk, the main requirements for UAV integration into normal airspace. The UAV technologies beyond this definition, featuring a greater autonomy, are also quite well developed and, while integration is not currently planned, it could plausibly follow a successful period of development in the UAS sector [1].

The UAV swarm is a group of vehicles that perform the flight in a group, communicating with each other and assisting other UAV in tasks' accomplishment. Many applications for UAV swarm use there are foreseen, they may be search, rescue and payload (nonhomogeneous UAVs) transportation. A swarm could cover a big area, especially if where only small UAVs could be used and would require only one operator. In order to model UAV swarm motion and control it is better to use the bottom-up modelling approach and use the decentralized method for UAVs coordination. The main advantages of such principle are flexible, adaptive and efficient group organization of system with low autonomy, where the intelligence is distributed through all swarm participants, even in case of one UAV loss. Currently, an important challenge is the reduction of the number of operators required for performing a multi-UAV mission. This challenge can be addressed by increasing the autonomy of fleets and providing capabilities of operators to the interfaces. This article presents a proposal of control method for UAV swarm flight performance, so the multi-UAV system will be able autonomously perform shaping and maintain the expected formation with desired flight parameters.

2 Related Work

The results of analysis show that most of known methods for multi-UAV control have a number of significant limitations that are connected with multiple conflicts resolution and group formation. Particularly, the main disadvantage of such methods is connected with pairwise way (between two vehicles) of potential conflicts resolution, when this issue needs to be done in a global way. For example, the system called Traffic Alert and Collision Avoidance System (TCAS) that is already installed aboard uses a range of measurements and range-rate estimates to determine if a conflict exists in the horizontal plane. In case potential conflict presence TCAS searches through a set of climb or descent maneuvers (Fig. 1) and choose the best one accounting flight performance characteristics and flight plan of only two aircraft to provide safe conflict resolution [2, 3]. Methods developed for a group control in robotics do not include such feature, so UAVs must deal with constant movement and limited turning ability, which makes collision avoidance much more complicated [4].

Multiple conflicts can be resolved in pairwise and global way, where pairwise means that all conflicts will be resolved sequentially in pairs and global means analysis of general traffic situation. The first way refers to TCAS principle of operation where one conflict induces a new one until conflict free trajectory will be found, but there is a big probability that it leads to "domino effect". That's why the second way will be used in the combination with algorithms.

It is possible to define four main classes of algorithms:

- geometric approach;
- stochastic approach;
- linear programming approach;
- potential fields approach.

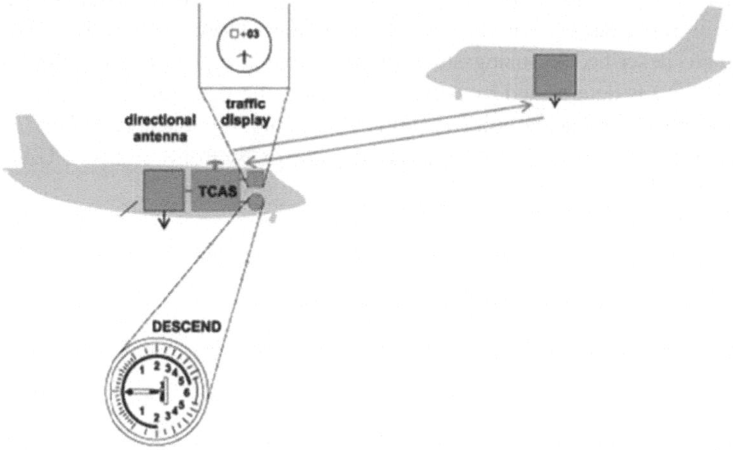

Fig. 1. TCAS Air part principal of operation

The classical approach is called geometric, where aircraft trajectory predictions are based on linear projections of the current vehicle states. Such projections can be computed efficiently, and prediction errors are negligible for short time periods, but it still cannot be used for multiple vehicle conflicts resolution due to computational complexity, so it requires time and space discretization [5]. In [6] presented idea for global conflict resolution with geometric approach by aggregation of vehicles in one artificial vehicle (Fig. 2) [6] with its center but for this more complicated algorithm is required, and from practical point of view the rate of updating surveillance information should be fast.

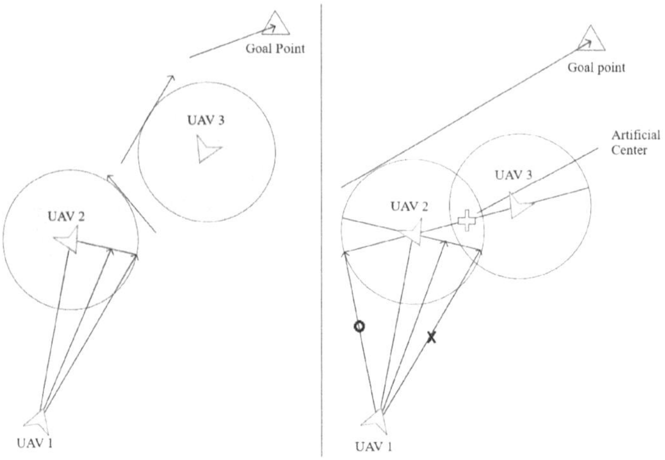

Fig. 2. Multiple conflicts resolution based on geometric approach

The class of stochastic approaches is related to the problem of probabilistic conflict detection in the presence of various uncertainties during the flight. The aircraft dynamics are described by using stochastic differential equations, and the future aircraft's trajectory is determined by solving the stochastic trajectory optimization task, it could be applied for the conflict definition at rather big distances (Fig. 3) [7], so stochastic approach can be hardly applied in order to control a group of UAVs flying close to each other.

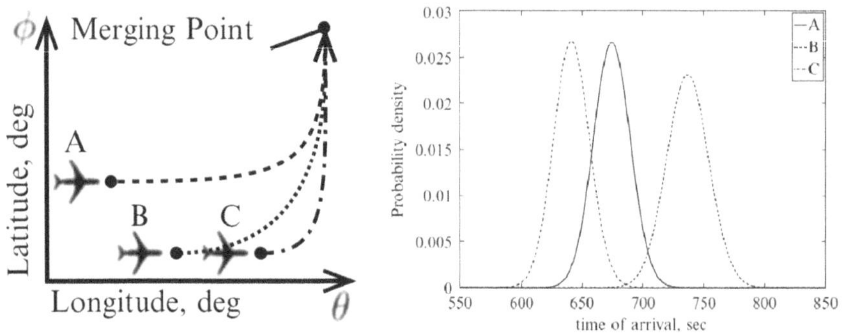

Fig. 3. Multiple conflicts resolution based on probabilistic approach

Linear programming is a mathematical method [8] where an optimal control problem lies in finding trajectories that minimize objective function. There distinguish two main approaches, where the first approach, the optimal control is converted to a finite dimensional Nonlinear Program (NLP) by using collocation on finite elements and by reformulating the disjunctions involved in modeling the protected zones by using continuous variables. In the second approach, the optimal control is converted to a finite dimensional Mixed Integer Linear Program (MILP) using Euler discretization and reformulating the disjunctions involved with the protected zones by using binary variables and Big-M techniques. The drawback of such approach is flyability of the optimal trajectories due to its safety and performance aspects. Also it's computationally expensive with the number of vehicles increasing and causing "the curse of dimensionality" (Fig. 4) [8].

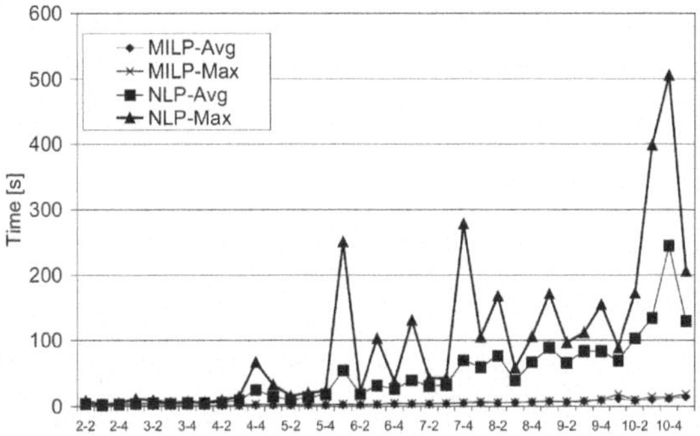

Fig. 4. Maximum and average computational times depending on number of UAVs for multiple conflicts resolution based on probabilistic approach

The common disadvantage of these methods is that they do not meet the main requirements with respect to autonomous UAVs: the absence of any communication links with the appropriate ground stations, with on-board computational and power sources being limited.

The summarized disadvantages of the analyzed methods make no possibility to simultaneously use a combination of such parameters as heading, speed and altitude change maneuvers to resolve multiple potential conflicts. Therefore, it is necessary to develop some new methods for multiple autonomous UAVs control in a group in a three-dimensional space. The method, developed in this article, is the evolution of potential field method proposed in article [9]. A potential fields approach is based on assigning magnetic or electrical charges of the same sign to UAVs, while the opposite charges are assigned to destinations, with the principle being based on the laws of physics according to which the like particles will repel each other, while the destinations having the opposite charges will attract them. The main feature of such approach is UAVs do not necessarily need to know the positions of all other aircraft, so artificial force generated by each UAV allows them to avoid each other spontaneously, at the same time keeping a group form [10]. According to [11], this approach is scalable and can be applied to a big number of UAVs, even in case of multiple conflicts without a ground control station (Fig. 5) [9].

The Artificial Potential Field (APF) approach was introduced in [12], it was used for collision avoidance where the robot is attracted by the destination position and is repulsed from obstacles or other objects. Last years, this method was extended and modified, in order to solve the task of autonomous vehicles path planning either in the stationary or dynamic environments. There are two methods called Formation Potential Field (FPF) and Modified Artificial Potential Field (MAPF). The first one is used for the multiple vehicles formation problem, combining multiple local attractive fields generated by other vehicles to keep the swarm shape with multiple local repulsive potential fields generated by obstacles and vehicles to prevent collisions. A global attractive potential is added to denote a destination area and virtual leader located in the formation center is introduced [13]. The second method - MAPF - is intended for multiple UAVs control and maintains a formation. It can deal with static and dynamic obstacles and differ from FPF by its ability to prevent destination area attractive potential field naturalization by other vehicles repulsive potential field [14].

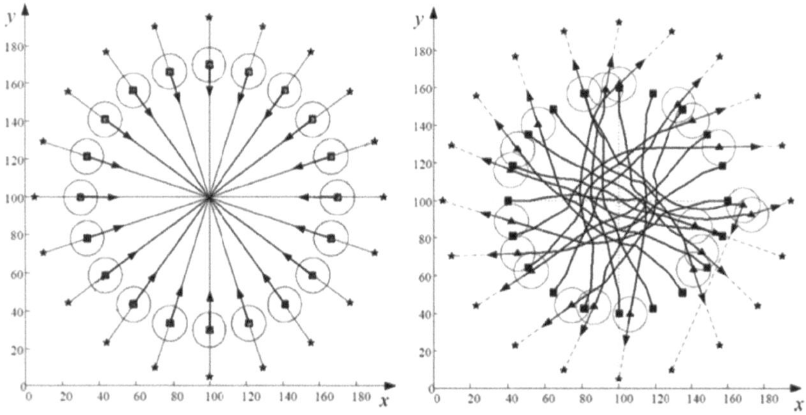

Fig. 5. Multiple conflicts resolution between dynamic objects based on potential fields approach

3 Problem Statement

To solve this problem, a potential field approach is used. This method uses the property of the real world charged particles generating a force field (electric or magnetic), caused attraction and repulsion forces when these particles interact. The matter itself is a typical example of the self-organization principle in our nature. UAVs are considered as the dynamic objects with the same sign, with the point of destination having the opposite sign, it is analogous to the free movement of the aircraft autonomous motion where they constantly have the potential conflicts, and it is required avoiding collisions with other dynamic objects or static/dynamic obstacles. In this case, the term ‘potential conflict’ is a situation, when the minimum separation standard between dynamic objects is violated. The protection zone of dynamic objects is generally defined as follows: the minimum allowed horizontal separation and the vertical separation requirement depending on the dynamic objects’ sizes. The dynamic objects collision is the process of interaction between the dynamic objects or obstacles at a distance in which the dynamic objects change their direction of motion and the speed module.

The dynamic objects interact similarly to the particles of substances that are found in other aggregate states of matter (solid, liquid). The forces act simultaneously. For the different dynamic objects, the general character of the gravity force from distance is qualitatively the same: the attraction force between dynamic objects dominates at large distance, while the repulsion force acts at a short distance. Figure 6 shows the qualitative dependence of forces interaction between two dynamic objects found at distance r between two dynamic objects is presented, where F^{+} and F^{-} - are the dependence of the attraction and repulsion forces respectively, and $F^{+}+F^{-}$ - is a resultant force. At a critical distance $r = r_{cr}$ the resultant force is equal to zero, i.e., the forces of attraction and repulsion are counterbalanced (Fig. 7). This distance r_{cr} corresponds to the equilibrium distance between the dynamic objects.

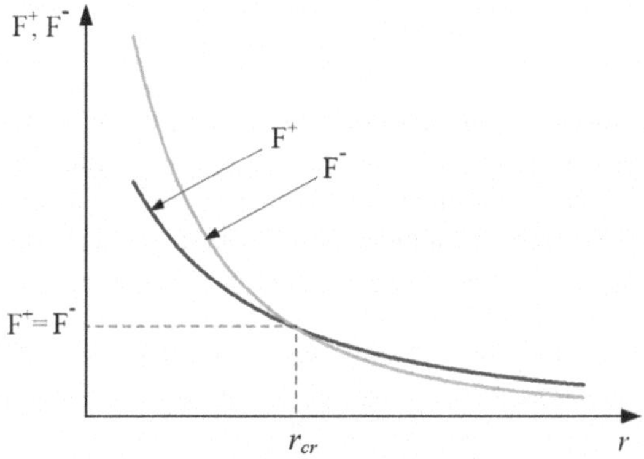

Fig. 6. The dependence attraction and repulsion forces between dynamic

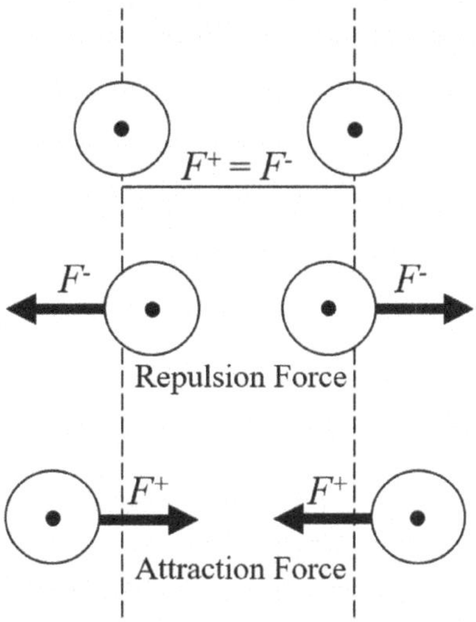

Fig. 7. Attraction and Repulsion forces action

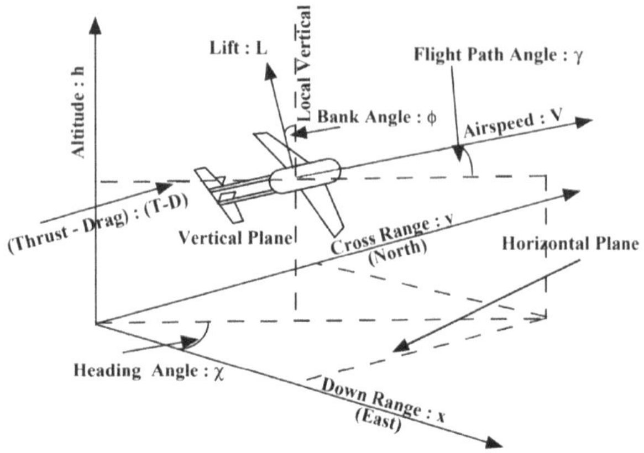

Fig. 8. UAV coordinate system

This article considers a group system consisting of n autonomous UAVs, with a point-mass model used to describe UAV formation movement. The related variables are defined with a respect to the inertial coordinate system and shown in Fig. 8.

The point-mass UAV model captures most of the dynamical effects encountered in the civil aviation aircraft. The point-mass equations of motion are formulated with a respect to a coordinate system shown in Fig. 8. The point-mass model assumes that the UAV thrust is directed along the velocity vector, and that the UAV always performs coordinated maneuvers. It further assumes a flat, non-rotating earth. These assumptions are reasonable for UAVs operating within different ranges, therefore, this method can be used in a conflict resolution between different types of UAVs, with the fidelity provided by the point-mass model being adequate for formulating these problems.

Point-mass models are applicable for the spherical earth approximations that can also be developed. The fuel expenditure is negligible, i.e., the center of mass is time-invariant [15]. Under these assumptions, the motion equations of the i-th UAV can be described as follows:

$$\begin{aligned} \dot{x}_i &= V_i \cos\gamma_i \cos\chi_i; \\ \dot{y}_i &= V_i \cos\gamma_i \sin\chi_i; \\ \dot{h} &= V_i \sin\gamma_i; \\ \dot{\gamma} &= \frac{L_i \cos\varphi_i - g m_i \cos\gamma_i}{V_i m_i}; \\ \dot{\chi} &= \frac{L_i \sin\varphi_i}{m_i V_i \cos\gamma_i}; \\ \dot{V} &= \frac{T_i - D_i}{m_i} - g \sin\gamma_i; \end{aligned} \tag{1}$$

where: $i = 1, 2, \ldots, n$ is the index of multiple UAVs under consideration. x_i, y_i, h_i denote the components of UAV gravity center position. For i-th UAV, x_i is down range; y_i is cross range; h_i is altitude; V_i is ground speed; γ_i is flight path angle; χ_i is heading angle; T_i is engine thrust; D_i is drag; m_i is mass; g is acceleration due to gravity; φ_i is bank angle; L_i is vehicle lift. Bank angle φ_i and engine thrust T_i are the control variables for an aircraft. A bank angle is commanded via combining rudder and aileron trims, thrust is commanded by engine throttle. The g-load $n_i = L_i/gm$ is controlled by elevator, though it refers only to UAV construction characteristics having higher limits due to the absence of crew on board an aircraft in comparison to traditional application. Throughout the UAV swarm control process, these control variables will be constrained to remain within their respective limits. The most common constraints considered are upper and lower bounds on ground speed (V_i), altitude (h_i), g-load (n_i), thrust (T_i), bank angle (φ_i) and climb or descent rates.

Heading angle χ_i and flight path angle γ_i are computed as:

$$\tan \chi_i = \frac{\dot{y}_i}{\dot{x}_i} \tag{2}$$

$$\tan \gamma_i = \frac{\dot{h}_i}{V_i} \tag{3}$$

In an air traffic, a conflict resolution is determined by separation constraints, forming the so-called conflict envelopes or 'protection zones' so that UAVs flight trajectories do not overlap during a flight. The conflict between two UAVs or an UAV with the above-mentioned obstacles implies that their altitude should differ in value h_{pr} given in UAV flight performance characteristics, or they should not get closer in the horizontal plane than indicated by value r_{pr}. The protection zone can be visualized for each UAV as shown in Fig. 9.

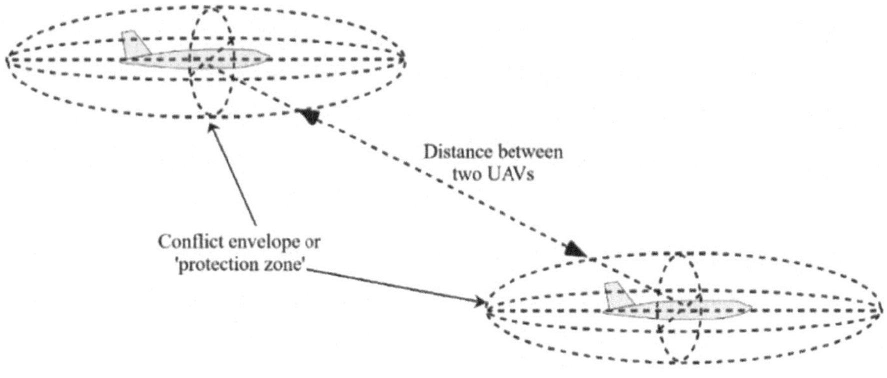

Fig. 9. Spheroidal conflict envelope or 'protection zone' and distance between two UAVs in the vertical plane

The model of UAV swarm formation and control is based on three characteristics: autonomy, which is provided by fully independent vehicle activity, localization, each vehicle is aware of local traffic situation and should not know about a whole air picture, decentralization, there is no any head of swarm.

It means that the UAV can receive information about another UAVs or vehicles by communication channels and it helps to estimate a range between them, and to use a different type of sensors to scan an environment for the obstacles presence. As a result, based on collected information, a decision about flight trajectory changes can be made. UAV collects the information about the coordinates and the flight parameters of other vehicles, the obstacles location and the shape. Then an autopilot system transfers it to the control command based on UAV flight dynamic model (Fig. 10). The difficulty of such model is connecting with the fixed-wing type vehicles considered in the paper, because they must keep a minimum velocity and have maneuverability limitations.

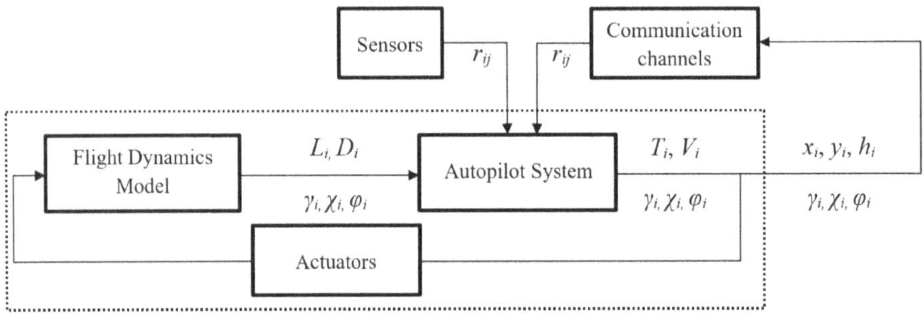

Fig. 10. Model of single UAV and parameters exchange mechanism

4 Method of the UAV Swarm Formation and Flight Control

In order to apply this approach, it is required to transfer the real world properties of UAVs and their position coordinates to the virtual world with its synergetic properties, with the potential conflicts that may occur on the flight path being taken into account [16].

This process includes the following steps:

- structural and parametric synthesis of the virtual world;
- structure formation and parameters of virtual measuring systems that provide conflict free trajectories calculation.

UAVs are transferred from the real to the virtual world as dynamic objects, with mass, attraction and repulsion potentials values being assigned to them [17]. So, the equilibrium state can be represented as:

$$F^{+}\left(m_i, m_j, G, r_{cr}^{\alpha}\right) = F^{-}\left(m_i, m_j, G, r_{cr}^{\beta}\right) \tag{4}$$

where m_i, m_j – masses of i-th and j-th dynamic bodies, G – gravitational constant,

Attraction and repulsion forces can be calculated as:

$$F_{ij}^{+} = \frac{Gm_im_j}{r_{ij}^{\alpha}}; \qquad \alpha \in \{2, 3, \ldots\}; \tag{5}$$

$$F_{ij}^{-} = \frac{Gm_i m_j r_{\kappa p}}{r_{ij}^{\beta}} \qquad \beta \in \{3, 4, \ldots\}; \tag{6}$$

Projections of attraction and repulsion forces between *i*-th and *j*-th bodies on axes *X* and *Y* are calculated by the formulas:

$$\mathrm{F}_{ijx}^{+} = \mathrm{F}_{ij}^{+}\frac{|\mathrm{x_i} - \mathrm{x_j}|}{\mathrm{r_{ij}}} \qquad \mathrm{F}_{ijx}^{-} = \mathrm{F}_{ij}^{-}\frac{|\mathrm{x_i} - \mathrm{x_j}|}{\mathrm{r_{ij}}} \tag{7}$$

$$\mathrm{F}_{ijy}^{+} = \mathrm{F}_{ij}^{+}\frac{|\mathrm{y_i} - \mathrm{y_j}|}{\mathrm{r_{ij}}} \qquad \mathrm{F}_{ijy}^{-} = \mathrm{F}_{ij}^{-}\frac{|\mathrm{y_i} - \mathrm{y_j}|}{\mathrm{r_{ij}}} \tag{8}$$

$$\mathrm{r_{ij}} = \sqrt{(\mathrm{x_i} - \mathrm{x_j})^2 + \left(\mathrm{y_i} - \mathrm{y_j}\right)^2} \tag{9}$$

In Eqs. (5) and (6), the aggregate state of the virtual world environment (solid, liquid, gas) is chosen by the ratio α/β, which characterizes the self-organization degree of the dynamic objects. The aggregate state analogy of a virtual environment can serve as an aggregate state of matter - gaseous, liquid, crystalline, etc.

The resultant vector at each point of dynamic object location consists of the attraction and repulsion forces sum, $F_{ij}^{+} + F_{ij}^{-}$, but can perform a group formation, so to produce dynamic objects movement there should be present one more force which takes into account thrust force P_{ijx}, P_{ijy} direction with projection on axes *X* and *Y* (Fig. 11):

$$\mathrm{F_{ijx}} = \mathrm{F}_{ijx}^{+} + \mathrm{F}_{ijx}^{-} + P_{ijx} \tag{10}$$

$$\mathrm{F_{ijy}} = \mathrm{F}_{ijy}^{+} + \mathrm{F}_{ijy}^{-} + P_{ijy} \tag{11}$$

$$\mathrm{F_{ij}} = \mathrm{F}_{ij}^{+} + \mathrm{F}_{ij}^{-} + P_i(\chi_i) \tag{12}$$

The main condition for dynamic object motion should be satisfied in the following way: $F_{ij}^{+} + F_{ij}^{-} < P(\chi_i)$. The group consists of *n* dynamic objects and each of them can be described by the system of equations:

$$\frac{d^2x_i}{dt^2} = \frac{1}{m_i}\sum\nolimits_{i \neq j}^{n}\left(\mathrm{F}_{ijx}^{+} - \mathrm{F}_{ijx}^{-} + P_{ijx}\right) \tag{13}$$

$$\frac{d^2y_i}{dt^2} = \frac{1}{m_i}\sum\nolimits_{i \neq j}^{n}\left(\mathrm{F}_{ijy}^{+} - \mathrm{F}_{ijy}^{-} + P_{ijy}\right) \quad i \in n, j \in n. \tag{14}$$

The main advantage of the formed virtual world is when the dynamic objects approach the critical distance r_{pr}, the resultant force acting on them is zero, i.e., the forces of attraction and repulsion balance each other. Thus, r_{pr} allows to set the size of the dynamic objects protection zone.

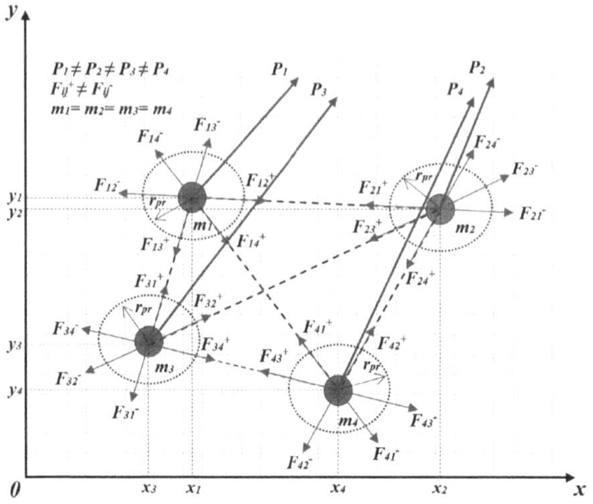

Fig. 11. The forces scheme with four dynamic objects in the original position

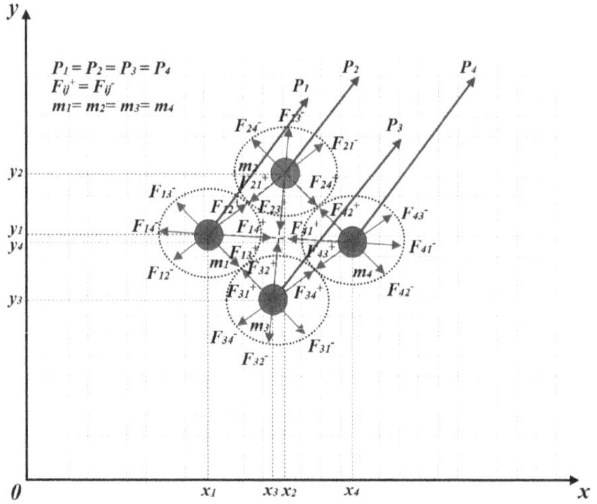

Fig. 12. The forces scheme with four dynamic objects after group formation

$$F_{ij}^{+} = F_{ij}^{-} \tag{15}$$

The absence of such zones intersections, taking into account the forecasted position of the dynamic objects uncertainty, allows maintaining a guaranteed level of traffic safety in the UAV swarm flight control (Fig. 12).

If a static obstacle occurs on a multi-UAV path, the group interacts with it through applying attraction F_O^{+} and repulsion F_O^{-} forces (Fig. 13). This type of maneuver can be conducted provided F_O^{-} is neglected, because the obstacle is static:

$$F_O^+ < \mathrm{F}_{\mathrm{ij}}^+ + \mathrm{F}_{\mathrm{ij}}^- + P_i(\chi_i) \tag{16}$$

In order to satisfy the condition (16) the resultant vector at each point of dynamic object location (12) should be changed according to the form (17), where a and b are user-defined weighting factors. In a such way, it is possible to regulate a virtual connection strength between dynamic objects and regulate their state from solid to liquid in case of collision avoidance with obstacles.

$$\mathrm{F}_{\mathrm{ij}} = \mathrm{aF}_{\mathrm{ij}}^+ + \mathrm{bF}_{\mathrm{ij}}^- + P_i(\chi_i) \tag{17}$$

The forces F_O^+ and F_O^- created by obstacles are directed from geometric centers (point O) (Fig. 13). It can lead to 'stop' effect occur, when attraction and repulsion forces vectors lies on one line with the opposite directions, which is not allowed in case of fixed-wing UAVs application.

To solve such issue, obstacle forces vectors should start at a boundary line and directed with tangent according to the rule, where δ point the angle of victor direction:

$$\mathrm{F}_{\mathrm{oj}}^+ = \mathrm{acos}(\delta)\mathrm{F}_{\mathrm{ij}}^+ \tag{18}$$

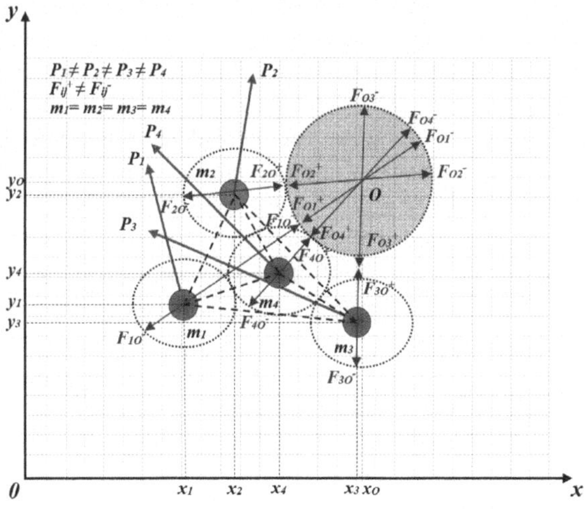

Fig. 13. The scheme of forces with four dynamic objects in a group obstacle avoiding

The values of heading angle χ_i and ground speed V_i may change depending on dynamic objects location relative to the obstacle and destination point.

5 The UAV Swarm Formation and Flight Control Simulation

In order to find out if the potential field approach can be applied for decentralized UAV swarm formation and flight control problem solution, Matlab simulators were used. All in all, 3 cases were simulated with 8 of dynamic objects, with UAV being referred to as a dynamic object.

The flight path was divided into 3 main stages of flight: (1) swarm aggregation; (2) obstacle avoidance; (3) straight line flight in a group to the destination. Figures represent dynamic objects movement trajectory in 2D (a), change in ground speed V_i (b), heading angle χ_i (c) and distance between moving dynamic objects, with dotted line showing protection zone with radius 3 m (d).

$$\tan \chi_i = \frac{\dot{y}_i}{\dot{x}_i} \text{ or } \tan \chi_\mathrm{i} = \frac{\mathrm{F_{ijy}}}{\mathrm{F_{ijx}}} \tag{19}$$

$$V_i = \sqrt{\dot{x}_i + \dot{y}_i} \tag{20}$$

The dynamic objects are in their original positions with the starting speed being equal to zero. At the first stage of modelling, due to the attraction action (5) and repulsion (6) forces the process of group formation begins, which depends on the distance between them (9). Heading angle χ_i has the same direction as vector F_{ij}, which is projected on axes X (10), Y (11) and is formed by their sum, including thrust force (12). Simultaneously, the shape of group formation is regulated be the equilibrium state (4), (15).

In Experiment 1, 8 dynamic objects were considered with the point-mass of 1 kg and protection radius 3 m, with two 6 m-radius obstacles to overcome and the swarm state was considered as liquid. On Fig. 14 shown projections of attraction and repulsion forces vectors with the condition that obstacles only repulse dynamic object $F_o^- = 0$ and destination zone only attract $F_d^+ = 0$, at the same time on dynamic objects act both forces in order to provide swarm aggregation and keep its shape during flight performance. On Figs. 15 and 16, 8 equal dynamic objects (considered as fixed-wing UAVs) move from the initial positions to destination area in swarm where they are being pushed away from each other and settle at their equilibrium distance of so called 'non-conflict'. This figure also demonstrates the swarm moving and accurately performing collision avoidance in three dimensions.

The UAVs were placed at the random initial positions, but under influence of attractive and repulsion forces UAVs have to move as a swarm to a destination area, thereby demonstrating velocity matching as well as swarm aggregation (Fig. 17) and avoiding two obstacles (Fig. 18) placed between the starting points and destination area. Result of simulation, destination area was reached in 1650 s.

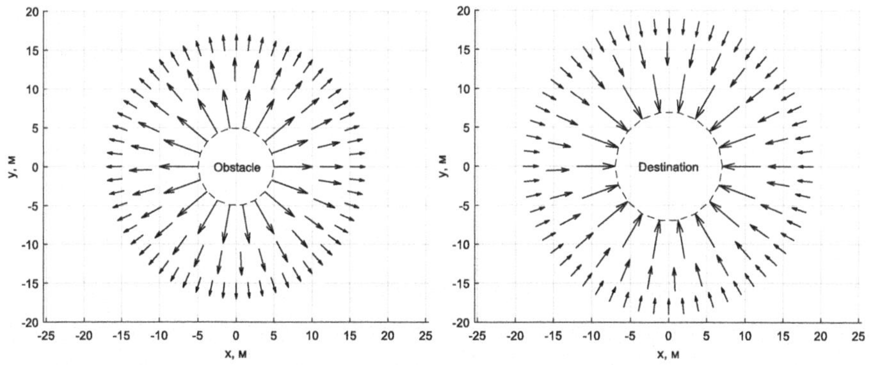

Fig. 14. Experiment 1. Attraction and Repulsion forces vectors projection for obstacles and destination area

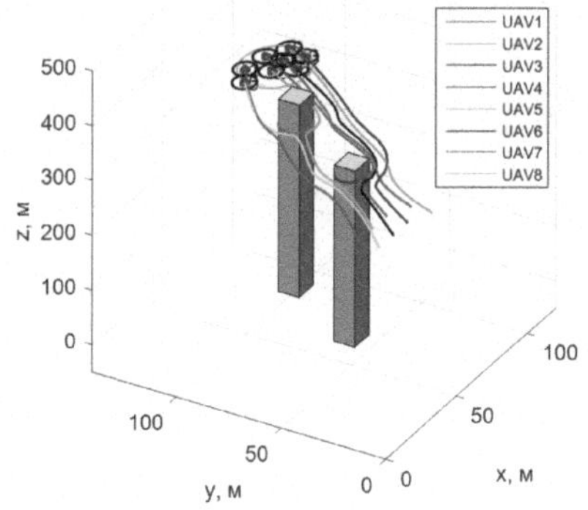

Fig. 15. Experiment 1. UAV swarm movement to destination area in 3D with t = 1650 s

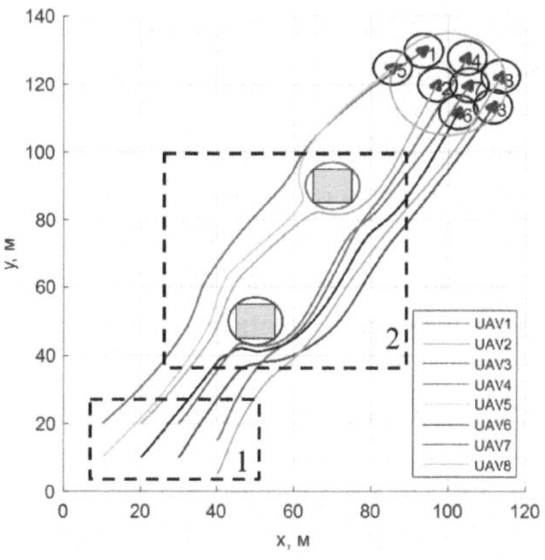

Fig. 16. Experiment 1. UAV swarm trajectory of movement in 2D

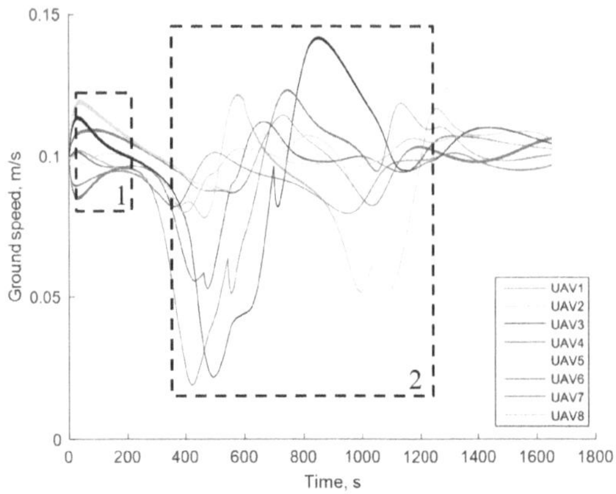

Fig. 17. Experiment 1. UAVs ground speed on time dependence

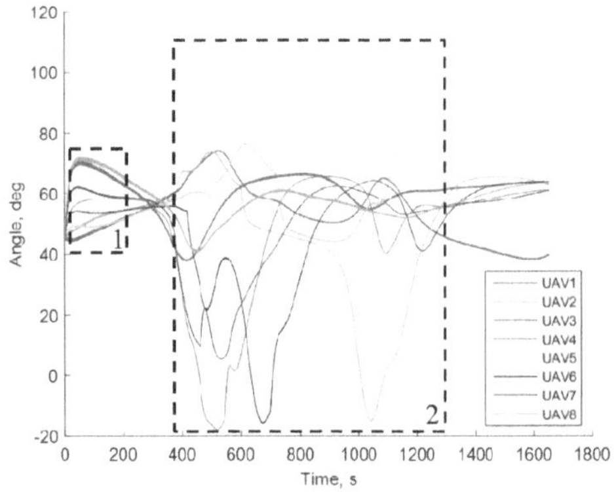

Fig. 18. Experiment 1. UAVs heading angles change

The Experiment 1 results prove the potential field method applicability for UAV swarm aggregation and control. Small differences were observed between theory and simulation results because it requires the mathematical model of concrete UAV type and implementation of stabilization control laws.

In Experiment 2, 8 dynamic objects were considered whose point-mass was 1 kg and protection radius was 3 m, with two obstacles in the way whose radii 6 m and the swarm state was considered as solid but left side rule was applied for conflict resolution. The experiment's aim is to show the swarm acting in case of non-standard obstacle shape and it will require group motion in one direction in order to keep the aggregate state. In comparison with previous one, obstacle repulsion force directed not perpendicular, but on tangent line clockwise (Fig. 19). Multi-UAV movement shown in 3D (Fig. 20) with various height obstacles and destination zone. UAVs trajectories represent the real dynamic objects behavior in a flight (Fig. 21) and verify the applicability. Verification is based on a ground speed (Fig. 22), heading angle (Fig. 23), and distance between UAVs taking into account protection zone. Intersection of protection zones leads to immediate change of heading angle and ground speed depending on the obstacle size and shape.

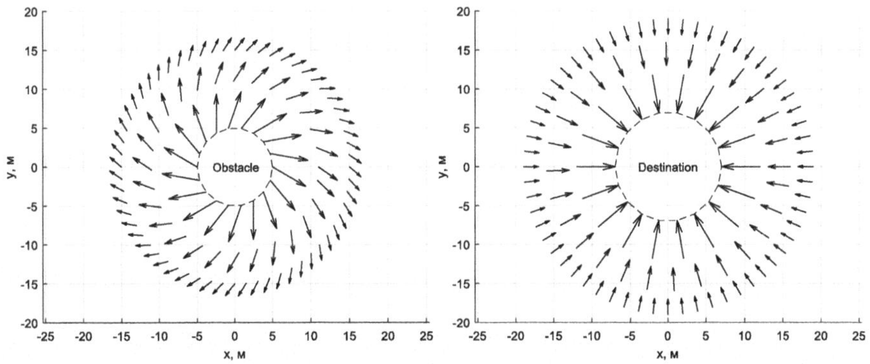

Fig. 19. Experiment 2. Attraction and Repulsion forces vectors projection for obstacles and destination area

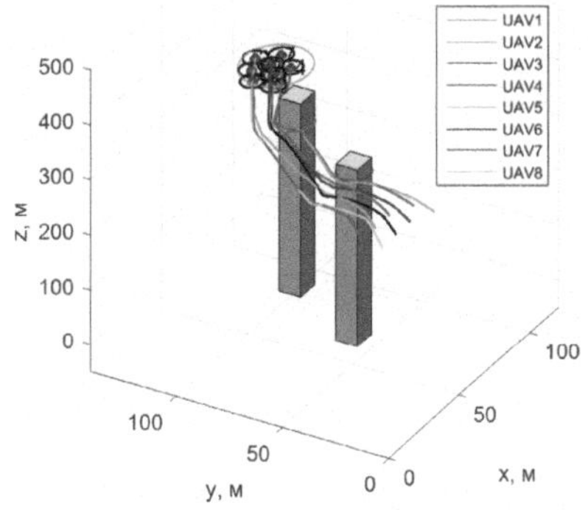

Fig. 20. Experiment 2. UAV swarm movement to destination area in 3D with t = 1550 s

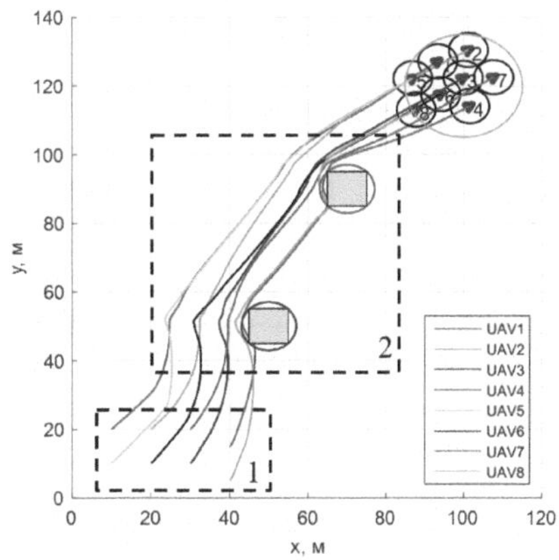

Fig. 21. Experiment 2. UAV swarm trajectory of movement in 2D

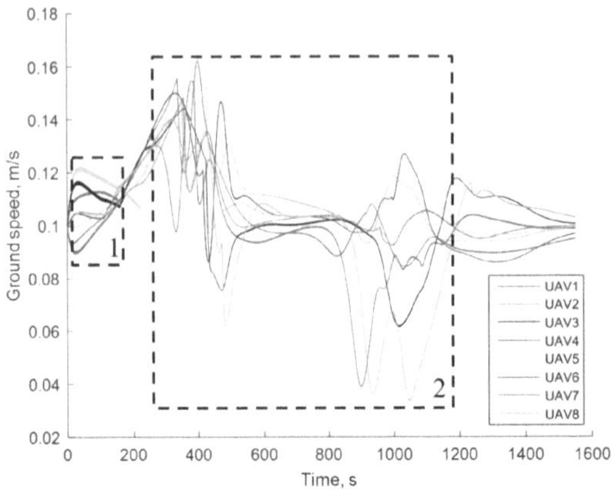

Fig. 22. Experiment 2. UAVs ground speed on time dependence

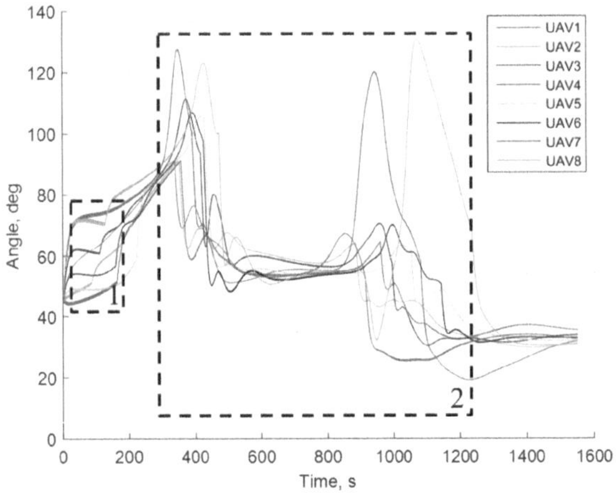

Fig. 23. Experiment 2. UAVs heading angles change

In Experiment 3, 8 dynamic objects were considered, whose point-mass was 1 kg and protection radius was 3 m, with two obstacles in the way, whose radii 6 m and the swarm state was considered as a solid, but a right side rule was applied for a conflict resolution. Obstacles repulsion force directed on a tangent line counterclockwise (Fig. 24). Simulation in 3D (Fig. 25) proves algorithm applicability and gives the opportunity to consider conflict resolution in vertical plane, even in case of low obstacles detection range, fixed-wing UAVs due to their aerodynamic characteristics will not have enough time and space for maneuver. UAVs in case of potential conflict detection will change both heading angle and flight altitude (Fig. 26).

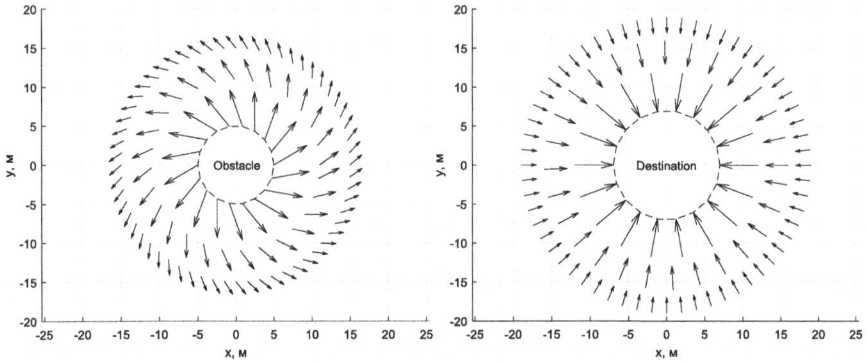

Fig. 24. Experiment 3. Attraction and Repulsion forces vectors projection for obstacles and destination area

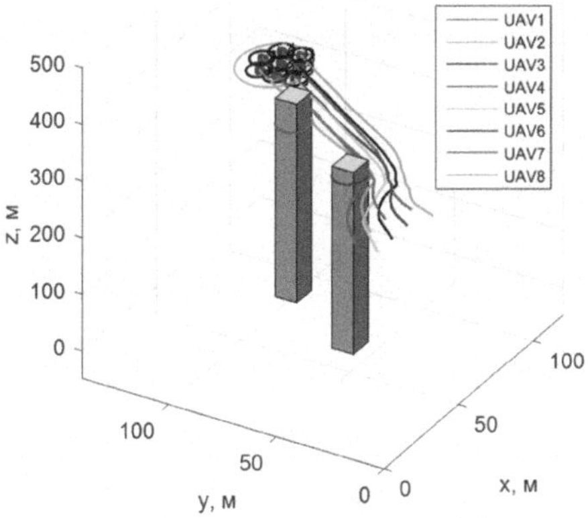

Fig. 25. Experiment 3. UAV swarm movement to the destination area in 3D with t = 1500 s

The Experiment 3 result is analogous to previous experiments, where all flight performance parameters are in an allowable range (Figs. 27 and 28). During such flight observed uneven power consumption, which will lead to decrease of flight time, to solve this issue may be considered the case of uniform motion with constant speed where heading angle changes only, or in the combination with altitude. It may require a bigger detection range, so to provide enough time and space for a conflict resolution. The time spent for destination zone achievement is the smallest in comparison to other experiments and alert about potential conflict detection appears on the time basis criterion because UAVs move with different speeds in order to achieve artificial protection zone of obstacle.

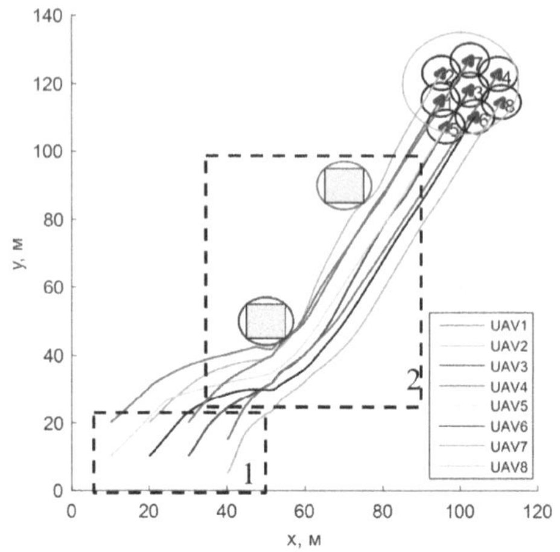

Fig. 26. Experiment 3. UAV swarm trajectory of movement in 2D

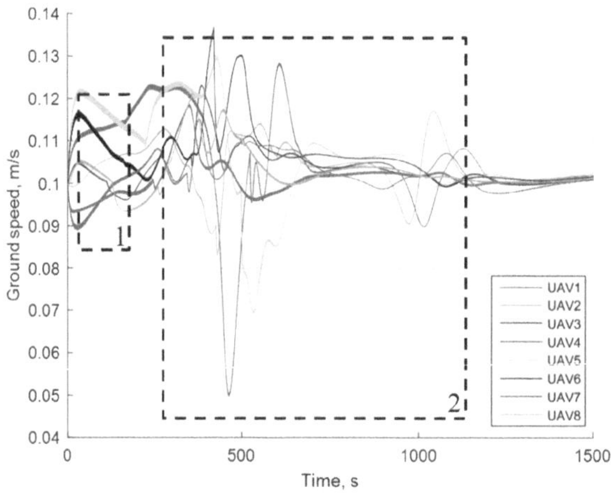

Fig. 27. Experiment 3. UAVs ground speed on time dependence

All experiments were performed with the same initial data: UAVs, obstacles, destination zone coordinates. The results of simulations shown the time required to reach destination zone in Experiment 1 t = 1650 s, Experiment 2 t = 1550 s, Experiment 3 t = 1500 s.

The artificial potential field based method can efficiently obtain the conflict-free results. They lead to small deviations from the nominal paths and the additional flight distances. The solution obtained by the UAV swarm control algorithm is time varying. Each UAV should keep on calculating conflict free results according to the outer environment. The distributed decentralized algorithm of autonomous UAVs control is also fast. The provided trajectories are smooth and low-cost results. The results show

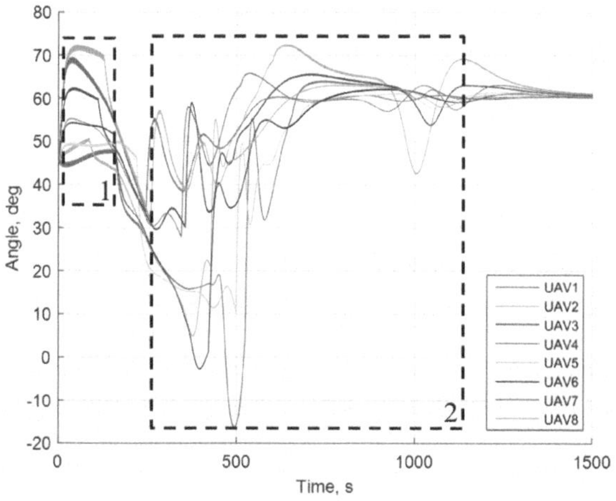

Fig. 28. Experiment 3. UAVs heading angles change

that proposed method is efficient in solving the multiple conflicts problem in the dynamic environment.

Artificial Potential Field (APF), Formation Potential Field (FPF), Modified Artificial Potential Field (MAPF) and Evolutionary Artificial Potential Field (EAPF) methods described in [12–14, 18] are solving the problem of multiple vehicles formation and control, mostly devoted to the robot path-planning problem. These methods are associated with robots or flying objects without maneuverability limitations and in some cases they require a central control station for trajectory optimization. APF allows only prevent collisions with obstacles, FPF devoted to multiple vehicles formation control, EAPF provides smooth and optimal paths and MAPF is decentralized and does not require a high computational capability. Proposed method composes all advantages of methods listed above. It allows in decentralizes way to control a big formation of UAVs that act independently, to modify a flight trajectory when the unexpected obstacles detected, to avoid collisions with different size obstacles without any loss of swarm shape, and to provide optimal path based on an environment. The main breakthrough of proposed method is a fixed wing UAV aerodynamic and the flight characteristics integration to MAPF with an excessive robustness in a dynamic environment and it can guarantee a mission performance.

6 Conclusions

UAVs are widely used in different areas of human activity, and UAV swarm performance has many advantages compared with the performance of an individual UAV. Research institutions and groups are currently developing an algorithm for a group of UAV autonomous control since manual control is not available.

For multi-UAV formation control, the artificial potential field approach is used, where UAVs are denoted as the interacting dynamic objects influenced by attraction and repulsion forces. The movement of each dynamic object is described by a system of equations, with the direction of movement coinciding with a thrust force angle projected on each of axes.

The algorithm allows to accurately steer UAV fixed-wing type swarm to user defined positions. The applied method has a number of limitations related to such flight performance parameters: ground speed, turn rate, bank angle, distance between UAVs in swarm and obstacles. UAV swarm control method is effective in creating stable multi-UAV group and it was checked using simulation. It includes 3 experiments with 8 UAVs and 2 obstacles located at the different positions with the same mass and protection zone around. The tasks were to form a group, avoid the obstacles, and continue a movement into the destination area with no change in the shape of the group. The results show that in this form the approach can be applied to a group aggregation and multi-UAV flight control. All dynamic objects moved within the allowable range determined by heading angle χ_i and ground speed V_i keeping within the protection zones.

The simulations have shown that potential field approach is an adequate method to control swarms of fixed-wing UAVs. The functionality of method can be extended by checking the maximum number of UAVs in swarm that could be controlled, loss of UAVs in swarm and its influence on the flight task performance, collision avoidance with dynamic obstacles, and collision avoidance between swarms by the conflict resolution in the horizontal and vertical planes.

References

1. Boucher, P.: Domesticating the drone: the demilitarisation of unmanned aircraft for civil markets. Sci. Eng. Ethics **21**, 1393–1412 (2015)
2. Debajit, D., Kumar, S.: A novel approach towards the designing of an antenna for aircraft collision avoidance system. AEU – Int. J. Electr. Commun. **71**, 53–71 (2017)
3. Kuchar, J.K., Yang, L.C.: A review of conflict detection and resolution modeling methods. IEEE Trans. Intell. Transp. Syst. **1**(4), 179–189 (2000)
4. Murray, R.M.: Recent research in cooperative control of multivehicle systems. ASME: J. Dyn. Syst. Meas. Control **129**(5), 571–583 (2007). https://doi.org/10.1115/1.2766721
5. Geser, A., Muñoz, C.: A geometric approach to strategic conflict detection and resolution. In: Proceedings of the 21st Digital Avionics Systems Conference, vol. 1, pp. 6B1-1–6B1-11 (2002)
6. Park, J.-W., Oh, H., Tahk, M.-J.: UAV collision avoidance based on geometric approach. In: Proceedings of the 2008 SICE Annual Conference, 20th-22nd August 2008, Tokyo, pp. 2122–2126 (2008)
7. Matsuno, Y., Tsuchiya, T.: Probabilistic conflict detection in the presence of uncertainty. Air Traffic Management and Systems. LNEE, vol. 290, pp. 17–33. Springer, Tokyo (2014). https://doi.org/10.1007/978-4-431-54475-3_2
8. Borrelli, F., Subramanian, D., Raghunathan, A., Biegler, L.: MILP and NLP techniques for centralized trajectory planning of multiple unmanned air vehicles. In: Proceedings American Control Conference, pp. 5763–5768 (2006)
9. Chepizhenko, V.I.: Energy-potential method of dynamic objects polyconflicts guaranteed collision resolution. In: Cybernetics and Computer Engineering, no. 168, pp. 80–87 (2012)

10. Leonard, N.E., Fiorelli, E.: Virtual leaders, artificial potentials and coordinated control of groups. In: Proceedings of the 40th IEEE Conference on Decision and Control, vol. 3, pp. 2968–2973 (2001)
11. Nguyen, B.Q., et al.: Virtual attractive-repulsive potentials for cooperative control of second order dynamic vehicles on the Caltech MVWT. In: Proceedings of the American Control Conference, vol. 2, pp. 1084–1089 (2005)
12. Khatib, O.: Real-time obstacle avoidance for manipulators and mobile robots. Int. J. Robot. Res. **5**(1), 90–98 (1986). https://doi.org/10.1177/027836498600500106
13. Liu, X., Ge, S.S., Goh, C.H.: Formation potential field for trajectory tracking control of multi-agents in constrained space. Int. J. Control **90**, 1–15 (2016)
14. Yin, H., Cam, L.L., Roy, U.: Formation control for multiple unmanned aerial vehicles in constrained space using modified artificial potential field. Math. Model. Eng. Probl. **4**(2), 100–105 (2017). https://doi.org/10.18280/mmep.040207
15. Ruibin, X., Gaohua, C.: Formation flight control of multi-UAV system with communication constraints. J. Aerosp. Technol. Manag. **8**(2), 203–210 (2016)
16. Pavlova, S.V., Pavlov, V.V., Chepizhenko, V.I.: Virtual Einstein force fields in synergy of navigation environment of difficult ergatic systems. In: Proceedings of the National Aviation University, no. 3, pp. 15–27 (2012)
17. Chepizhenko, V.I.: Synthesis of artificial gravitational fields virtual meters for the polyconflicts resolution in the aeronavigation environment. In: Proceedings of the National Aviation University, no. 2, pp. 60–69 (2012)
18. Vadakkepat, P., Tan, K.C., Ming-Liang, W.: Evolutionary artificial potential fields and their application in real time robot path planning. In: Proceedings of the 2000 Congress on Evolutionary Computation, vol. 1, pp. 256–263 (2000). https://doi.org/10.1109/cec.2000.870304

Availability Models of the Healthcare Internet of Things System Taking into Account Countermeasures Selection

Anastasiia Strielkina(✉), Vyacheslav Kharchenko, and Dmytro Uzun

National Aerospace University "KhAI", Kharkiv, Ukraine
{a.strielkina, v.kharchenko, d.uzun}@csn.khai.edu

Abstract. An active infiltration of information technology in the healthcare sector has led to a fundamental change in people's quality of life. Networked medical and healthcare devices and their applications are already creating an Internet of Medical Things which is aimed at better health monitoring and preventive care. But the new concepts and applying of new technologies bring certain risks including failures of devices, infrastructure which may lead to the worst outcome. In this regard, the security and safety problems of this technology using increase rapidly. This paper touches upon the issue of the healthcare Internet of Things (IoT) infrastructure failures and attacks on components and complete system. The purpose of the paper is to develop and research the availability models of a healthcare IoT system regarding failures and attacks on components. A detailed analysis of an architecture of healthcare IoT infrastructure is given. The main causes of the healthcare IoT based system failures are considered. This paper presents an approach to develop a Markov models set for a healthcare IoT infrastructure that allows considering safety and security issues. Much attention is given to developing and research of the Markov model of a healthcare IoT system considering failures of components. The analysis of obtained simulation results showed the rates that have the greatest influence on the availability function of the healthcare IoT system. In addition, it is presented a case study with a game theoretical approach to select countermeasure tools.

Keywords: Attack · Cloud · Countermeasure · Failure · Game theory · Insulin pump · Internet of Things · Markov model · Security · Vulnerability

1 Introduction

1.1 Motivation

The paradigm of the Internet of Things (IoT) implies the possibility of massively and inexpensively connecting to an information network (for example, the Internet) any physical object and control systems for these objects. IoT in general promises textually to every citizen and every company, regardless of the industry - its own set of benefits and improvements, savings and growth, the release of time and new opportunities. On

V. Ermolayev et al. (Eds.): ICTERI 2018, CCIS 1007, pp. 220–242, 2019.
https://doi.org/10.1007/978-3-030-13929-2_11

the basis of these statements, the IoT has already found applying in many industries. According to predictive forecasts [1, 2], the number of networked and connected devices will increase to 25.6 billion. In 2017 IoT has been ranked as the first among the eight breakthrough technologies that can change the business model of companies or entire industries, advancing artificial intelligence, augmented reality, technology related to the creation and management of the drones, blockchain etc. [3]. The IoT has already a great impact in many economical areas [4] as transport, energy, healthcare, industry, agriculture, wearables, smart retails, smart homes, etc.

One of the most promising and already most advanced industries are medicine and healthcare. Networked medical and healthcare devices and their applications are already creating an Internet of Medical Things which is aimed at better health monitoring and preventive care for creating better conditions for patients who require constant medical supervision and/or preventive intervention. Healthcare and medical organizations (providers) also attempt to collect and analyze data that generate the IoT devices that are essential for prospective innovations.

And one of the most sought-after fields in healthcare and medicine treatment, monitoring and prognosis is Diabetes. According to [5] an estimated 422 million adults were living with diabetes in 2014, compared to 108 million in 1980, the global prevalence of diabetes has nearly doubled since 1980, rising from 4.7% to 8.5% in the adult population, it caused 1.5 million deaths in 2012, and higher-than-optimal blood glucose caused an additional 2.2 million deaths and they predict that Diabetes will be the 7th leading cause of death in 2030. But the new concepts and applying of new technologies bring certain risks including failures of devices, infrastructure which may lead to the worst outcome - the death of the user (patient).

Nevertheless, with all the benefits of using such networked devices, the security, safety and reliability risks are increasing. Thus, the security, safety and reliability assessment of such systems is a complex process. Such systems are characterized by a large number of failures due to the dynamism, multicomponence and multilevelness. For reducing this issues the fragmentedness of the models being developed should be used in some cases to describe repeated parts of models which have similar structure and differ only values of some parameters. It concerns fragmentedness caused by changing of design faults and attacked vulnerabilities number and the corresponding failure rates.

1.2 State-of-the-Art

For today there are a lot of papers that describe opportunities and benefits of the smart and intellectual technologies using in the field of healthcare and medicine and at the same time they describe the security and safety problems of this technology using.

One of the most famous and almost all covering paper is [6]. The authors tried to show all the healthcare IoT trends, solutions, platforms, services and applications. They outlined main problems during development and using of such devices related mostly to standardization and regulatory issues. In addition, that paper analysed healthcare IoT security and privacy features, including requirements, threat models, and attack taxonomies and proposed an intelligent collaborative security model to minimize security risk. But the authors did not address the issues of reliability and safety analysis, did not consider the possible failures of the healthcare IoT system and its particular components and the influence on performance.

The authors of [7] presented three use cases for quality requirements for IoT in healthcare applications. One of them is for safety and violence. They gave a simple construct for a patient or caregiver safety use case. Also, they refer to the US Underwriters Laboratories [8] and as well recommended using "traditional techniques for defining misuse and abuse cases".

Goševa-Popstojanova and Trivedi in [9] provided an overview of the approach to reliability assessment of systems. The architecture of system could be modelled as a discrete time Markov chain, continuous time Markov chain, or semi-Markov process.

The Markov model that considers the technical conditions of typical network components of the IoT-based smart business centre was presented in [10].

A Markov Queuing approach to analysing the Internet of Things reliability with some experimental results was proposed in [11].

The paper [12] describes an approach to developing a Markov models' set for a healthcare IoT infrastructure that allows considering safety and security issues. It details the models sets for the healthcare IoT system based on Markov process approach.

The existing standards and regulatory acts in the field of healthcare [13–15] provide different approaches for risk management to determine safety and security of medical and healthcare devices.

Attacks on vulnerabilities of IoT based systems can be simulated using Markov's modelling. The authors of [16] was presented and explained how Markov modelling can be used to evaluate the reliability of the complex systems parallel redundant system. The paper [17] shows that dependability consists of many measures (as reliability, availability, safety, performability and security and its attributes). Authors presented a good state-of-the-art how Markov models can be applied to the dependability and security analysis.

To choose an effective protection tool against various types of attacks, it is possible to use the methods of game theory. Game theory is a formal approach designed to analyse interactions between several participants in a process that have different interests and make decisions. Obviously, there are two sides in each system from point of view the cybersecurity procuring: the side of the attack and the defense side (the information protection tool) with opposing interests.

Researches of [18] showed the applicability of the game theoretical approach to address the network issues and presented a taxonomy of games classification.

An approach for modeling the decision-making process of cyber security monitoring using a game-theoretic approach was presented in [19].

Nevertheless, despite a large number of researches regarding healthcare IoT, there are no papers that consider safety and reliability issues of healthcare IoT systems taking into account failures of hardware and software components and system failures.

1.3 Objectives, Approach and Structure of the Paper

The goal of the paper is to analyze and develop a model that describes the healthcare IoT system failures and attacks, and their influence of availability indicators. Our approach is based on review of the variety of existing techniques and mathematical models for similar systems and step by step development of a set of states and transitions caused by failures and attacks on the healthcare IoT system components.

In this context, this paper proposes the availability Markov models that includes possible failures and attacks on the healthcare IoT system and method to select countermeasures. The remainder of this paper is organized as follows. The second section describes an architecture of healthcare IoT infrastructure and possible failures and attacks during its operation, a Markov models set for the healthcare IoT infrastructure that allows taking into account the specificity of end user devices, communication channels, technologies of data flows and safety and security issues of these components. The third is devoted to the development of a Markov model of a healthcare IoT system considering failures of components and a Markov model considering attacks on vulnerabilities and analysing simulation results of the models. In the fourth section the game theoretical approach to select the countermeasure is presented. The last section concludes and discusses future research steps.

2 Analysis of Healthcare IoT Behaviour

2.1 The Architecture of Healthcare IoT Infrastructure

Analysis of the latest publications related to this topic [6, 11, 20] allows us to present a generalized architecture of the healthcare IoT infrastructure that can be seen in Fig. 1.

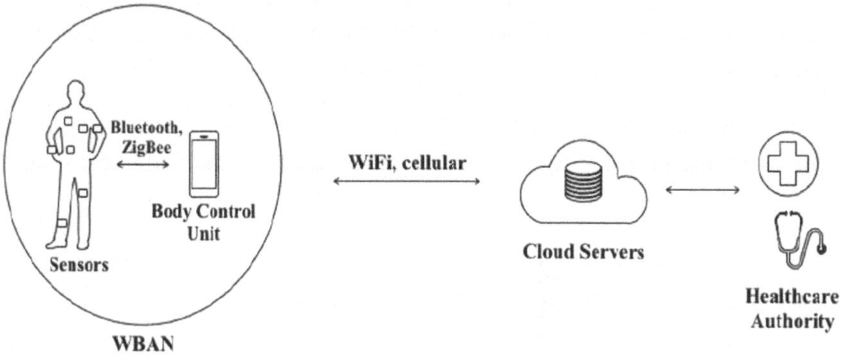

Fig. 1. The general architecture of the healthcare IoT infrastructure.

Thereby it is possible to identify the main components and subcomponents of healthcare IoT system. They are:

- Wireless body area network (WBAN) consists of different sensors located in different parts of human's body and body control unit. Sensors are used to record physiological processes and convert the received data into a format convenient for perception and analysis. There are different kinds of medical sensors and first of all they are classified as consumer products for health monitoring, wearable external, internally embedded and stationary [21]. These sensors or even devices can capture such data as blood pressure, temperature, electrocardiogram (ECG), electroencephalogram (EEG), accelerometer, the global positioning system (GPS), electromyography (EMG), etc. Data collected by sensors are transmitted to the body control unit using e.g. Bluetooth or ZigBee protocols. The control unit is designed to read reports, monitor status, change settings, and update the device's firmware. It can directly connect to Cloud servers if it has WiFi or cellular interfaces or through monitoring unit using Bluetooth or WiFi;

- Cloud servers provide easy access to servers, storage, databases and a wide range of software services on the Internet. The main purposes of the cloud are storage, analytics, and visualization. Clouds provide reception of telemetry data in the required volume from the devices and determination of the way of processing and storing the obtained data, allow healthcare telemetry analysis to provide valuable information both in real time and later and send commands from the cloud or gateway device to a specific healthcare device. Also, the server part of the Internet of things' cloud should provide the device registration capabilities that allow preparation of the device and control which devices are allowed to connect to the infrastructure and device management for monitoring the status of devices and monitor their actions. Using cloud services, it is possible to effectively store and dynamically process data, interact and integrate data;
- A healthcare authority pulls an analytical report for each patient to check the patient's illness status. He evaluates the data and sends a notification. The patient receives a notification that advises whether to consult a doctor.

In this paper, the main subject of the study is a general insulin pump operating in the infrastructure of IoT. An activity diagram for the insulin pump operating independently without interaction with other devices or Internet was described in [22].

The author illustrated how the software transforms an input blood sugar level to a sequence of commands that drive the insulin pump. Figure 2 shows an improved version for the insulin pump operating in the infrastructure of IoT and interaction with other components.

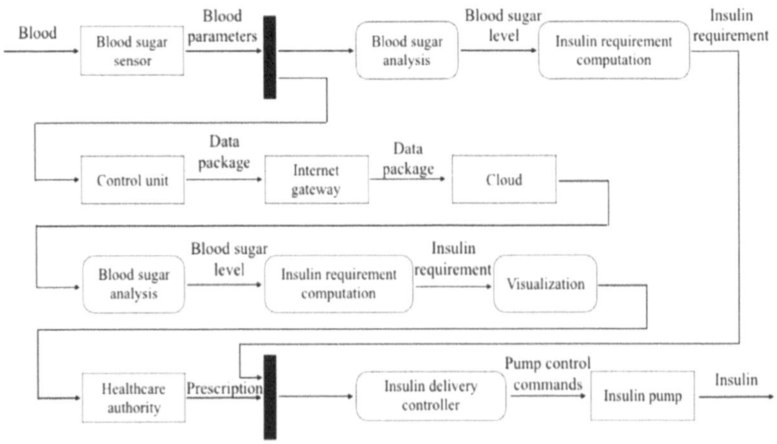

Fig. 2. High-level processes for the insulin pump in the context of IoT.

The data from the blood sugar sensor send to the blood sugar analysis and insulin requirement computation what is carried out by integrated technical possibilities and tools of the insulin pump and/or sends to the Cloud servers via the Internet gateway for further processing, storage, and visualization. The patient's data can be analysed using e.g. artificial intelligence tools in the Cloud. The decision made by artificial intelligence tools sends to the healthcare authority for the conclusive prescription and finally to the patient or insulin pump user. In more details, decisions that were made by the healthcare authority are also loaded into the Cloud, and then insulin pump user (control unit) downloads prescriptions.

2.2 Analysis of Failures and Attacks on the Healthcare IoT Systems

It is clear that the healthcare IoT based system is a safety-critical system. If the insulin pump or any other significant element of the infrastructure fails to operate or does not operate correctly, then the patients' health may be damaged, or they may fall into a coma because their blood sugar levels are too high or too low, or the doctor's prescription is not received by the patient in time, etc. Consequently, the healthcare IoT system must meet availability characteristics and provide round-the-clock service with no exceptions. In this way, there are some vital high-level requirements that such IoT system must meet:

1. The system shall be available to deliver insulin when required (system availability).
2. The system shall perform reliably and deliver the correct amount of insulin to counteract the current level of blood sugar [22].
3. Any component of the IoT system shall interact with any other when required.
4. The system shall be able to scale.
5. The Cloud component shall be able to process, storage and visualize all patients' data when required.
6. The healthcare authority component shall be able to respond to all patients' requests when required, etc.

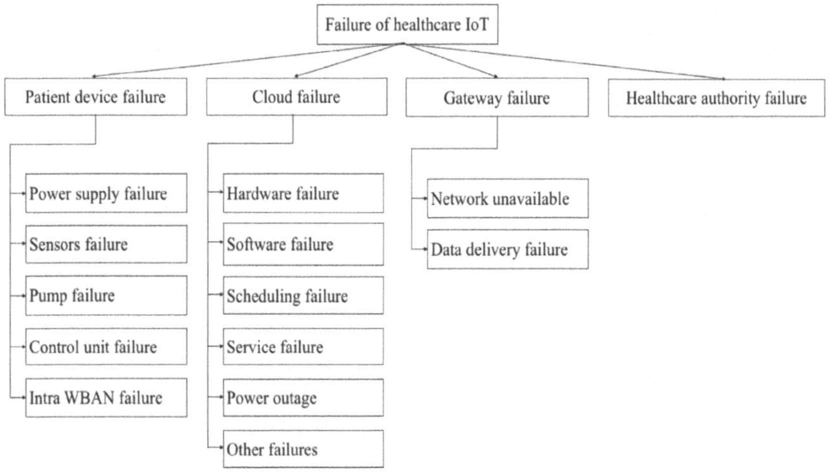

Fig. 3. Classification of the healthcare IoT failures.

Thereby as in any other information and technology systems, failures also may occur in the IoT based systems. Figure 3 depicts in outline the main causes of healthcare IoT based system failures.

In papers [21–27] were described failures of insulin pumps that were caused by different reasons (e.g. sensors failure, control unit failure due to hardware and/or software, etc.). Analysis of papers [28–31] shows the possible failures of Cloud servers. These failures are caused due to software failure, hardware failure, scheduling, service failure, power outage, denser system packaging, etc. Accordingly, it is possible to assert that the reasons of failures may be variable and depend on failures of healthcare IoT infrastructure each component.

Since attacks are the possible consequences of the threat implementation the existing vulnerabilities. Therefore it is necessary to consider the attack as a malicious action affecting the healthcare IoT system's performance.

About 250 cyber attacks were targeted on health sector (only publicly disclosed incidents) in 2016–2017 [32]. There are several types of attacks on IoT that were discussed in many papers [33–35]. The security issues of insulin pump in the context of cyber-security systems were shown in [36]. The authors of these papers presented attacks' targets, weaknesses, and technique of the security attacks. The main categories of IoT attacks are aimed for control, data, controllers (end-nodes) and networks. Attacks on data are very devastating in the healthcare field due to the physician–patient privilege and a patient privacy and confidentiality. Attacks on control involve imply an intruder's intention to gain access to the management of both the entire healthcare IoT system and individual components. Attacks on controllers are aimed at end-nodes (patients' devices) to gain access to control them and make a physical damage. Attacks on networks are aimed to sniffing out, copying the confidential information or any other data flowing in the networks.

After analysing classification of attacks according the main aims and focus is presented in Fig. 4.

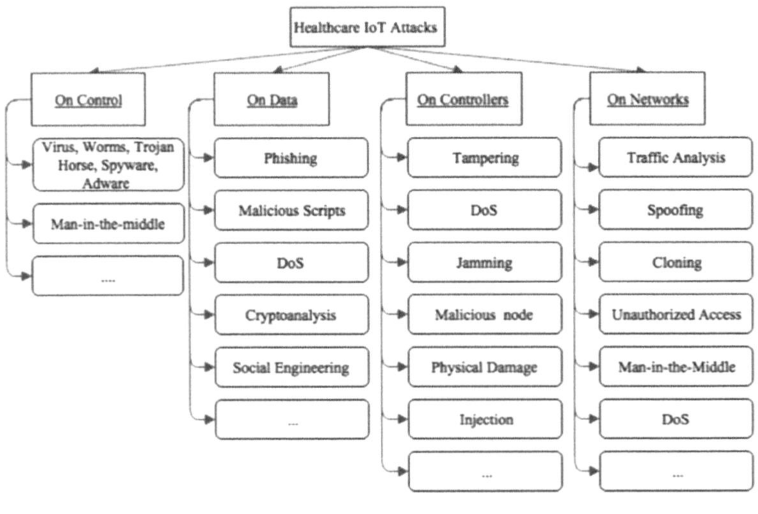

Fig. 4. Classification of the healthcare IoT attacks.

Such attacks on vulnerabilities can prevent the devices and infrastructure to communicate correctly and without failures. According to the presented classifications (Figs. 3 and 4) the availability Markov models of failures and attacks on the healthcare IoT infrastructure respectively will be presented in the next section.

2.3 The Model Set for the Healthcare IoT System Behaviour

For the formalization of the healthcare IoT system behaviour, it was proposed to apply the set-theoretic model that represents the structure of the system and the cause-effect relationships between its components in [12]. The model that describes a set of models' behaviour of the healthcare IoT infrastructure is presented in Fig. 5.

M_{00}^{10}			M_{00}^{F0}		
$M_{R_I0}^{10}$	$M_{0R_D}^{10}$	$M_{R_IR_D}^{10}$	$M_{R_I0}^{F0}$	$M_{0R_D}^{F0}$	$M_{R_IR_D}^{F0}$
M_{00}^{1V}			M_{00}^{FV}		
$M_{R_I0}^{1V}$	$M_{0R_D}^{1V}$	$M_{R_IR_D}^{1V}$	$M_{R_I0}^{FV}$	$M_{0R_D}^{FV}$	$M_{R_IR_D}^{FV}$

Fig. 5. Model range for healthcare IoT systems behaviour [12].

The basic models were described, in details, simple cases with a few models of healthcare IoT system based on the queueing theory in [11]. These models describe streams of the requests and attacks on vulnerabilities and procedure of recovery by a restart and eliminating of ones.

According to the presented model range, $M_{R_IR_D}^{FV}$ is a model that takes into account all possible states of the healthcare IoT system (functional states, vulnerabilities and reliability issues), where F – is the cardinality of a functional states set $F = \{S_{F_0}, S_{F_1}, \ldots, S_{F_m}\}$; V – is the cardinality of the security vulnerabilities set $V = \{S_{V_0}, S_{V_1}, \ldots, S_{V_p}\}$, where p is a number of vulnerabilities; R_I – is the cardinality of the degradation levels of the healthcare IoT infrastructure set $R_I = \left\{S_{R_{I_0}}, S_{R_{I_1}}, \ldots, S_{R_{I_k}}\right\}$, i.e. ability to serve applications from device nodes; R_D – is the cardinality of the devices (for this paper, networked insulin pump) failures set $R_D = \left\{S_{R_{D_o}}, S_{R_{D_1}}, \ldots, S_{R_{D_1}}\right\}$.

In general, in the healthcare IoT system, the failures of single (or attacks on) subcomponents are possible. These failures may lead to the failures of the main components of infrastructure (i.e. insulin pump, cloud, etc.). In its turn, the failures of main components may lead to failure of the whole healthcare IoT system. Figure 6 shows the dependence of the healthcare IoT system failures, where state 0 corresponds to condition when there is no any failure in the system, state 1 – there is one failure (of subcomponent), state 2 – there are two failures (subcomponent and main element), state 3 – there are three failures (the failure of the whole healthcare IoT system).

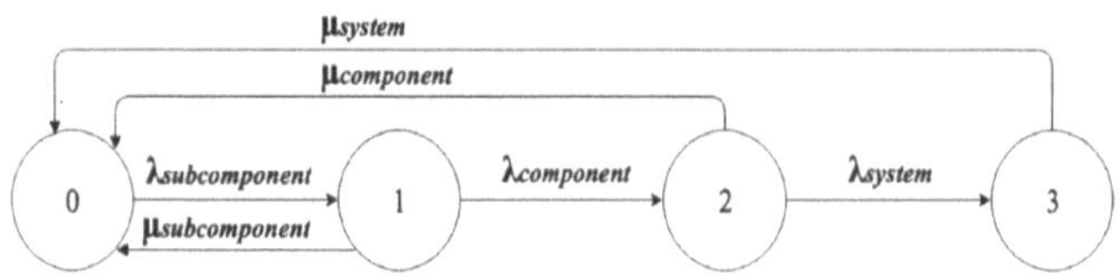

Fig. 6. Dependence of the healthcare IoT failures.

According with this model and the approach discussed above (see Fig. 5) the availability models of failures and attacks (M_{00}^{1V}) on the healthcare IoT infrastructure will be presented in the next section.

3 Availability Markov Models of Healthcare IoT System

Based on the approach of the set-theoretic model discussed above, it is necessary to develop mathematical models for an availability function definition. In this paper, the assumptions considered to the development of the availability models are the following:

- The rates of failures and attacks are constant, the flows of failures and attacks' rates obey the Poisson distribution law.
- The models do not consider eliminating of any reasons because of what failures caused, the system provides just standard protection tools.
- The tools of control and diagnostics are triggered ideally, it means that they detect correctly and in time all appearing failures and faults.
- The occurring process in the healthcare IoT system is a process without after-affects, the probability of each event in the future depends just on the state of the IoT system in the present time and does not depend how the system arrived at this state previously. Thus, it satisfies the Markov property.

3.1 A Markov Model of Healthcare IoT Considering Component Failures

Development of the Model. In more details Fig. 7 shows a Markov graph of the functioning of the main components of healthcare IoT system if failures occur, where λ - the failure rate, μ - the recovery rate. Thereby, the basic states of the healthcare IoT system are: 1 - normal condition (upstate) system; 2 - failure due to the power supply (battery) pump causes discharge, recharging and/or causing dam-age; 3 - failure of any one and/or more sensors of the insulin pump due to the out-of-order, does not deliver any output to inputs, delivers null output values and/or no meaningful values and/or impurity etc.; 4 - pump failure (inaccurate size/rate of insulin dose) due to the components defects, improper position of pump, ambient temperature, air pressure and/or design errors etc.; 5 - software of insulin pump control module failure due to buffer overflow or underflow, incorrect libraries, wrong algorithms or programming, threshold setting error etc.; 6 - hardware of insulin pump control module failure due to overheating, short or open circuit, high leakage current, high or low impedance, missed alarm, false alarm, fail to read/write data and/or de-sign error etc.; 7 - intra wireless body area network (WBAN) communication failure due to the packet loss, isolation, a communication module failure (e.g., L2CAP, BNEP etc.), header corruption and/or length mismatch and/or payload corruption etc.; 8 - insulin pump (as the patient's complex) failure due to the failure of any one or more main components; 9 - extra gateway communication partial failure due to data delivery failures; 10 - extra gateway communication partial failure due to Blue-tooth/cellular/WiFi network unavailable; 11 - partial failure due to the refusal of the mobile application of the reader (control unit); 12 - cloud software failure due to a planned or unplanned reboot, software updates and/or complex design; 13 - cloud hardware failure due to hard disk failures, RAID controller, memory and/or other devices; 14 - cloud scheduling failure due to overflow and/or timeout; 15 - cloud service failure due to request stage and/or execution stage; 16 - cloud failure due to power outage; 17 - cloud failure due to the failure of any one and/or more cloud components; 18 - failure due to incorrect assignment or programming of the device by a healthcare authority related to device functions or lack of functions; 19 - failure of the IoT healthcare system.

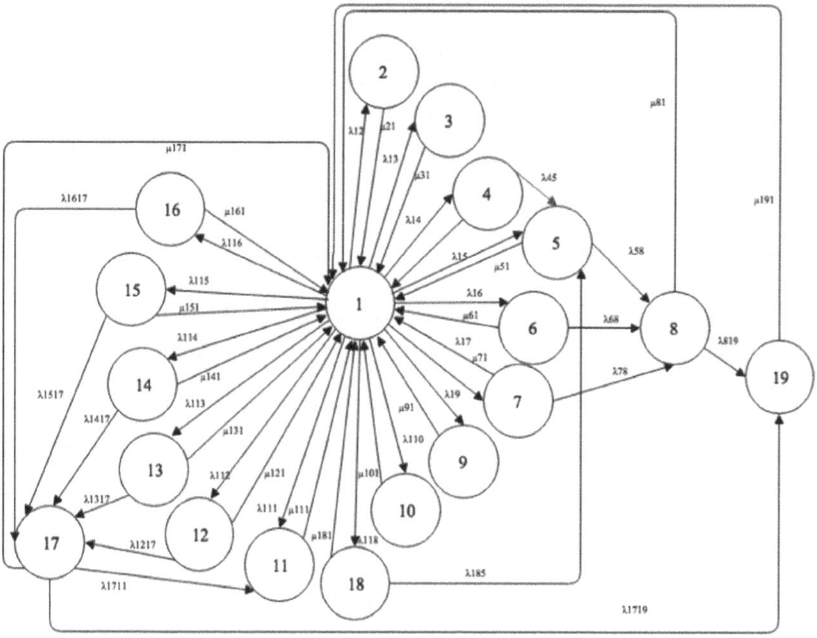

Fig. 7. A Markov's graph of the healthcare IoT failures.

A system of the Kolmogorov differential equations for presented Markov model is:

$dP_1/dt = -(\lambda_{12} + \lambda_{13} + \lambda_{14} + \lambda_{15} + \lambda_{16} + \lambda_{17} + \lambda_{19} + \lambda_{110} + \lambda_{111} + \lambda_{112} + \lambda_{113} + \lambda_{114} +$
$+ \lambda_{115} + \lambda_{116} + \lambda_{118})P_1(t) + \mu_{21}P_2(t) + \mu_{31}P_3(t) + \mu_{41}P_4(t) + \mu_{51}P_5(t) + \mu_{61}P_6(t) +$
$+ \mu_{71}P_7(t) + \mu_{81}P_8(t) + \mu_{91}P_9(t) + \mu_{101}P_{10}(t) + \mu_{111}P_{11}(t) + \mu_{121}P_{12}(t) + \mu_{131}P_{13}(t) +$
$+ \mu_{141}P_{14}(t) + \mu_{151}P_{15}(t) + \mu_{161}P_{16}(t) + \mu_{171}P_{17}(t) + \mu_{181}P_{18}(t) + \mu_{191}P_{19}(t);$
$dP_2/dt = -\mu_{21}P_2(t) + \lambda_{12}P_1(t);$
$dP_3/dt = -\mu_{13}P_3(t) + \lambda_{13}P_1(t);$
$dP_4/dt = -(\mu_{41} + \lambda_{45})P_4(t) + \lambda_{14}P_1(t);$
$dP_5/dt = -(\mu_{51} + \lambda_{58})P_6(t) + \lambda_{16}P_1(t) + \lambda_{45}P_4(t) + \lambda_{185}P_{18}(t);$
$dP_6/dt = -(\mu_{61} + \lambda_{68})P_6(t) + \lambda_{16}P_1(t);$
$dP_7/dt = -(\mu_{71} + \lambda_{78})P_7(t) + \lambda_{17}P_1(t);$
$dP_8/dt = -(\mu_{81} + \lambda_{819})P_8(t) + \lambda_{58}P_5(t) + \lambda_{68}P_6(t) + \lambda_{78}P_7(t);$
$dP_9/dt = -\mu_9P_9(t) + \lambda_{19}P_1(t);$
$dP_{10}/dt = -\mu_{101}P_{10}(t) + \lambda_{110}P_1(t);$
$dP_{11}/dt = -\mu_{111}P_{11}(t) + \lambda_{111}P_1(t) + \lambda_{1711}P_{17}(t);$
$dP_{12}/dt = -(\mu_{121} + \lambda_{1217})P_{12}(t) + \lambda_{112}P_1(t);$
$dP_{13}/dt = -(\mu_{131} + \lambda_{1317})P_{13}(t) + \lambda_{113}P_1(t);$
$dP_{14}/dt = -(\mu_{141} + \lambda_{1417})P_{14}(t) + \lambda_{114}P_1(t);$
$dP_{15}/dt = -(\mu_{151} + \lambda_{1517})P_{15}(t) + \lambda_{115}P_1(t);$
$dP_{16}/dt = -(\mu_{161} + \lambda_{1617})P_{16}(t) + \lambda_{116}P_1(t);$
$dP_{17}/dt = -(\lambda_{1711} + \lambda_{1719} + \mu_{171})P_{17}(t) + \lambda_{1217}P_{12}(t) + \lambda_{1317}P_{13}(t) + \lambda_{1417}P_{14}(t) +$
$+ \lambda_{1517}P_{15}(t) + \lambda_{1617}P_{17}(t);$
$dP_{18}/dt = -(\mu_{181} + \lambda_{185})P_{18}(t) + \lambda_{118}P_1(t);$
$dP_{19}/dt = -\mu_{191}P_{19}(t) + \lambda_{919}P_8(t) + \lambda_{1719}P_{17}(t).$

Initial values are:

$$P_1(0) = 1, Pi(0) = 0, i = 2, 3, \ldots, 19.$$

To solve a system of the linear Kolmogorov differential equations it is necessary to carry out the collection and analysis of statistics on failures of healthcare IoT systems.

Simulation of the Model. Hence the initial data for Markov model simulating were taken from [16, 18–20] for the insulin pump failures, for the Cloud failures [29–31] and experts' assessments. Due to the heterogeneous nature and complexity of statistical data, and not to overflow with excess information, the sequence of rates' calculations and the rates are not given in this paper.

The working state is state 1, and eighteen others are states with failures of different components and parts of the healthcare IoT system. The obtained probabilities of finding the healthcare IoT system in each state of Markov model are shown below (stationary values):

Pf1 = 0.9853745;	Pf2 = 0.000103622;	Pf3 = 0.003330566;
Pf4 = 0.000251795;	Pf5 = 0.001162896;	Pf6 = 0.0003859747;
Pf7 = 0.006145591;	Pf8 = 0.0009757008;	Pf9 = 0.001486934;
Pf10 = 0.0006328081;	Pf11 = 3.207395e−05;	Pf12 = 3.070724e−05;
Pf13 = 1.056492e−05;	Pf14 = 2.90451e−05;	Pf15 = 2.596815e−05;
Pf16 = 5.734457e−06;	Pf17 = 3.038937e−08;	Pf18 = 1.545724e−05;
Pf19 = 2.957985e−08.		

Hence A(t) = P1(t), Fig. 8 shows the availability function value changing before a transition to the stationary value ($A_{stationary}$ = 0.9853745). According to the simulation results the function gets a qua approximately at step 2300 h, i.e. 3 months later after beginning of work.

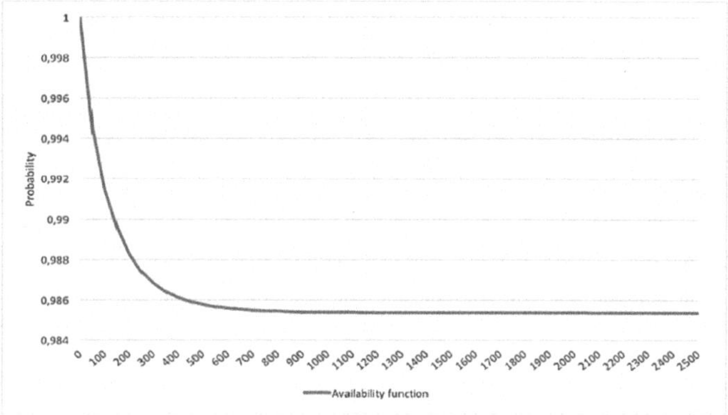

Fig. 8. Availability function changing before a transition to the stationary value.

Figures 9, 10, 11 and 12 show the decreases of the availability function value depending on the different types of failures of the healthcare IoT systems rates:

- the transition rate λ13 from the upstate of the system to a state 3 of the failure of any one and/or more sensors of the insulin pump due to the out-of-order, does not deliver any output to inputs, delivers null output values and/or no meaningful values and/or impurity etc. (Fig. 9);
- the transition rate λ15 from the upstate of the system to a state 5 of the software of insulin pump control module failure due to buffer overflow or underflow, incorrect libraries, wrong algorithms or programming, threshold setting error etc. (Fig. 10);
- the transition rate λ110 from the upstate of the system to a state 10 of the extra gateway communication partial failure due to Bluetooth/cellular/WiFi network unavailable failure (Fig. 11);
- the transition rate λ112 from the upstate of the system to a state 12 of the cloud software failure due to a planned or unplanned reboot, software updates and/or complex design (Fig. 12).

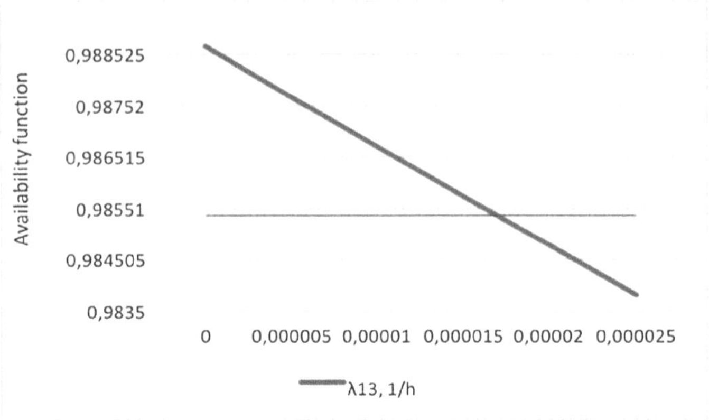

Fig. 9. Dependence of the availability function value depending on the changing λ13 rate.

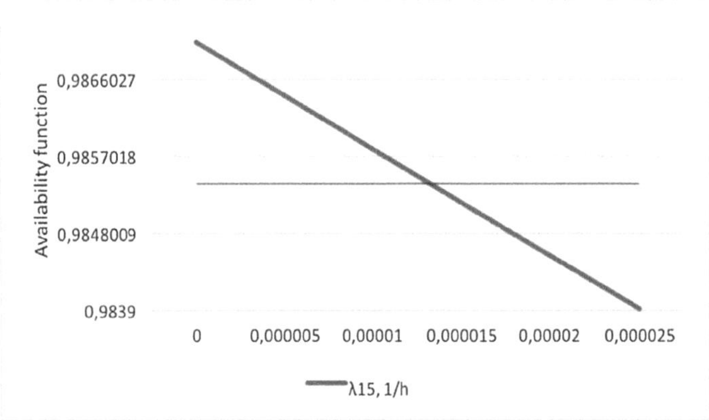

Fig. 10. Dependence of the availability function changing depending on the changing λ15 rate.

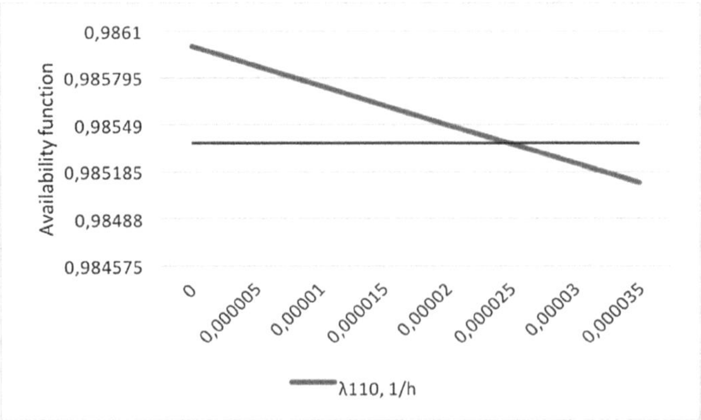

Fig. 11. Dependence of the availability function value depending on the changing λ110 rate.

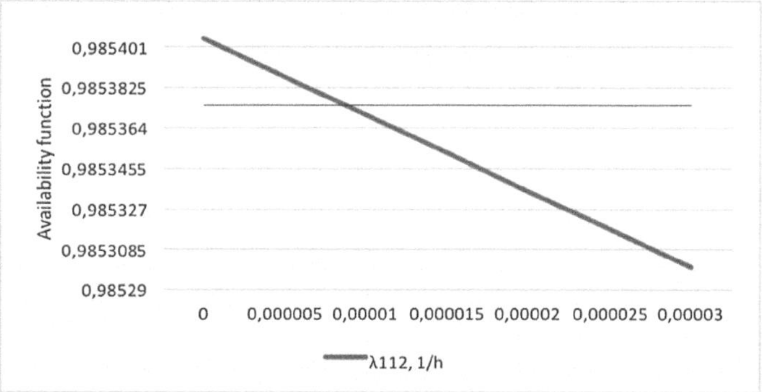

Fig. 12. Dependence of the availability function value depending on the changing λ112 rate.

The reducing of the availability function value from the upstate of the healthcare IoT system the states with failures occurs due to the different reasons, e.g. design defects in hardware/software components, impact of external influences (attacks) etc. Accordingly to Figs. 9, 10, 11 and 12, with changing of the rate λ13 = 0 … 0.000025 1/h the availability value decreases from 0.9886673 to 0.983804; for the rate λ15 = 0 … 0.000025 1/h the availability value decreases from 0.9870318 to 0.9839211; for the rate λ110 = 0 … 0.000035 1/h the availability value decreases from 0.9859984 to 0.9851145; for the rate λ112 = 0 … 0.00003 1/h the availability value decreases from 0.9854048 to 0.9853004. The obtained results' analysis shows that the greatest influence on the change in the availability function is the λ15 rate and the next is the λ13 rate (i.e. different components of the patient device (for our case of insulin pump) failures). The least influence has the failures of cloud components due to the rapid recovery time [29–31]. These results are confirmed by the statistical data (of availability, reliability and accessibility) of AWS, Google, Azure, IBM, etc. cloud providers and end-nodes manufactures (insulin pumps) [16, 18–20, 29–31].

The analysis of obtained results shows that the complete failure of the healthcare IoT system does not happen too often (one case on the analysed time interval due to the complete failure of the Cloud). Nevertheless, failures of constituent elements of the system arise quite often that may affect the performance of mission-critical functions of the healthcare IoT system and in the worst case, lead to the death of the patient. The most often failures are due to the failure of the insulin pump and its particular elements and components and some components of the Cloud.

Availability of the system can be improved by more fast recovery (repair) of the equipment and system resources and application of more reliable devices.

3.2 A Markov Model of the Healthcare IoT Considering the Attacks on Vulnerabilities

Development of the Model. According to the approach described above (Fig. 5) for the case M_{00}^{1V}, a simplified Markov's graph was presented in [12] and showed below.

The model does not take into account eliminating of vulnerabilities and design faults. The failure and/or attack, recovery and/or repulse rates are constant.

Figure 13 shows a Markov's graph of the main components functioning of the healthcare IoT system during attacks, λ - the failure and/or attack rate, μ - the recovery and/or reflection rate. Thereby, the basic states of the healthcare IoT system are: 1 – normal condition (upstate) system; 2 – traffic analysis attack; 3 – spoofing attack; 4 – cloning attack; 5 – unauthorized access to the network or database; 6 – failure of the network; 7 – failure due the data leakage; 8 – man-in-the-middle attack; 9 – DoS/DDoS attack; 10 – failure due the loss of control; 11 – attacks on software (i.e., viruses, worms, Trojan horses, spyware, adware, etc.); 12 – phishing attack; 13 – malicious scripts injection attacks; 14 – social engineering; 15 – failure or controllers (hardware, end-nodes); 16 – tampering of the end-nodes attack; 17 – jamming attack; 18 – malicious node injection attack; 19 – physical damage; 20 – complete failure of healthcare IoT system.

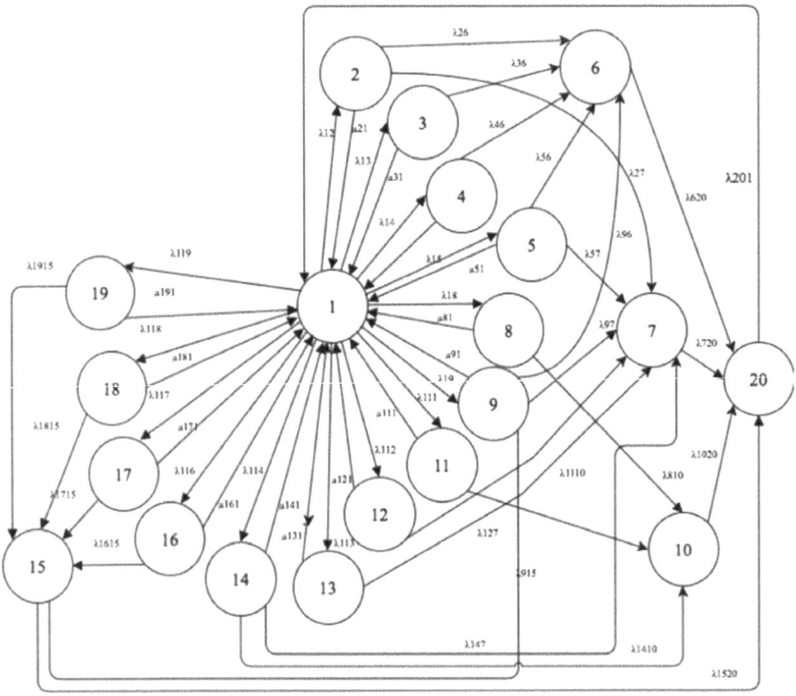

Fig. 13. Markov's graph of attacks on the healthcare IoT infrastructure.

Such Markov model for the healthcare IoT system can be divided into four levels: the first level is the upstate (state 1), the second level are the states state the transition to which occurred due to attacks (states 2–5, 8, 9, 11–14, 16–19), the third level implies states with different failures (states 6, 7, 10, 15); the fourth level is a failure of a whole system (state 20). So, there is only one up-state in the considering system.

Simulation of the Developed Markov Model. After solving the system of Kolmogorov-Chapmen equations, it is possible to obtain the availability function value of the healthcare IoT system, the number of system failures due to cyber attacks.

To solve a system of linear Kolmogorov differential equations it is necessary to collect and analyse statistics on failures and attacks on the healthcare IoT infrastructure. The data for model simulation was taken from [37, 38] (for the healthcare cyber attacks [39–41]) and experts' assessments.

The obtained stationary value probabilities for the considered Markov model are:

Pal = 0.9200121;	Pa2 = 5.73576e−05;	Pa3 = 4.294692e−05;
Pa4 = 0.0001696415;	Pa5 = 0.0002171216;	Pa6 = 0.01081887;
Pa7 = 0.04191969;	Pa8 = 4.88764e−05;	Pa9 = 0.0001009014;
Pa10 = 0.01365914;	Pa11 = 5.696207e−05;	Pa12 = 0.0001208568;
Pa13 = 1.517664e−05;	Pa14 = 0.001280493;	Pa15 = 0.00138444;
Pa16 = 0.000618885;	Pa17 = 0.002611121;	Pa18 = 0.0005009447;
Pa19 = 0.005841538;	Pa20 = 0.0005229814.	

The dependence on the changing of the availability function value depending on the changing of the different types of the attacks on the healthcare IoT system rates was shown in [12].

4 Game Theoretical Model for Choosing the Protection Tool and Assure of the Healthcare IoT Availability

4.1 Development of the Game Theoretical Model for Choosing the Countermeasures

Analysing security issues against various threats it is advisable to consider the actions of the two sides: the parties to the protect (of the healthcare IoT system) and the parties to the offender (an attacker). The relationship between these players is determined as a payoff matrix. The condition for the effective protection is a rule: the cost of the protection tools should be less than the cost of the losses incurred in the successful implementation of attack.

An approach how to calculate an effective protection factor was presented in [42]:

$$\lambda_{ij} = \frac{S_i}{\left(1 - p_{ij}^{(p)}\right) \times p_{ij}^{(a)} \times D}, \quad i = \overline{1,n},\ j = \overline{1,m}, \tag{1}$$

where S_j - is a cost of protection tool; $p_{ij}^{(p)}$ – is an attack reflection probability; $p_{ij}^{(a)}$ – is an attack probability; D – the value of the average damage of the healthcare IoT system.

Due to the attacks' classification presented in Fig. 4 the protection tools were divided according to attacks classification. Tables 1, 2, 3 and 4 define the matrix of attacks reflection probabilities. These probabilities were determined as the expert's assessments.

Table 1. The pay-off matrix for the network attacks

Attack/Protection tool	Attack probability	Protection tools			
		IDS	Firewall	Authentication	Cryptographic
		Attack reflection probability			
Traffic analysis	5.74E−05	0.7	0.85	0.001	0.3
Spoofing	4.29E−05	0.7	0.8	0.1	0.001
Cloning	0.000169642	0.01	0.9	0.5	0.001
Unauthorized access	0.01081887	0.01	0.9	0.9	0.6
MITM	4.89E−05	0.01	0.6	0.5	0.4
DoS/DDoS	0.000100901	0.4	0.8	0.001	0.001

Table 2. The pay-off matrix for the controllers attacks

Attack/Protection tool	Attack probability	Protection tools			
		IDS	Firewall	Cryptographic	Tamper proofing and self destruction
		Attack reflection probability			
Tampering	0.000618885	0.01	0.01	0.3	0.9
DoS/DDoS	0.000100901	0.4	0.8	0.001	0.6
Jamming	0.002611121	0.01	0.4	0.01	0.91
Malicious node	0.000500945	0.01	0.7	0.01	0.7

Table 3. The pay-off matrix for the data attacks

Attack/Protection tool	Attack probability	Protection tools				
		IDS	Firewall	Authentication	Cryptographic	Training
		Attack reflection probability				
Phishing	0.000120857	0.001	0.3	0.5	0.8	0.001
Malicious scripts	1.51E−05	0.5	0.5	0.001	0.8	0.1
DoS/DDoS	0.000100901	0.4	0.8	0.001	0.001	0.0001
Traffic analysis	5.74E−05	0.7	0.85	0.001	0.3	0.001
Unauthorized access	0.01081887	0.01	0.9	0.9	0.6	0.2
Social engineering	0.001280493	0.001	0.001	0.6	0.2	0.7

Table 4. The pay-off matrix for the control attacks

Attack/Protection tool	Attack probability	Protection tools			
		IDS	Firewall	Authentication	Cryptographic
		Attack reflection probability			
Software attacks	5.70E−05	0.2	0.95	0.1	0.2
MITM	4.89E−05	0.4	0.5	0.001	0.4
Social engineering	0.001280493	0.2	0.6	0.7	0.2

According to the statistical data the average damage of the attack *D* on the healthcare IoT system is about USD 10000 (e.g., insurance payment). The cost of protection tools (*Sj*) at a proportion is presented by the corresponding values as in Table 5.

Table 5. Protection tools cost (at a proportion)

IDS	Firewall	Authentication	Cryptographic	Tamper proofing and self destruction	Training
50	10	4	7	2	2

The Wald's maximin model was used for selection the optimal protection tools (it helps to choose the best of the worst protection tool). Accordingly, using (1) for each attack classification group the protection factors were determined. They are: for network attacks was chosen Firewall and optimal protection factor is $\gamma_{network} = 1.924311$, for attacks on controllers were chosen Tamper Proofing and Self Destruction and optimal protection factor is $\gamma_{controllers} = 1.851060607$, for attacks on data was chosen Firewall and optimal protection factor is $\gamma_{data} = 1.924311$, and for attacks on control was chosen Authentication and optimal protection factor is $\gamma_{control} = 1.78094921$.

Thus, after defending and analyzing the game matrix values, it is possible to assess the cost of each solution for protecting the IoT healthcare systems and choose the most effective tool for all range of attacks. In addition, the proposed method allows to choose the set of protection countermeasure tools.

4.2 Discussion of the Results of Rectification of Markov Modelling Considering Game Theoretical Outcomes

In accordance with the game theoretical approach described above the results were obtained which made it possible to correct the initial parameters of the Markov model. The simulation results of Markov model depicted in Fig. 13 using the modificated initial date according with game theoretical outcomes are presented below:

Pa1 = 0.969398;	Pa2 = 1.633232e−05;	Pa3 = 1.203168e−05;
Pa4 = 5.160693e−05;	Pa5 = 5.949243e−05;	Pa6 = 0.003040368;
Pa7 = 0.01373675;	Pa8 = 1.516303e−05;	Pa9 = 2.945473e−05;
Pa10 = 0.004921064;	Pa11 = 1.685261e−05;	Pa12 = 3.745716e−05;
Pa13 = 4.155224e−06;	Pa14 = 0.000466583;	Pa15 = 0.0007246805;
Pa16 = 0.0001917184;	Pa17 = 0.0007433931;	Pa18 = 0.0001546808;
Pa19 = 0.006155206;	Pa20 = 0.0002249595.	

The availability functions without and with the using of countermeasures values changing are presented in Fig. 14.

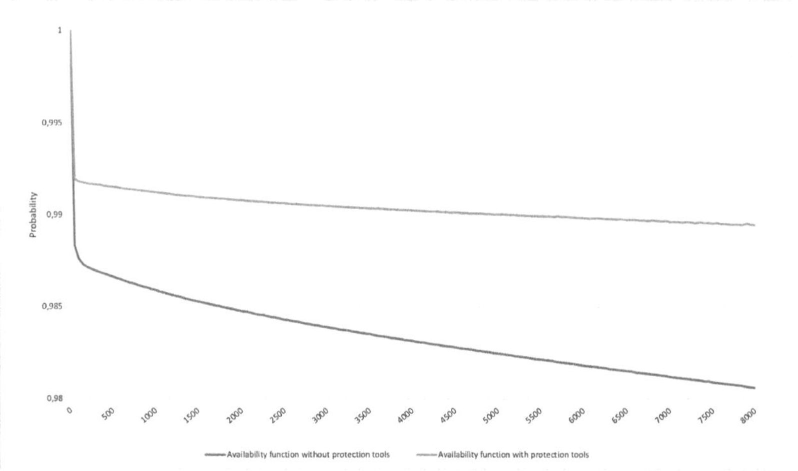

Fig. 14. The availability functions' values changing.

Accordingly, the obtained results show that the availability function value (stationary value) for the healthcare IoT system without using of protection tools was $A_{stationary}$ = 0.9200121, ad after implementing of countermeasures tool into the system the availability function value changed to $A_{stationary}$ = 0.969398. In this way, with the game theoretical results' implementation and by choosing of countermeasures the availability value can be increased up to 4%. On the other hand, when it comes to the healthcare IoT system unavailability, which was about 100 − 92 = 8%, and after a rectification is 100 − 96 = 4%, it can be concluded that there is a potential for reducing the healthcare IoT system's unavailability by 200% or 2 times.

5 Conclusions and Future Work

Due to the use of the IoT technologies, the interaction of objects, environment, and people will be extremely active, and it is making it possible to hope that the world will be "smart" and a well-appointed for a person. However, at the same time, the IoT faces a number of problems that can prevent us from taking power of its potential advantages.

In this paper, the overview of the healthcare IoT system failures and attacks is presented. A Markov models set for the healthcare IoT infrastructure that allows taking into account the specificity of end user devices, communication channels, technologies of data flows and safety and security issues of these components has been developed. Based on the conducted analysis and classification of the main possible failures and attacks on healthcare IoT infrastructure the Markov models considering failures and attacks on components are constructed. For the developed models the probabilities of finding IoT system in each state of the Markov model are shown. The obtained results show possible most frequent failures and attacks on the healthcare IoT system. The game theoretical approach to select the countermeasure tool was presented.

Next steps of research will be dedicated to development of more general dependability models for healthcare IoT systems and combining results of this paper and models considering both the reliability, safety and security requirements and issues and a dynamical nature of the failures and attacks rates. Besides, we plan to develop technique to support decision making on choosing countermeasures according with criteria "availability/safety-cost".

Acknowledgements. This paper implies results obtained during involvement in the Erasmus+ programme educational project ALIOT «Internet of Things: Emerging Curriculum for Industry and Human Applications» (reference number 573818-EPP-1-2016-1-UK-EPPKA2-CBHE-JP, web-site http://aliot.eu.org) in which the appropriate course is under development (ITM4 - IoT for health systems). Within its framework, the teaching modules related to IoT systems modelling were developed. The authors would like to thank colleagues on this project, within the framework of which the results of this work were discussed.

The authors also would like to show deep gratitude to colleagues from Department of Computer Systems, Networks and Cybersecurity of National Aerospace University n. a. M. Ye. Zhukovsky «KhAI» for their patient guidance, enthusiastic encouragement and useful critiques of this paper.

This research is also supported by the project STARC (Methodology of SusTAinable Development and InfoRmation Technologies of Green Computing and Communication) funded by Department of Education and Science of Ukraine.

References

1. Understanding the Internet of Things (IoT), p. 14. GSM Association, London (2014)
2. Press Release: Global Internet of Things market to grow to 27 billion devices, generating USD3 trillion revenue in 2025. https://machinaresearch.com/news/press-release-global-internet-of-things-market-to-grow-to-27-billion-devices-generating-usd3-trillion-revenue-in-2025/. Accessed 02 Mar 2018

3. A Decade of Digital: Keeping Pace with Transformation, 10th edn, p. 30. PwC's Digital IQ Research (2017)
4. Vermesan, O., Friess, P.: Internet of Things: Converging Technologies for Smart Environments and Integrated Ecosystems. River Publishers, Delft (2013)
5. World Healthcare Organization: Global Report on Diabetes (2016)
6. Islam, S.M.R., Kwak, D., Kabir, M.D.H., Hossain, M., Kwak, K.-S.: The internet of things for health care: a comprehensive survey. IEEE Access **3**, 678–708 (2015). https://doi.org/10.1109/ACCESS.2015.2437951
7. Laplante, P.A., Kassab, M., Laplante, N.L., Voas, J.M.: Building caring healthcare systems in the internet of things. IEEE Syst. J. 1–8 (2017). https://doi.org/10.1109/jsyst.2017.2662602
8. Applied Safety Science and Engineering Techniques. Taking Hazard Based Safety Engineering (HBSE) to the Next Level, p. 11. IEEE (2010)
9. Goševa-Popstojanova, K., Trivedi, K.S.: Architecture-based approach to reliability assessment of software systems. Perform. Eval. **45**(2–3), 179–204 (2001). https://doi.org/10.1016/S0166-5316(01)00034-7
10. Kharchenko, V., Kolisnyk, M., Piskachova, I., Bardis, N.: Reliability and security issues for IoT-based smart business center: architecture and Markov model. In: 2016 Third International Conference on Mathematics and Computers in Sciences and in Industry (MCSI), Chania, pp. 313–318 (2016). https://doi.org/10.1109/mcsi.2016.064
11. Strielkina, A., Uzun, D., Kharchenko, V.: Modelling of healthcare IoT using the queueing theory. In: 2017 9th IEEE International Conference on Intelligent Data Acquisition and Advanced Computing Systems: Technology and Applications (IDAACS), Bucharest, pp. 313–318 (2016). https://doi.org/10.1109/idaacs.2017.8095207
12. Strielkina, A., Kharchenko, V., Uzun, D.: Availability models for healthcare IoT systems: classification and research considering attacks on vulnerabilities. In: 2018 9th IEEE International Conference on Dependable Systems, Services and Technologies, DESSERT'2018, Kyiv, pp. 48–65 (2004). https://doi.org/10.1109/dessert.2018.8409099
13. U.S. Department of Health and Human Services Food and Drug Administration Center for Devices and Radiological Health Office of Device Evaluation: Applying Human Factors and Usability Engineering to Medical Devices: Guidance for Industry and Food and Drug Administration Staff (2016)
14. U.S. Department of Health and Human Services Food and Drug Administration Center for Drug Evaluation and Research Center for Biologics Evaluation and Research: Guidance for Industry: Q9 Quality Risk Management (2006)
15. ISO 14971:2007 Medical Devices – Application of Risk Management to Medical Devices (2007)
16. Trivedi, K.S., Selvamuthu, D.D.: Markov modeling in reliability. In: Encyclopedia of Quantitative Risk Analysis and Assessment. Wiley, Hoboken (2008). https://doi.org/10.1002/9781118445112.stat03635
17. Nicol, D.M., Sanders, W.H., Trivedi, K.S.: Model-based evaluation: from dependability to security. IEEE Trans. Depend. Secure Comput. **01**(1), 48–65 (2004). https://doi.org/10.1109/tdsc.2004.11
18. Roy, S., Ellis, C., Shiva, S., Dasgupta, D., Shandilya, V., Wu, Q.: A survey of game theory as applied to network security. In: 2010 43rd Hawaii International Conference on System Sciences (2010). https://doi.org/10.1109/hicss.2010.35
19. Chung, K., Kamhoua, C.A., Kwiat, K.A., Kalbarczyk, Z.T., Iyer, R.K.: Game theory with learning for cyber security monitoring. In: 2016 IEEE 17th International Symposium on High Assurance Systems Engineering (HASE) (2016). https://doi.org/10.1109/hase.2016.48

20. Maksimović, M.V., Vujović, V., Perišić, B.: A custom internet of things healthcare system. In: 2015 10th Iberian Conference on Information Systems and Technologies (CISTI), Aveiro, pp. 1–6 (2015). https://doi.org/10.1109/cisti.2015.7170415
21. Malik, M.A.: Internet of Things (IoT) Healthcare Market by Component (Implantable Sensor Devices Wearable Sensor Devices System and Software) Application (Patient Monitoring Clinical Operation and Workflow Optimization Clinical Imaging in Fitness and Wellness Measurement) - Global Opportunity Analysis and Industry Forecast 2014–2021. Allied Market Research, p. 124 (2016)
22. Sommerville, I.: Software Engineering, 9th edn. Pearson, London (2010)
23. Zhang, Y., Jones, P.L., Jetley, R.: A hazard analysis for a generic insulin infusion pump. J. Diab. Sci. Technol. **4**(2), 263–283 (2010). https://doi.org/10.1177/193229681000400207
24. Wetterneck, T.B., et al.: Using failure mode and effects analysis to plan implementation of smart i.v. pump technology. Smart i.v. Pump Technol. **63**, 1528–1538 (2006)
25. Rafeh, R., Rabiee, A.: Towards the design of safety-critical software. J. Appl. Res. Technol. **11**(5), 683–694 (2013). https://doi.org/10.1016/S1665-6423(13)71576-1
26. Klonoff, D.C., Reyes, J.S.: Insulin pump safety meeting: summary report. J. Diab. Sci. Technol. **3**(2), 396–402 (2009). https://doi.org/10.1177/193229680900300224
27. Guenego, A., Bouzillé, G., Breitel, S., Esvant, A., Poirier, J.-Y., et al.: Insulin pump failures: has there been an improvement? Update of a prospective observational study. Diab. Technol. Ther. **18**(12), 820–824 (2016). https://doi.org/10.1089/dia.2016.0265
28. Sharma, Y., Javadi, B., Si, W., Sunb, D.: Reliability and energy efficiency in cloud computing systems: survey and taxonomy. J. Netw. Comput. Appl. **74**, 66–85 (2016). https://doi.org/10.1016/j.jnca.2016.08.010
29. Reliability Pillar: AWS Well-Architected Framework. Amazon Web Services, p. 45 (2018)
30. Yanovsky, M., Yanovskaya, O., Kharchenko, V.: Analysis of methods for providing availability and accessibility of cloud services. In: 2016 12th International Conference on ICT in Education, Research and Industrial Applications. Integration, Harmonization and Knowledge Transfer, Kyiv, pp. 414–426 (2016)
31. Cloud Computing Vulnerability Incidents: A Statistical Overview, p. 21. Cloud Vulnerabilities Working Group (2013)
32. McAfee Labs Threat Report, 13 p. (2017)
33. Abomhara, M., Kien, G.M.: Cyber security and the Internet of Things: vulnerabilities, threats, intruders and attacks. J. Cyber Secur. Mob. **4**(1), 65–88 (2014). https://doi.org/10.13052/jcsm2245-1439.414
34. Farooq, M., Waseem, M., Khairi, A., Mazhar, S.: A critical analysis on the security concerns of Internet of Things (loT). In: International Journal of Computer Applications, vol. 111, 6 p. (2015). https://doi.org/10.5120/19547-1280
35. Nawir, M., et al: Internet of Things (IoT): taxonomy of security attacks. In: 3rd International Conference on Electronic Design, pp. 321–326 (2016). https://doi.org/10.1109/iced.2016.7804660
36. Humayed, A., Lin, J., Li, F., Luo, B.: Cyber-physical systems security—a survey. IEEE Internet Things J. **4**(6), 1802–1831 (2017). https://doi.org/10.1109/JIOT.2017.2703172
37. Kaspersky Security Bulletin: Overall statistics for 2017 Kaspersky Lab, 29 p. (2017)
38. The State of Industrial Cybersecurity 2017. The Business Advantage Group Limited, 23 p. (2017)
39. Bell, G., Ebert, M.: Health care and cyber security: increasing threats require increased capabilities. In: KPMG International, USA, 7 p. (2015)

40. Le Bris, A., El Asri, W.: State of Cybersecurity & Cyber Threats in Healthcare Organizations. ESSEC Business School, 12 p. (2016)
41. The State of Cybersecurity in Healthcare Organizations in 2016. Ponemone Institute, 32 p. (2016)
42. Strielkina, A., Tetskyi, A., Selin, B., Solovyov, O., Uzun, D.: Service for vulnerabilities analysis and security assessment of open source systems. CERes J. **1**(2), 53–64 (2015)

Geographic Information Systems: Should They Be Used in Public Finance Reform Development?

Tetiana Paientko(✉)

Kyiv National Economic University named after Vadym Hetman, Peremohy Avenue 54/1, Kyiv 02000, Ukraine
tpayentko109@gmail.com

Abstract. Public finance reform is one of the most complex areas of decision-making. It requires proper implementing methodology and tools for visualizing possible results of reform in a society. The purpose of this article is to show how geographic information systems (GIS) can be used in the development of reform in the sphere of public finance. GIS could provide a wide range of analysis and better support for ideas of reforms. GIS is useful in cases of public finance reform because it makes it possible to combine statistical, demographical, and geographical analyses. Also, GIS provides necessary visualization that helps ordinary citizens to understand a proposed reform and its aftereffects. GIS can increase transparency and accountability of government, because it is quite difficult to manipulate open map data. The article provides examples of how GIS can be used to justify reforms in public finance, namely, optimizing funding for formal school education and health care. Data from Ukraine was used for the analysis. This choice is due to the presence of radical reforms in this country, citizens' distrust of reforms, and lack of transparency of information about the activities of the public sector. GIS was used to test whether government reforms would comply with European standards of access to schooling and health care for all citizens. The results of the study show that such standards could not be achieved because the government did not take into account the low quality of infrastructure when developing reform programs. GIS is not a perfect tool and several challenges should be also considered. Firstly, the software for GIS must be revised quite often. Secondly, GIS software continues to change and improve over time and there are now several GIS applications that range from being relatively free (having limited tools) to being very expensive (for example, ArcGIS 10.). Thirdly, to follow the idea of increasing transparency, all GIS software should have no conflicts between each other. This means that data from one GIS software can be easily exported into different GIS applications.

Keywords: Geographic information systems · Public finance · Reform · Health care reform · Education reform · Taxpayer funds

V. Ermolayev et al. (Eds.): ICTERI 2018, CCIS 1007, pp. 243–261, 2019.
https://doi.org/10.1007/978-3-030-13929-2_12

1 Introduction

Public finance reform is one of the most complex areas of decision-making. At first glance, the problem is related exclusively to the redistribution of taxpayer money. However, for most post-socialist countries, this problem is not easy to solve. This is due to several reasons, one of which is limited financial resources, so it is often difficult for a government to balance the interests of different members of society in the process of financial resource allocation. The second problem is closely related to the former and lies in the institutional immaturity of society, which makes public opinion relatively easy to be manipulated. Institutional immaturity of society has several characteristics. First is the appearance of democracy in form, but without substance. This means that democracy theoretically exists, there is an electoral system in the country, elections are held, but in fact, power belongs to a small group of people who successfully manipulate the elections. Second, there is a merger of business and political power, a result of which is a class of oligarchs that captures not only power over the distribution of financial flows, but also control over state regulators. Third, the society's passivity; people do not believe that change is possible in the country. This is a source of a crisis of trust, where people become skeptical and cynical about any reform in the country. The most sensitive area for crisis of trust is public finance.

This problem is not formed in a single moment, but over time, therefore it cannot be solved quickly. Having carefully studied the process of reform in the public finances of developed countries, it can be concluded that there are several key success factors. First, it is respect for the taxpayers. This means that the government does not hold to a concept of "government money", but rather of "taxpayer money". Expenditures are made as an expression of the will of the taxpayers, not based on the desires of a small group of powerful people. Secondly, there is maximum transparency of reforms meaning that taxpayers can track the use of their money [1]. To achieve a required level of transparency is not a major problem at present. Using modern information technology, accurate information can be available to every member of society. Thirdly, there is an individual approach to the development of a reform mechanism, which ensures high efficiency. In this case modern technology is very useful too.

One of the mistakes of making reform in many developing countries is simply copying the experience of neighboring countries, which can lead to serious errors and irreparable losses of financial resources. The Ukrainian government has been working over the past few years to introduce ideas of fiscal decentralization, but at the same time it is trying to replicate the experience of countries whose decentralization can be considered successful. However, the peculiarities of the Ukrainian economy, the unevenness of regional economic development, and the large physical territory require the development of approaches specific to Ukraine. One such approach can be considered as being the use of geographic information systems (GIS) in support of the economic feasibility of reforms in public finance and the mechanisms for their implementation. Also, GIS could be used to increase the level of transparency and accountability, because the performance of GIS is easily understandable even for lay people. Furthermore, it is not very easy to manipulate GIS results after being published, because people understand their own geographic areas.

The most complex areas of reform in public finance are government funding for education and health care. In Ukraine, the share of these in the government has always been quite high (more than 20%). Funds for these areas have not always been used effectively. This situation has led to a deterioration in the quality of education and health care and should be regarded as an inefficient use of taxpayer funds. The government is proposing a reduction in spending on education and health care as one of the solutions of the current problem. This decision is causing confusion among the majority of citizens, since the income level of the majority of the population is very low. People are concerned that they simply will not have access to education and health care. I see the solution to this problem in a different way. There is no need to reduce government expenditures for education and health care, but funds must be used more efficiently. Taxpayers not only need to understand the essence of the reform, they should be able to track the changes in effectiveness. Since the allocation of funds to finance education and healthcare is related to the territorial aspect, the use of GIS will improve the quality of analytical calculations and minimize the risk of errors.

The paper is organized as follows. The next section explores the theoretical background of GIS use in public finance reform. The third part describes the methodology of the research. The forth part is divided into two subsections. The first presents the possibilities of GIS for optimizing government financing of health care in Ukraine. The second part presents the results of assessing the feasibility of using GIS to justify reforms in the government funding of education. This is followed by a brief discussion on how GIS could help to increase transparency and accountability in public finance reform. The purpose of the article is to show the possibilities of using geographic information systems (GIS) in the development of reform in the sphere of public finance.

2 Theoretical Background

Different aspects of public finance reforms are represented in recent publications by prominent authors. A careful study of western economists discussing public finance reform shows that there are two main focuses in this field. The first one is how public finance reform can help to fight corruption through increasing transparency and accountability of the government [2–4]. Economists argue that many reforms in public finance failed because of corruption and lack of transparency and accountability. Furthermore, economists have proven that new approaches in public finance reform are needed to increase efficiency. B. Dressel stated, “Citizen participation and a commitment to accountability and transparency have become common in the ‘good governance’ discourse globally, but the extent of the changes the Philippine government has initiated in terms of how it manages and spends its money is remarkable by any standard” [5]. This means that one of the key factors in increasing the efficiency of public finance reform is citizen involvement in the process of their implementation.

The second focus of recent publications is concerning recognizing the signals of poor quality of reforms and their failures and how the quality of reforms in public finance can be improved [1, 6–11]. It is believed that the quality of public finance reform could be increased by considering the causes that prompted reform in the first place. For example, fiscal crises (in Tanzania, the UK, Canada, Ukraine, Asian economies), political changes (post-socialistic countries), changes in public expectations (Canada, the UK, post-socialistic countries), and post-conflict situations (Rwanda, Burundi, Liberia) all contributed to the shape of specific reforms. The specific cause features are very different, but all of them are intrinsically tied to the quality of peoples' lives in those countries. It means that people should be involved in the reform process and they must have access to the all information on how their funds are being used. The words "their funds" must be used, because those funds are collected as taxes, and as such belong to the taxpayers. Even if funds are raised as government borrowings they are also belong to taxpayers because taxpayers will pay those debts in the future.

Many countries have already started the process of making government activity transparent to the people. A significant impact on this process was made by information technology (IT) development. For example, e-procurement is now standard practice, and this helps to prevent corruption and increase transparency of public fund expenditures. IT could be used in different areas of public finance reform. This is why some economists think that GIS could also be useful in this area. GIS is not completely new in economic science. UK and USA universities started promoting GIS-based economic research in the 1990's [12–14].

At the beginning of this century, GIS became a part of econometric methodology [15, 16]. Later GIS became useful in research related to demography problems [17, 18].

New horizons for GIS are represented in the articles of Anselin and Rey (eds.) [19], Goodchild [20, 21], Sianko and Small [22]. They mentioned that "GIS has been helpful in answering questions relate access to social and health services…. GIS can also be used in demography to study issues related to migration and migration related health problems" [22]. GIS is often used in measuring distances and evaluating access of different groups of people to some facilities [23, 24], including public schools [25].

GIS is defined most generally as technology for processing a specific class of information – geographic information. Processing is understood to encompass creation, acquisition, storage, editing, transformation, analysis, visualization, sharing, and any other functions amenable to execution in a digital domain [14]. GIS is a very good tool not only for visualization of the Earth's surface information with specific properties, but also as a proper tool for measuring distances considering the quality of roads and characteristics of a particular region (flat fields or mountains). Also, GIS could help to prevent fraud in public expenditures in cases of national disasters, for example floods. The information about the number of houses and their characteristics was in the GIS before a flood, so it easy to calculate how much is needed to compensate people who lost their homes.

Furthermore, GIS can help to create data visualization. GIS data could be easily shared with other researchers and with people who are interested in the results of research. It could help to increase transparency in the public finance reform process, because every citizen could have access to the data, which are presented in an understandable way. GIS helps to create different maps which are good tools to show the interaction between different variables. GIS can help to improve analysis, and this will help to avoid mistakes in developing ideas for reform.

3 Methodology

As an example of how GIS can be used in the public finance reform process, health care and education reform in Ukraine has been chosen. The methodology of research was as follows:

1. Create a questionnaire on social networking sites to ask people to express their opinions. Healthcare reform questionnaires were posted on social networks and sent to e-mail addresses of potential respondents. Taking into account that the reform of school education will more affect rural areas, teachers from rural schools were involved in the survey. They have direct contact with parents of students studying in rural schools, this is why they were involved in interviewing work.
2. Analyze peoples' opinion and mapping. Summary tables were created at this stage.

Mapping was done using special software. It allowed the creation of maps to calculate the best locations of schools and healthcare facilities taking into account public opinion. Mapping allows differences in population density, age distribution, disease prevalence, poverty and the ability to access health care facilities or schools to be considered.

The GIS is composed of a regional geographical data base and a collection of spatial models. The regional geographical data base contains a number of thematic map layers including administrative boundaries, roads, bridges; remote sensed data; digital terrain models; and a comprehensive collection of statistical records. The process of mapping consists of several stages. First of all, specific healthcare and education variables were defined. For health care facilities those variables are population, age distribution, levels of income, access to public transport, and quality of roads. For schools, those variables are the number of children, their ages, access to public transport, and quality of roads. Secondly, clustering was done to define target groups and possible risk. Thirdly, the mapping was applied. To build a map, the "Open street map" tool was used [26]. This means that only real data was used for analysis and cannot be falsified. The whole map is divided into several sectors. For further analysis buffering must be used. Buffering allows an understanding of how sensitive infrastructure quality and population density impact the study. Various types of buffering could be used (Fig. 1).

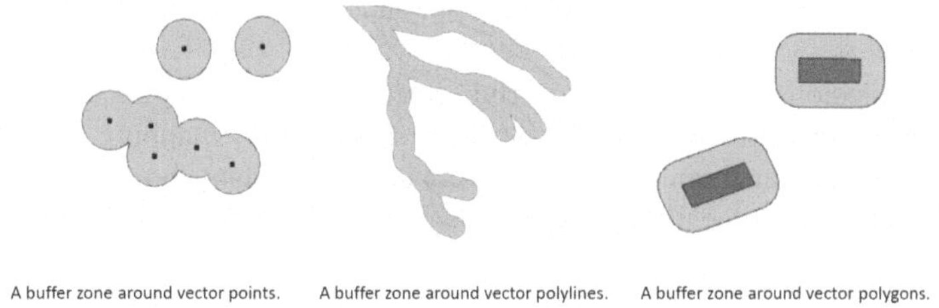

Fig. 1. Buffering in GIS

Multi ring buffering was used here (Fig. 2). Multi ring buffering helps eliminate facilities that have similar accessibility factors and are located close to each other, as shown on the right of Fig. 2. Also this is a good tool for visualizing data, because people are able to see where they live and where a hospital could be located and how fast they could reach it. If such data is kept open to the public, it cannot be manipulated.

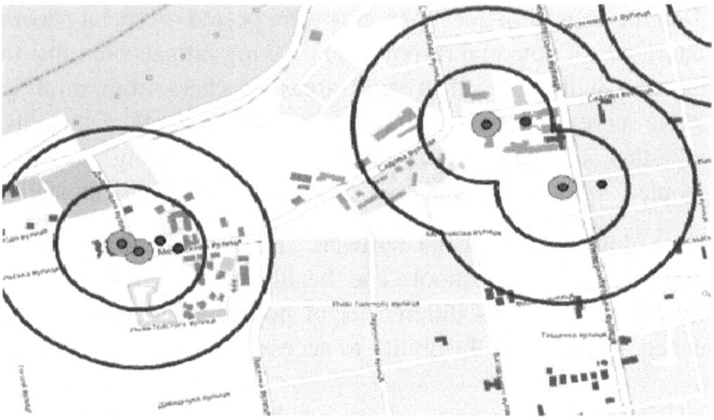

Fig. 2. Multi ring buffering

Also, every interested citizen can view the map and understand if the governmental decision about facility location (health care or educational) was reasonable or not from their own point of view. If people disagree, they can easily voice their opinion (for example, for reasons of lack of public transportation or bad quality of roads requiring more time to get to a certain facility). Overlay analysis and summary statistics were then used. These tools help calculate optimal distances between hospitals and residences. Overlay and vector analysis also include characteristics of the earth's surface and quality of roads in its calculations.

3. Analyze government proposals for health care and education reform and mapping. Mapping was done considering government decisions on the locations of schools and health care facilities. GIS can be used to identify observations by both characteristics and location and then perform simple statistical operations. Differences between mapping based on public opinion and government decisions were analyzed.
4. Draw conclusions. The most powerful aspect of GIS is arguably its ability to quickly analyze spatial data. This analysis is supported by mapping, and helps with data visualization and can be seen as a way of making data about possible impacts of health care and education reforms transparent and understandable for people.

4 Efficiency Estimation Procedure

4.1 GIS for Health Care Reform

The first step of the research was the creation of a questionnaire for the social network survey and choose the target group. It should be noted that in Ukraine the age of active users of professional social networks is between twenty and forty-five years old. The target group consisted of five thousand people, 60% women and 40% men, with 32% having a background in economic science, and 68% with a background in medical care, education, farm production, social services, etc. Because of Ukraine's size, a representative scope can be made using target groups chosen from four oblasts (Kyiv, Zhytomyr, Cherkasy and Chernihiv). The results of the survey are presented in Table 1.

Table 1. Results of survey on public finance reform, %

Question	Strongly disagree	Disagree	Agree	Strongly agree	Cannot decide
Do you think that information about public finance in Ukraine is not transparent enough?	2	3	77	14	4
Do you think that information about public finance reforms is not fair?	1	3	56	26	14
Do you think that information about public finance reforms is not easy to understand?	3	4	63	29	1
Do you think that citizens should be more involved in public finance reforms?	3	2	53	15	28
Do you think that citizens can prevent bad public finance management?	12	29	24	11	24
Do you think that government can manipulate information about public finance?	4	3	57	21	15
Do you think that citizens should be able to track each UAH paid as a tax?	1	3	83	11	2

As can be seen from Table 1, many people are interested in tracking public finance reform, but at the same time roughly 20% people cannot decide. This group of people cannot make a decision about whether they are interested or not in how government uses their money. It proves the existence of a group of people who are "passive", because they do not understand information about public finance or they do not trust government, or because of other reasons. This is a large percentage for a society that wants to follow democratic ideas. It should be noticed that many people think that information about public finance in Ukraine is not transparent enough and information about reforms in public finance is not fair or not understandable.

The second poll was about the establishment of hospital districts in Ukraine (Table 2).

Table 2. Results of survey on health care reform, %

Question	Strongly disagree	Disagree	Agree	Strongly agree	Cannot decide
Do you think that the establishment of hospital districts will improve health care in Ukraine?	46	36	10	3	5
What kind of risks do you see in establishing of hospital districts					
Bad access to the health care facilities	2	2	71	19	6
Absence of medical care in villages	3	3	67	20	7
Lack of doctors, because they will leave Ukraine	4	6	45	44	1

According to the project on health care reform in Ukraine, a district hospital center must be located in a settlement with 40,000 inhabitants or more and serve a region with at least 200,000 inhabitants. Taking into account the amount of population served, five hospitals should be located in Chernihiv oblast, but according to the requirement that the hospital must be located in a town with more than 40,000 inhabitants, this means only one hospital will be funded by the government, and any others will be deprived of funding. In contrast, in the Zhytomyr region, six hospitals should be located there, but according to the minimum population requirement, only three hospitals will be funded by the government, again meaning the remaining hospitals will lose funding. The calculations that were done for Kyiv oblast do not include the population of Kyiv. At least nine hospitals are needed there, but according to the minimum population requirement, only five hospitals will be funded by the government, leaving the remainder unfunded. For the Cherkasy oblast those indicators are six and three respectively (Table 3).

Table 3. Main characteristics for hospital districts in selected oblasts

Indicators	Zhytomyr oblast	Kyiv oblast	Chernihiv oblast	Cherkasy oblast
Territory, square kilometers	29832	28131	31865	20900
Population, mln people	1.240	1.735	1.056	1.231
Number of towns with a population of more than 40,000 (equal to the number of hospital districts in the oblast)	4	5	2	3

It appears that Ukrainian citizens may have reasons to be unhappy with the coming healthcare reform. The district councils involved will decide for themselves how much they need hospitals. The idea of reform is based on the amount of people and does not take into account how people are able to reach the hospitals. Another problem is related to a district council deciding how many hospitals are needed. Because of a lack of funding and differences in the economic development of different districts, this decision becomes almost impossible. The idea of creating hospital districts is not wrong, but its implementation must be based not on the number of people living in a town, but on equal access to the hospital for every citizen. This becomes critical because Ukrainian infrastructure is in very poor condition and public transportation is not universally available.

According to general healthcare requirements, in case of emergency, travel time to a hospital should be within 15–20 min. This is not possible to provide if only one or two hospitals are located in a territory of 30,000 km^2. This is a case in which GIS could be helpful. One of the most common tools of GIS is spatial analysis. Spatial analysis comprises a set of techniques and tools designed to analyze data in a spatial context. A GIS database captures not only links between properties at the same place, but also such spatial concepts as proximity, containment, overlap, adjacency, and connectedness. Visualization in spatial context (commonly in the form of a map) is an obvious and powerful way of detecting pattern, anomaly, and even causation Goodchild (2011). It means that by using GIS, the planning of new hospital districts could be improved by better analysis and data visualization. Proper analysis and better visualization will help prove that health care reform is reasonable. This will help decrease tension between government and society and help to build a relationship based on trust, responsibility and accountability.

Today, several software products are available. The best one is ArcGIS 10 [27]. ArcGIS 10 provides spatial querying, attribute querying, tabular visualization, statistical analysis, advanced mapping, map publishing in pdf format and map printing, etc.

The idea of using GIS in establishing hospital districts is as follows:

1. An analysis of the population in a district and population density is made. At this stage, the rate of population growth, the proportion of children and pensioners, is analyzed. It is important to determine how long the current demand for health care services will remain. It is important to determine the potential need for public health care services for low income people.
2. An analysis of infrastructure, its quality and availability for the population of certain district is made. It is necessary at this stage of the analysis to calculate the average distance from each citizen's home to the potential location of the hospital. Then the average length of time that is necessary for the emergency car to reach the patient must be calculated. The next step is to calculate the average length of time needed to get to the hospital by public transport. It should be noted that GIS allows relatively accurate calculation of time intervals, since it is possible to take into account the terrain and the quality of roads. To prevent data manipulation at the information input stage (for example, the quality of roads may be overestimated), the original data should be publicly available so that any citizen can get acquainted with it. The openness of information at this stage will help prevent the appearance of unreasonable statements about the poor quality of reforms in Ukraine.
3. Visualization could be done through mapping (Fig. 3).

Fig. 3. Results of an analysis of infrastructure quality

The use of spatial analysis allows not only to combine the processing of geographic and economic data, but also to visually demonstrate the result of calculations. Visual demonstration of reform progress in the public domain helps better explain the expected result of the reforms.

In order to determine the optimal number of hospital districts, the analyzed territories were divided into polygons. The polygons were built taking into account the size of an area, and population numbers and density. Then, using a buffer analysis, the optimal locations for hospitals and the creation of hospital districts around them were determined.

This analysis was completed for four oblasts in Ukraine (Table 4).

Table 4. Results of GIS analysis for Chernihiv, Zhytomyr, Kyiv and Cherkasy oblasts

Indicators	Chernihiv oblast	Zhytomyr oblast	Kyiv oblast	Cherkasy oblast
Area square kilometers	31865	29832	28131	20900
Population, mln people	1.056	1.240	1.735	1.231
Number of hospital districts	5	6	10	7

As can be seen in Table 4, the results of GIS analysis show the necessity of establishing a larger number of hospital districts than would be possible according to government requirements. This is because the government project did not include the poor condition of roads in its analysis.

The idea of reforming the health care system is not inappropriate. However, if the government manipulates input data on infrastructure quality, and does not disclose information on the development of the reform and its consequences, it can discredit the actions of the government, and the reform itself.

4.2 GIS for Formal Education Reform

The reform of the educational sector is extremely important. The training of the younger generation determines the future of the state. Obviously, Ukrainian education does not correspond today to the modern needs of the individual nor society, the needs of the economy nor to world standards.

That is why the reform envisages a systemic transformation of the education sector, the main objective of which is new high-quality education at all levels: from elementary school to higher education institutions. The main task of education should be the formation of conscious, socially active citizens, capable of ensuring economic growth and cultural development of the country.

Education reform should cover all areas of education, but formal school education reform is the most problematic. The main problem is the consolidation of schools, and hence the risk of worsening access to education for children who live in rural areas.

Technically, the consolidation of schools is voluntary and aims to improve the quality of education. The closure of small schools is necessary primarily due to the poor quality of education in small schools and the optimization of the educational network. According to official statistics, as of 2016, more than half of the schools in Ukraine were located in rural areas, and 613 schools had an enrollment of up to twenty-five students, which meant that one teacher at a school taught several courses. Also, those schools have a deficiency of material needs.

According to a study by CEDOS, 15% of rural teachers teach more than three courses, 8% in small cities, and 5% in medium and large cities. Almost 30% of rural schools do not have any computer connected to the Internet, whereas in large cities almost all computers are connected to the Internet.

Politicians claim that the reform offers the united territorial communities (UTCs) the establishment of support schools and their branches. In accordance with decentralization reform, schools are transferred to a UTC, which entails funding from the community budget. That is why the community now has the right to establish schools and determine which school or branch will receive financial support from the community. Accordingly, the decision on the liquidation or reorganization of a school is made by the community on its own.

However, a rapid increase in the number of UTCs without a proportional increase in the transfer of funding from the central government for infrastructure development reduces the motivation of communities to unite and develop education.

In 2016, the transfer of funding from the central government for community infrastructure was one billion UAH and it was distributed among the budgets of 159 UTCs in proportion to community size and population. In 2017, the amount of the transfer of funding from the central government was set at 1.5 billion UAH. It should be noted that financing from the central budget was mainly carried out at the expense of a special fund, and this source of financing is not stable. In addition, the indicated funds were distributed among 366 UTCs, it means that each community received less money in 2017 than in 2016 [28].

Thus, the greatest risk of resistance to the implementation of formal school education reform may arise among villagers and small towns. As was mentioned in the methodology section, the first stage of the research is creating a questionnaire on social networking sites to ask people to express their opinions. Taking into account that the reform of formal school education will more affect rural areas and small towns, teachers from rural schools and from small towns were involved in the survey. They asked parents to answer questions from the questionnaire. It should be noted, that the response from the parents was not very high, only 52% people filled out the questionnaire. Summary results of the survey are presented in Table 5. As can be seen from the table, parents are seriously concerned about the consequences of school reform. They have good reasons for this.

Table 5. Results of the survey about formal school education reform, %

Question	Strongly disagree	Disagree	Agree	Strongly agree	Cannot decide
The merger of schools will destroy the Ukrainian villages	2	5	48	33	12
Many teachers will lose their jobs as a result of school education reform	3	6	33	33	25
Students from villages will not have good access to schools	1	2	45	43	9
Schoolchildren will not be able to get a quality education, because parents will not be able afford textbooks	3	5	46	38	8
Schoolchildren from rural areas and small towns will not be able to get quality education in schools, so they will not be able to enter universities	2	4	48	39	7

Nowadays the management system of school education is constructed in a way that any school in a village or small town receives funding from the local government (raion). The advantages of the present system include the presence of a single monitoring center, the possibility of coordinating the work of school in a certain region, the possibility of staff rotation between schools or redistribution of government financing in cases of unpredicted circumstances requiring urgent financial support (urgent repair of equipment or liquidation of emergency).

The disadvantages of the planned management system are a lack of correlation between the quality of education services and funding. Also, there are cases when the schools that have equal sizes can receive different funding. According to the government's plan, these problems should be solved after the completion of the decentralization reform, when full school management will be transferred to the community level. Under such conditions, the principle of subsidiarity, which means the provision of public services by the government body that is closest to the citizen, must be fully implemented. In our case, it means that the general management and financing of a particular school should be carried out not from the district center (raion), but directly by the executive bodies of the territorial community on which territory a school is located. This is the universally accepted practice of civilized democratic societies. This idea may not work now in Ukraine. First, Ukraine is a democratic state only nominally. In fact, public administration is centralized, and citizens are in a passive position and do not take responsibility for themselves. Secondly, even after the completion of decentralization reform, the problem of financing school education in depressed regions will not be solved.

Long-term work is needed to solve these problems. First, the availability of information that every citizen can understand regarding reform effects must be improved. For reforms that are related to territorial interests, GIS is a very important tool. GIS helps to prevent unsound resistance from citizens, and also prevents information manipulation by politicians. Secondly, the merits of reform cannot only be copies of the successes of developed countries. Each country has its own historical and cultural peculiarities of development. These must necessarily be taken into account while developing a reform strategy.

In order to implement the reform of formal school education in Ukraine it is necessary to take into account the availability of school education for children. This means the physical accessibility of the school for children of all ages. It must be considered that long travelling by bus to school is harder for small children, and it is harder for them to get up early. Therefore, the possibility of establishing elementary schools for children from six to ten years of age should be considered, and enlargement should be done among secondary schools.

The optimal number of elementary schools for a given locality can be determined from the vector analysis of GIS (Fig. 4).

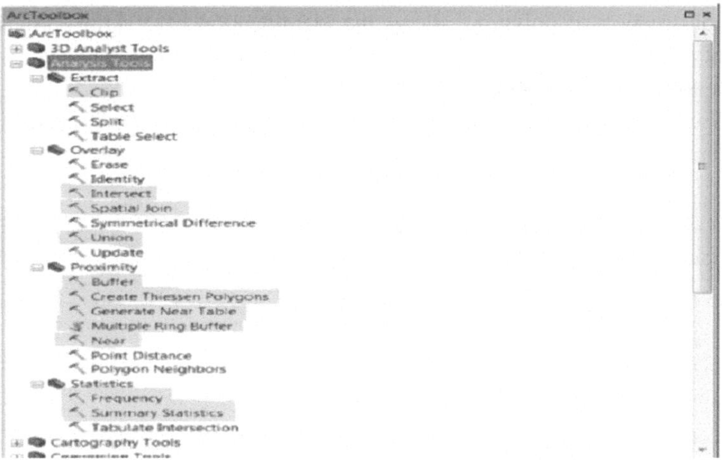

Fig. 4. Tools that were used for vector analysis in GIS

The buffering distance is one of the common tools to determine the best location for a facility. The buffer distance can vary according to numeric values provided in the vector layer attribute table. In our case, the buffer distance depends on the age of school children, quality of roads, and the availability of busing. It should be noted that school bus service is not available in every region of Ukraine. This is why two calculations were done. One is for where busing is available, the other where unavailable.

For each region buffers were merged into a single geometric object to avoid overlapping areas. Then multiple buffers at specified distances were created. The next step is calculating distance and additional proximity information between the input features and the closest feature in another layer. To calculate a distance, the "GEODESIC" method was used, because it takes into account the curvature of the spheroid and correctly deals with data near the dateline. Then Thiessen Polygons were created (Fig. 5).

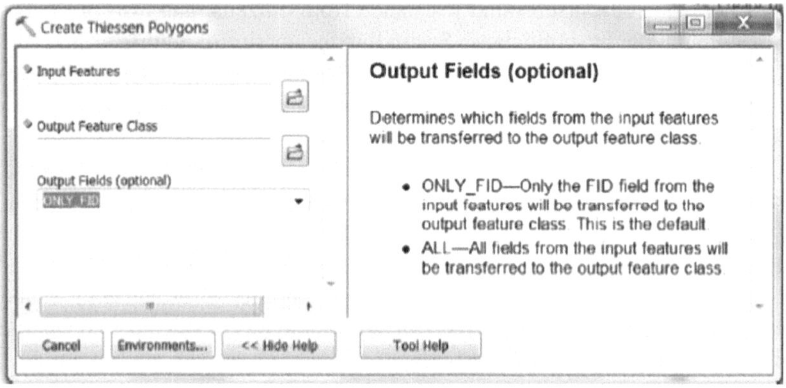

Fig. 5. Creating Thiessen Polygons

The construction of polygons allows the identification of several options for the location of primary schools, taking into account student accessibility. Then a summary of statistic tools were used (Fig. 6).

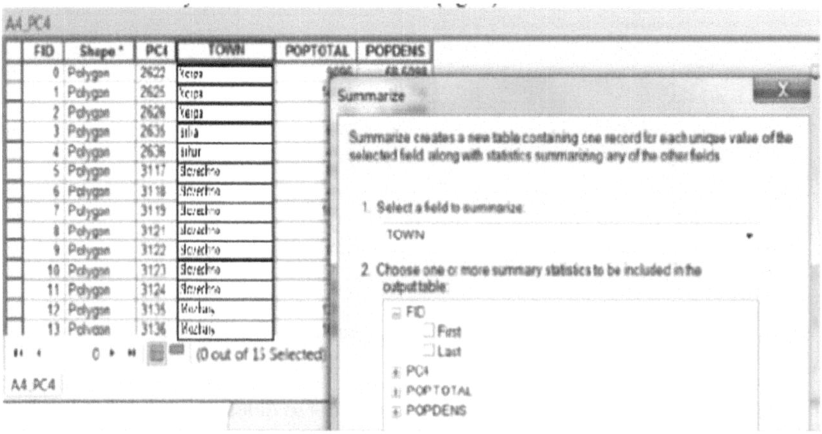

Fig. 6. Summary statistic

Statistical GIS tools allow the optimal distance from home to school to be calculated, as well as the travel time that is needed. Analysis of frequency distribution (Fig. 7) shows that for the number of school children from 1 to 100 at least one primary school is needed, for the number from 101 to 1000 at least two primary schools are needed, for the number of children from 1001 to 4000 six schools are needed, and for number of children from 4001 to 12000 at least twelve schools are needed.

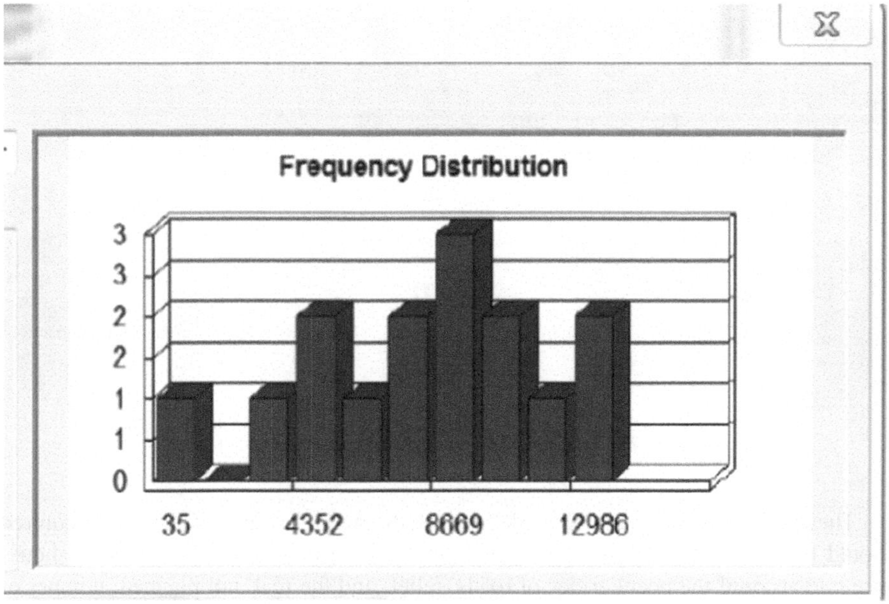

Fig. 7. Frequency distribution

A calculation was done for Ovruch raion of Zhytomyr oblast. This raion was chosen as an example because it has 144 villages. Raion infrastructure and busing levels can be described as average. The actual number of schools and their location was compared with a GIS calculation. According to the GIS calculations, there was a 17% discrepancy in the number of needed schools compared to government reform figures. Corrections were made considering the age distribution of school children. As a result of school reform, thirty-six schools will be located in Ovruch raion, with one being primary. According to GIS calculations, forty-two schools are actually needed for Ovruch raion, fourteen being primary schools. The location of twenty-four schools is shown in Fig. 8. It can be seen that three schools are located a distance from the larger group of schools. This is explained by the quality of infrastructure; travel is more difficult to the larger grouping.

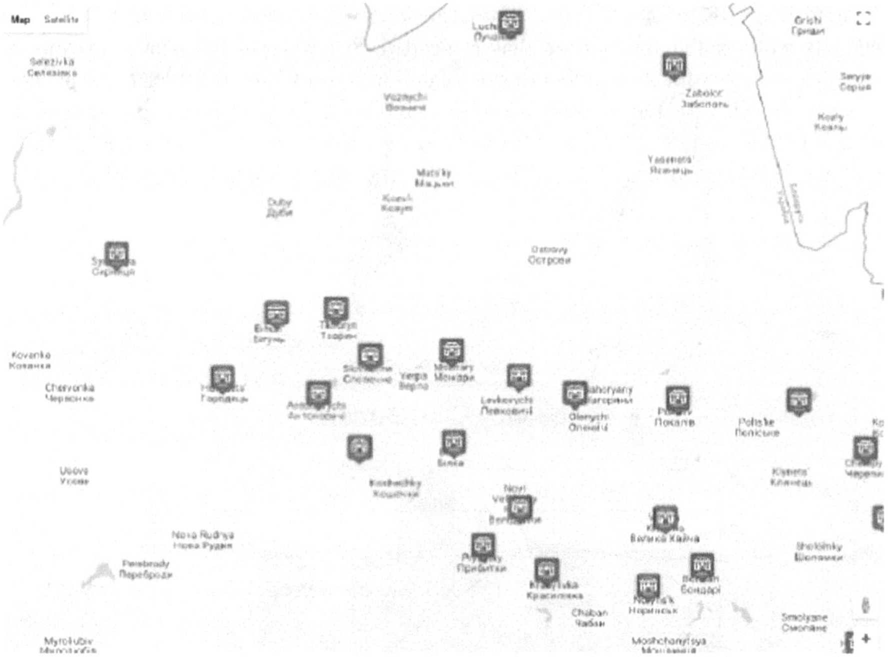

Fig. 8. School location in Ovruch rayon (extract)

These calculations should be made for each district of Ukraine. The results acquired should be used to justify the reform of formal school education. Calculations should take into account the peculiarities of roads, relief, and the real, not planned, number of school buses. It should be noted, that school buses are not available for each region. According to the report of Accounting Chamber of Ukraine, only 791 buses were bought from the planned number of 1600 in 2008–2011. In 2012–2016, 1200 school buses were planned to be bought, but only 326 were bought because of economic and political problems in Ukraine. When the quality of roads is improved, and each region is provided with a sufficient number of school buses, the number of primary schools may be reduced. The decision to reduce the number of primary schools should also be taken based on GIS calculations. These results should be available to local citizens for discussion. Mapping visualization would further help avoid misunderstandings and manipulations in the process of decision making.

5 Conclusions

GIS can be a very useful tool in public finance reform development. GIS could provide a wide range of analysis and better support for reform concepts. It also makes possible a combination of statistical, demographical and geographical analyses. Moreover, GIS provides very good visualization that helps ordinary citizens understand how reforms would affect them. GIS can increase government transparency and accountability, because it is quite difficult to be manipulated by open source map data.

Reforms in public finance are often linked to the distribution of funding or the tax burden between districts. The complexity of reform justification is caused by the need to work simultaneously with spatial and statistical data. The article provides examples of how GIS can be used to justify reforms in public finance, namely, optimizing funding for formal school education and health care.

The basic idea of health care reform is to improve the quality of medical services and reduce public funding for the maintenance of health facilities. Medical reform involves the creation of hospital districts, the location of which is tied to population size in a certain area, and in the city in which the central hospital will be located. The disadvantage of the reform is that the need to ensure equal access to health care facilities for all citizens was not taken into account. Research conducted with GIS allowed calculating the optimal number of hospital districts for four regions (oblasts) of Ukraine. The GIS-estimated amount is larger than that proposed by the government of Ukraine. This means it is not possible to reduce public expenditures in health care in a short-term. Furthermore, if the government follows trough on its plans in this area, then that result maybe increased resistance from the citizens. Moreover, equal access to health care facilities will be reduced.

The use of open street maps for spatial analysis reduces the risk of manipulation by input data. The presentation of the results of the analysis in the form of maps showing hospital districts reached within a reasonable time from any home in the district would allow reasonable citizen feedback potentially reducing further resistance to the government reforms.

The main goal of education reform is to merge schools, thereby reducing cost. This idea is negatively perceived in villages and small towns. According to the results of the questionnaire, many people have fears that accessibility to education will become more difficult for children living in remote villages. GIS calculations showed that the people's fears are reasonable. The government's calculations did not take into account the time that children must spend on the road to get to school. Also, it was assumed in the government's calculations that all regions were provided with school buses, which is not the case at present.

The results of the calculations presented in the article were made in GIS for one district, taking into account the distribution of children by age, the availability of school buses and the quality of roads. The analysis showed that more schools are needed, some of which are elementary schools. A subsequent decrease in the number of schools, including their merger, should be carried out after improving the quality of roads and providing better access to schools. As in the case of health care reform, the use of GIS makes finding the best solutions between reducing funding for formal school education and ensuring that it is accessible to all children possible. Map visualization allows not only a clearer illustration of reform results, but also more people with the ability to comment intelligently on this problem, since such representation of information is understandable for the majority of population.

GIS is not a perfect tool and several challenges should be also considered. Firstly, the software for GIS must be revised quite often. Secondly, GIS software continues to change and improve over time and there are now several GIS applications that range from being relatively free (having limited tools) to being very expensive (for example, ArcGIS 10.). Thirdly, to follow the idea of increasing transparency, all GIS software should have no conflicts between each other. This means that data from one GIS software can be easily exported into different GIS software.

References

1. Fedosov, V., Paientko, T.: Ukrainian government bureaucracy: benefits and costs for the society. Bus. Manag. Stud. **3**(2), 8–19 (2017)
2. Allen, R., Schiavo-Campo, S., Garrity, T.: Assessing and Reforming Public Financial Management: A New Approach. World Bank, Washington (2004)
3. Gomez, P., Friedman, J., Shapiro, I. Opening Budgets to Public Understanding and Debate. IBP, Washington. Mode of access. http://www.transparency.cz/pdf/tsr_dstudie_02.pdf
4. Hedger, E., Kizilbash, A.Z.: Reforming Public Financial Management when the Politics aren't right: A proposal. ODI, London. Mode of access (2007). http://www.odi.org.uk/publications/opinions/89_PFM_politics_Nov07.pdf
5. Fjeldstad, O.: Anti-corruption reforms: challenges, effects and limits of world bank support. Background paper to public sector reform: what works and why? In: Fjeldstad, O.-H., Isaksen, J. (eds.) IEG Evaluation of World Bank Support. World Bank (2008)
6. Dressel, B.: Targeting the public purse: advocacy coalitions and public finance in the Philippines. Adm. Soc. **44**(6), 65S–84S (2012)
7. Dorotinsky, W., Pradhan, S.: Exploring corruption in public financial management. In: Campos, J.E., Pradhan, S. (eds.) The Many Faces of Corruption. World Bank, Washington (2007)
8. De Renzio, P., Dorotinsky, W.: Tracking Progress in the Quality of PFM Systems in HIPCs. PEFA Secretariat, Washington. Mode of access. http://www.pefa.org/report_file/HIPCPEFA%0Tracking%20Progress%20Paper%20FINAL_1207863932.pdf
9. Andrews, M.: PFM reform: signal failure. Mode of access. http://opinion.publicfinanceinternational.org/2013/03/pfmreform-signal-failure/
10. Fjeldstad, O.: Taxation and development: a review of donor support to strengthen tax systems in developing countries. WIDER Working Paper No. 2013/010 (2013)
11. Paientko, T.: Behavioral aspects of financial anomalies in Ukraine. In: CEUR Workshop Proceedings, vol. 1356, pp. 214–224 (2015). (Indexed by: Sci Verse Scopus, DBLP, Google Scholar). CEUR-WS.org/Vol-1356/ICTERI-2015-CEUR-WS-Volume.pdf
12. Langran, G.: Time in Geographic Information Systems. Taylor and Francis, London (1992)
13. Laudan, L.: Beyond Positivism and Relativism: Theory, Method, and Evidence. Westview Press, Boulder (1996)
14. Longley, P.A., Goodchild, M.F., Maguire, D.J., Rhind, D.W.: Geographical Information Systems: Principles, Techniques, Management and Applications. Wiley, New York (1999)
15. Bialynicki-Birula, I.: Modeling Reality: How Computers Mirror Life. Oxford University Press, New York (2004)
16. Anselin, L., Florax, R.J., Rey, S.J. (eds.): Advances in Spatial Econometrics: Methodology, Tools and Applications. Springer, Berlin (2004). https://doi.org/10.1007/978-3-662-05617-2
17. Castro, M.C.: Spatial demography: an opportunity to improve policy making at diverse decision levels. Popul. Res. Policy Rev. **26**, 477–509 (2007). Special Issue on Spatial Demography
18. Voss, P.R.: Demography as a spatial social science. Popul. Res. Policy Rev. **26**, 457–476 (2007). Special Issue on Spatial Demography

19. Anselin, L., Rey, S.J. (eds.): Perspectives on Spatial Data Analysis. Springer, Berlin (2010). https://doi.org/10.1007/978-3-642-01976-0
20. Goodchild, M.: New Horizons for the Social Sciences: Geographic Information Systems. Mode of access. http://www.geog.ucsb.edu/~good/papers/334.pdf
21. Goodchild, M.F.: A GIScience perspective on the uncertainty of context. Ann. Am. Assoc. Geogr. (2018)
22. Sianko, N., Small, M.: The Future of GIS in Social Sciences. Elsevier, New York City (2017)
23. Church, R., Murray, A.: Business Site Selection, Location Analysis and GIS. Wiley, Hoboken (2009)
24. Combes, P.P., Lafourcade, M.: Transport costs decline and regional inequalities: evidence from France. CEPR discussion paper 2894 (2001)
25. Rothstein, J.: Does competition among public schools benefit students and taxpayers? Comment. Am. Econ. Rev. **97**, 2026–2037 (2007)
26. Open street map [Electronic source]. https://www.openstreetmap.org/#map=6/48.537/31.168
27. Manual for working with ArcGIS 10 [Electronic source]. ftp://ftp.puce.edu.ec/Facultades/CienciasExactas/Geoinformatica/MANUALES%20PARA%20LA%20OPTATIVA/MANUAL%20ARCGIS%2010/Manual%20ArcGis%2010.pdf
28. Ministry of Finance of Ukraine [Electronic source]. https://www.minfin.gov.ua

ANN-Based Electricity Price Forecasting Under Special Consideration of Time Series Properties

Jan-Hendrik Meier(✉), Stephan Schneider, Iwana Schmidt, Philip Schüller, Thies Schönfeldt, and Bastian Wanke

Kiel University of Applied Sciences, Sokratesplatz 2, 24149 Kiel, Germany
{jan-hendrik.meier, stephan.schneider}@fh-kiel.de

Abstract. If one examines the spot price series of electrical power over the course of time, it is striking that the electricity price across the day takes a course that is determined by power consumption following a day and night rhythm. This daily course changes in its height and temporal extent in both, the course of the week, as well as with the course of the year. This study deals methodologically with non-linear correlative and autocorrelative time series properties of the electricity spot price. We contribute the usage of non-fully connectionist networks in relation to fully connectionist networks to decompose non-linear correlative time series properties. Additionally, we contribute the usage of long short-term-memory network (LSTM) to discover and to deal with autocorrelation effects.

Keywords: Electricity prices · Artificial neural network · LSTM · ARIMAX

1 Introduction

Despite all criticism of this approach, the random walk process has established itself for the modeling of stock prices. Pricing on electricity markets deviates significantly from the pricing on stock markets, as the underlying Markov property cannot be assumed for electricity markets as well. Produced electricity cannot be stored without significant losses and, accordingly, temporal arbitrage turns out to be highly inefficient. If one examines the spot price series of electrical power over the course of time, it is striking that the electricity price across the day takes a course that is determined by power consumption following a day and night rhythm. This daily course changes in its height and temporal extent in both, the course of the week, as well as with the course of the year. Accordingly, it can be concluded that the univariate time series shows non-linear correlative effects between daily, weekly, and yearly seasonal patterns as well as autocorrelative effects even without taking other explanatory variables into account.

The present study deals methodologically with non-linear correlative and autocorrelative time series properties of the electricity spot price. Correlation effects are adequately represented in classical fully connectionist networks but they cannot be meaningfully analyzed due to the high complexity of these networks. Therefore, we contribute the usage of non-fully connectionist networks in relation to fully

V. Ermolayev et al. (Eds.): ICTERI 2018, CCIS 1007, pp. 262–275, 2019.
https://doi.org/10.1007/978-3-030-13929-2_13

connectionist networks to decompose non-linear correlative time series properties. Hence, we use (i) different ANN architectures with non-fully and fully connectionist networks to discover and to deal with correlation effects on exogenous side/input layer, (ii) using a long short-term-memory network (LSTM) to discover and to deal with autocorrelation effects, and (iii) an ARIMAX model with daily, weekly, and yearly seasonal patterns reflected as binary coded variables as a benchmark for the aforementioned models.

The paper is organized as follows: In section two, the current state of the literature is presented, and the research gap is identified. In section three, sample and methodology are introduced. In section four, the results are presented and discussed. The study closes with a conclusion.

2 Literature Review and Research Gap

The number of electricity price forecasting articles has increased significantly in recent years. A particularly good overview can be obtained by Weron (2014). The author could identify 30 publications with a focus on ARIMA and its extensions. We could not identify further more recent articles in this special field of ARIMA-modelling of electricity prices. More recent electricity price forecasting literature is focused mainly on probabilistic forecasting and artificial intelligence. With regard to ANN, Weron could identify 56 publications. Subsequently, two further articles were published on electricity price forecasting using ANN that were not included in Weron's review (Dudek 2016; Marcjasz et al. 2018).

Comparing ARIMA(X) models of the Spanish and the Californian market with and without additional explanatory variables, Contreras et al. (2003) recognize that additional explanatory variables, such as hydropower, are only required in months of a high correlation between the explanatory variable and the price, while in months of low correlations these variables do not show significant predictive power. The authors were able to show average daily mean errors between 5% and 10% with and without explanatory variables.

When forecasting with ARIMA, Conejo et al. (2005b) argue that it could be necessary to use a different notation of the model for nearly every week. Accordingly, ARIMA-models turn out to be very unstable in their predictive power over time. Especially in spring and summer where the volatility was very high the ARIMA forecast provided poor results. The authors also introduce several other techniques, e.g. an ANN with a multilayer perceptron and one hidden layer. The ARIMA model outperforms the ANN in every period except for the September. The mean week errors with ARIMA are between 6% and 27% whereas the ANN shows errors between 8% and 32%.

Garcia et al. (2005) claimed that ARIMA-GARCH models show a better accuracy than seasonal ARIMA models. The authors present mean weekly errors of around 10% for relatively calm weeks. Misiorek et al. (2006) compare some linear and non-linear time series models. In contrast to the aforementioned authors, the simple ARX model - the exogenous variable is the day-ahead load forecast - shows a better result than a model with an additional GARCH component.

Conejo et al. (2005a) contributed a specified ARIMA model including wavelet transformation which was more accurate than the simple one. The wavelet transformation is applied to decompose the time series before predicting the electricity prices with ARIMA. This model outperforms the benchmark with a weekly error of 5% in winter and spring and 11% in summer and fall.

Applying a seasonal ARMA(X) process with three different explanatory variables of the temperature, Knittel and Roberts (2005) identified an inverse leverage effect with positive price reactions increasing the volatility more than negative ones. The authors further show that a higher order autocorrelation in the models is important to improve the results. The authors were able to show root mean squared errors for the out-of-sample week between 25.5 and 49.4 in the pre-crisis period and between 66.6 and 88.6 during the crises period. It is mentioned in the article that the data has a high frequency of large price deviations, which leads to these high forecast errors.

Zareipour et al. (2006) built an ARMAX and an ARX model with an average error in the 24-hour-ahead forecast of 8.1 and of 8.4 respectively, which is slightly better than the basic ARIMA model with an average of 8.8. With these models, it could be shown that market information in low-demand periods is not as useful as during high-demand periods. In general, the results have confirmed the contribution of the authors that market data improves the forecast results. Nevertheless, none of the models could forecast the extreme prices which increasingly occur in times of high-demand periods adequately.

Zhou et al. (2006) suggested that including error correction will lead to a more accurate result in forecasting with ARIMA. Therefore, they developed an ARIMA approach which is extended by an error series. This novel method turned out to show quite good forecasting accuracies with an average error of 2% and lower despite of periods with a high price volatility.

Koopman et al. (2007) were using an ARFIMA model, which is an ARIMA model with seasonal periodic regressions, and combined it with a GARCH analysis. The authors pointed out the importance of day-of-the-week periodicity in the autocovariance function when forecasting electricity prices. Beneath the implementation of the day of the week, binaries were included for the holiday effect to consider demand variations.

With the increase in available computational power in recent years, ANN became more and more popular in forecasting and forecasting research. Both, classical multilayer perceptron (MLP) and recurrent networks (Hopfield 1982; Haykin 2009), especially long short-term memory (LSTM) (Hochreiter and Schmidhuber 1997) networks, are used for forecasting purposes of time series data. Typically, all ANN architectures are composed of an input layer, a hidden layer with differing number of units, and an output layer. In fully connectionist networks, typically, lags and partially residuals are passed into the propagation function (Zhang et al. 1998; Adebiyi et al. 2014). Each node of a layer is usually fully connected to the units of the subsequent layer. MLP as well as LSTM networks are fed with differentiated time series data. The reason is the underlying characteristics of a time series itself. If time is the explanatory factor for the values of the endogenous variable, in our case the electricity price, the time series must be made stationary by differentiation to avoid spurious correlations.

Recurrent ANN have the possibility to incorporate the output of latter layer units again into earlier layer units, which is not possible in MLP networks. Commonly, the units of all hidden layers of recurrent networks are in a chain-like informational loop. A hidden unit can use its output as input (direct feedback), or it is connected to a hidden unit of the preceding layer (indirect feedback), or it is connected to a unit of the same layer (lateral feedback), or it is connected to all other hidden units (fully recurrent). The recurrent type of LSTM is typically direct feedback (Malhotra et al. 2015). The LSTM network, with regard to its inherent properties of "[…] maintaining its state over time in a memory cell […]" (Greff et al. 2017), is predestined for usage in time series analysis. In opposite to other recurrent network types a LSTM network solves the vanishing gradient problem (Hochreiter and Schmidhuber 1997).

Fully connectionist MLP are the most used type of ANN for electricity price forecasting (Weron 2014; Dudek 2016). They differ in usage of different explanatory variables, e.g. power consumption, weather, wind conditions, in addition to lag variables. Furthermore, the results of an MLP serve as benchmarks in comparison with the results of other forecasting models like ARIMA. Additionally, MLP is often used as the nonlinear part within a hybrid model, e.g. in combination with ARIMA. A further type of ANN, occasionally seen in the extent literature, is a recurrent network (Weron 2014), especially a nonlinear autoregressive exogenous model (NARX), a descendant of a recurrent network (Marcjasz et al. 2018).

The following Table 1 provides an author-based and criteria-based overview of electricity price forecasting with ANN in literature. The date criteria covers also higher aggregated levels of date-related factors like week, month, or season. The market criteria also covers different market-related factors such as exchange rate. The criteria others on the explanatory variables side subsumes a couple of non-typical factors especially used in electricity price forecasting with ANN.

Weron (2014) concluded that forecasting with univariate time series models is well known in the extent literature. Accordingly, including the right external input factors into the models, as well as dealing with nonlinear dependencies between endogenous and exogenous variables and among exogenous variables will become more important. In contrast to the author, we do not see that the time for univariate time series analysis of electricity prices is already over, as we still cannot see a satisfactory approach to meaningfully deal with the time-series characteristics of electricity prices. Although the Bayes-approach offers possibilities, it is rather unsuitable for practical use due to the high load of computer capacities during simulation operations. Hence, we see a research gap in handling the non-linear correlative effects between the exogenously modelled daily, weekly, and yearly seasonal patterns as well as autocorrelative effects within the time series and among the exogenously modelled variables. Our contribution is to close this research gap by using an ANN-based methodology. We perform a time series analysis for the German EEX "Phelix" Data using (i) different ANN architectures with non-fully and fully connectionist networks to discover and to deal with correlation effects on exogenous side/input layer, (ii) using a long short-term-memory network (LSTM) to discover and to deal with autocorrelation effects, and (iii) an ARIMAX model with time series features as binary coded variables as a benchmark for the aforementioned models.

Table 1. Author(s) - and criteria-based overview of electricity price forecasting with ANN

	Topology		Propagation		Activation				Explanatory Variables							
Author(s)	MLP	Recurrent	Linear-Weighted	Others	Linear	Logistic	Tangens-Hyperb.	Others	Lagged Variables	Power Consumption	Weather	Water	Time	Date	Market	Others
Adebiyi/Adewumi/Ayo (2014)	●		◑				●								●	
Barrow/Kourentzes (2018)	●		◑			◑	◑		●				●	●		
Conejo et al. (2005)	●		◑				●								●	
Dudek (2016)			◑		●	●									●	
Filho/Affonso/de Oliveira (2014)	●		◑			●									●	
Gajowniczek/Ząbkowskia (2014)	●		◑			●				●						
Ghiassi/Saidane/Zimbra (2005)		●	◑			●			●		●	●	●			
Graves/Schmidhuber (2005)		●	◑			●										●
Greff et al. (2017)		●	◑			●	●									●
Hu et al. (2008)	●		◑		◑	◑	◑								●	
Isa et al. (2009)	●		◑			●									●	
Kong et al. (2017)		●	◑			●	●			●				●		
Koutroumandis/Ioannou/Arabatzis (2009)	●		◑			●									●	
Krzemien et al. (2015)	●		◑		◑	◑	◑								●	
Lagoa et al. (2018)	●		◑		●					●				●		
Maciejowska/Nowotarski/Weron (2016)	●		◑		●	●			●					●		
Mandal/Senjyu/Funabashi (2006)	●		◑		◑	◑	◑				●		●	●		
Maniatis (2017)	●		◑		●									●		
Marcjasz/Uniejewski/Weron (2018)		●	◑		●		●		●				●			
Marín/Orozco/Velilla (2018)		●	◑			●			●					●		
Mirakyan/Meyer-Renschhausen/Koch (2017	●		◑			●			●				●	●		
Panapakidis/Dagoumas (2016)	●		◑		◑	◑	◑						●	●	●	
Psaradellis/Sermpinis (2016)		●	◑			●			◑	◑	◑	◑	◑	◑	◑	◑
Yamashita et al. (2008)	●		◑		◑	◑	◑			●	●			●		
Yamin/Shahidehpour/Li (2004)	●		◑		◑	◑	◑						●	●	●	
Zhang (2003)	●		◑			●			◑		●	●			●	

Legend ● Used ◑ Assumed

3 Sample and Methodology

At the European Energy Exchange (EEX), electricity spot prices (EPEX Spot), as well as future contracts are traded. The vast number of German municipal utility companies, but also large industrial consumers on the demand side, and European electricity suppliers on the supply side take part at the electricity trading at the EEX. The electricity volumes can be traded on the same day (intraday) or for the following day (day-ahead). Purchase and sale orders can be placed on an hourly basis as well as for the time blocks "baseload" (0.00 am–12.00 pm) or "peakload" (8.00 am–8.00 pm). These orders can be placed until 12.00 pm of every trading day for the next calendar day and will be processed primarily over the internet. A computer system ensures the automatic settlement of the purchase and sale orders and the fixing of the exchange price. Finally, around 12.40 pm the prices for the next day will be published via the internet and other data agencies.

The sample data used for this analysis is the EEX Phelix-DE day-ahead spot rate. It has established itself as a benchmark contract for European electricity. We considered time series data from January 1st 2015 until January 1st 2018. Each individual day has got 24 hourly price observations. The data underlying this analysis is complete.

Since the storage of electrical power is not possible without significant efficiency losses, the price shows daily, weekly, and yearly seasonality patterns. The seasonality of the time series certainly has its origin in the electricity demand over a day and night rhythm. Due to the daily, weekly and yearly seasonality patterns binary variables ("dummies") for these categories were introduced. To capture the seasonality, our models contain 23 hour-dummies for the daily seasonality, 6 weekday-dummies for the weekly seasonality and 11 month-dummies for the yearly seasonality.

Beneath seasonal and calendar day effects, the effects of wind power and solar energy increase the volatility of the time series which is particularly challenging in the prediction of the spot prices (Bierbrauer et al. 2007). More and more often, even negative electricity prices are documented at the EEX, which is mainly observable in times of weak demand combined with sunlight or strong wind. Since the present study focusses on seasonality patterns, other explanatory variables (e.g. wind or temperature) were not included into the models.

In this study, our models are trained on a training data set of two years prior the predicted months. We predict the months March, June, September, and December 2017. The mean squared error (MSE) and the root mean squared error (RMSE) – calculated on the differenced time series - is selected to assess and compare the different models. In most of the extent papers, this is the standard forecasting accuracy measure (Weron 2014). We apply this measure on the differenced time series because the accuracy measure should not depend on the electricity price itself. Furthermore, in the daily practice, the forecast would regularly be adopted to the latest realized prices by adding the predicted differences.

The ARIMAX-model used in this study is an extension of the classical ARIMA-model, introduced by Box and Jenkins (1971). To include seasonality into the model, the binary variables for hour, weekday, and month are applied in the X-term of the model, which means, that these variables are supplemented as additional regressor in the AR-Term. We used the Hyndman-Khandakar algorithm to find the best notation for the ARIMAX model (Hyndman and Khandakar 2008). The algorithm is using the KPSS tests to determine the number of differences (d) for the training dataset first. In a second step, the values of (p) and (q) are chosen for the training time series by minimizing the Akaike Information Criterion (AIC) out of every probable combination of these two parameters. As a result of this procedure, an ARIMAX(3, 1, 3) model with 40 binary coded variables is used for the analysis.

The ARIMAX model, which is used as a benchmark model at this point, is able to recognize autocorrelation of the time series using the lag variables. The binary-coded seasonal variables control the seasonality as additive constants for certain hours, certain days of the week, and certain months via the ARX-term. Relationships between these seasonal components cannot be recognized by this type of model. The ARIMAX-model is a fully linear model and cannot reflect non-linear behavior.

As described earlier, ANN are designed as fully connectionist networks. We suggest a different approach to discover information about the correlation of exogenous variables. We choose a step-by-step extension of the neural networks to better assess the influence of additional variables gradually introduced and connected to more and more units. Accordingly, we apply non-fully connectionist networks for the analysis. In particular, we chose network architectures that are comparable to classical statistical methods such as the ARIMAX model. The used network types are described in Table 2.

Table 2. Overview of applied artificial neural networks

Model	SLP/MLP	FC/NFC	Units Hidden Layer	Units Output Layer	Variables included				Description	
					Hour	Day	Month	Lags		
1a	SLP	FC		1	•	•	•	○	"ANOVA"-alike	Output unit receives hours, weekdays, and months
1b	SLP	FC		1	○	○	○	•	"ARI"-alike	Output unit receives lagged price changes
1c	SLP	FC		1	•	•	•	•	"ARIX"-alike	Output unit receives hours, weekdays, months, and lagged price changes
2a	MLP	NFC	3	1	•	•	•	○	no-interaction between the dummys	Hidden unit #1 receives hours Hidden unit #2 receives weekdays Hidden unit #3 receives months
2b	MLP	NFC	3	1	•	•	•	○	interaction between certain dummys	Hidden unit #1 receives hours & days Hidden unit #2 receives hours & months Hidden unit #3 receives months & days
2c	MLP	NFC	4	1	•	•	•	•	interaction between certain dummys no interaction with lags	Hidden unit #1 receives hours & days Hidden unit #2 receives hours & months Hidden unit #3 receives months & days Hidden unit #4 receives lags
2d	MLP	NFC	4	1	•	•	•	•	interaction between certain dummys and lags	Hidden unit #1 receives hours, days & lags Hidden unit #2 receives hours, months & lags Hidden unit #3 receives months, days & lags Hidden unit #4 receives lags
3	MLP	FC	4	1	•	•	•	•	all interactions	
4	LSTM	FC	4	1	•	•	•	•	LSTM all interactions	

• included ○ not included
FC = fully connected NFC = non fully connected

The input layer of the ANNs is composed of units for the binary coded hours, weekdays, and months as well as of units for the lags. To be in line with the ARIMAX-model, lags from 1 to 3 h are chosen for the lag units. Nevertheless, not in all networks all of these variables are used.

Recurrent neural networks are networks with loops, allowing information to persist in the network. Their ability to memorize data gives them particularly good features in the analysis of sequential data and time series. Long Short-Term Memory (LSTM) networks (Hochreiter and Schmidhuber 1997) are recurrent neural networks, that are capable of learning long-term dependencies. Thus, these LSTM networks should be able to reflect autocorrelative effects even better than standard MLP. For comparison, a fully connectionist LSTM-network is provided, too.

4 Results

The model forecasting accuracies in terms of the root mean squared error (RMSE) for all models as well as for all tested cases can be found in Table 3. In light of a monthly forecast horizon, the forecasting accuracies are in line with the expectation. In the extent literature, winter season is known to be more volatile and the price is more influenced by exogenous correlation effects, e.g. wind power. Accordingly, the comparatively poorer forecasting result in the winter season is also in line with expectations from literature. It is striking, that the ARIMAX benchmark performs nearly as well as the SLPs, the MLPs, and the LSTM. However, a clear improvement can be observed, the more interactions between variables are allowed inside the non-fully connectionist architectures. Accordingly, it must be assumed that these variables show more complex interactions than purely additive-linear models can reflect.

It is striking, that the LSTM-network of model 4 does not perform better than the fully connectionist MLP-network of model 3. This result can be explained with the fact that both models are already provided with lagged data, so that the LSTM-network cannot fully exploit its advantages to memorize sequential data.

The coefficients and model statistics of the ARIMAX model as well as the weights of all Single Layer Perceptron networks are given in Table 4. The color coding of this table is linearized between the highest and lowest value separately for lags, hours, weekdays, and month. The coefficients as well as the weights show well pronounced daily, weekly and yearly seasonality. Although the results are in line with expectation, the annual seasonality is based on only a few observations, resulting in high standard errors. The daily and weekly seasonality, on the other hand, can be described as stable, as inference is based on a large number of observations. By visual examination of the forecast, it can be seen that the ARIMAX model as well as the SLP predict repetitive daily patterns that oscillate across the course of the week. It is equally clear that the coefficients of the ARIMAX model show a slightly different behavior compared to all weights of the SLP or MLP networks which cannot be solely attributed to the different mathematical methods used.

The importance of individual inputs can be determined by their weights to the output layer in SLP networks or to the hidden layer in MLP networks, respectively. It is striking that in the SLP networks 1a and 1c the months receive almost no weight. The monthly patterns are therefore almost irrelevant to the model result. While the weekdays also show relatively small weights, the main driver of the model results are allocated in the lags (models 1b and 1c) and in the hourly patterns (model 1a and 1c).

Table 3. Model accuracies of the applied artificial neural networks

	Number of Units		Training Period		Testing Period			
Model	*Hidden Layer*	*Ouput Layer*	*Start*	*End*	*Start*	*End*	*MSE*	*RMSE*
ARIMAX			01.03.2015	28.02.2017	01.03.2017	31.03.2017	121.503	11.023
			01.06.2015	31.05.2017	01.06.2017	30.06.2017	7.181	2.680
			01.09.2015	31.08.2017	01.09.2017	30.09.2017	8.236	2.870
			01.12.2015	30.11.2017	01.12.2017	31.12.2017	33.412	5.780
1a		1	01.03.2015	28.02.2017	01.03.2017	31.03.2017	7.823	2.797
		1	01.06.2015	31.05.2017	01.06.2017	30.06.2017	7.157	2.675
		1	01.09.2015	31.08.2017	01.09.2017	30.09.2017	8.216	2.866
		1	01.12.2015	30.11.2017	01.12.2017	31.12.2017	33.420	5.781
1b		1	01.03.2015	28.02.2017	01.03.2017	31.03.2017	12.597	3.549
		1	01.06.2015	31.05.2017	01.06.2017	30.06.2017	9.907	3.148
		1	01.09.2015	31.08.2017	01.09.2017	30.09.2017	12.170	3.489
		1	01.12.2015	30.11.2017	01.12.2017	31.12.2017	42.956	6.554
1c		1	01.03.2015	28.02.2017	01.03.2017	31.03.2017	7.632	2.763
		1	01.06.2015	31.05.2017	01.06.2017	30.06.2017	6.633	2.576
		1	01.09.2015	31.08.2017	01.09.2017	30.09.2017	7.805	2.794
		1	01.12.2015	30.11.2017	01.12.2017	31.12.2017	34.285	5.855
2a	3	1	01.03.2015	28.02.2017	01.03.2017	31.03.2017	7.859	2.803
	3	1	01.06.2015	31.05.2017	01.06.2017	30.06.2017	7.098	2.664
	3	1	01.09.2015	31.08.2017	01.09.2017	30.09.2017	8.203	2.864
	3	1	01.12.2015	30.11.2017	01.12.2017	31.12.2017	33.405	5.780
2b	3	1	01.03.2015	28.02.2017	01.03.2017	31.03.2017	6.885	2.624
	3	1	01.06.2015	31.05.2017	01.06.2017	30.06.2017	6.399	2.530
	3	1	01.09.2015	31.08.2017	01.09.2017	30.09.2017	7.355	2.712
	3	1	01.12.2015	30.11.2017	01.12.2017	31.12.2017	32.539	5.704
2c	4	1	01.03.2015	28.02.2017	01.03.2017	31.03.2017	6.922	2.631
	4	1	01.06.2015	31.05.2017	01.06.2017	30.06.2017	5.294	2.301
	4	1	01.09.2015	31.08.2017	01.09.2017	30.09.2017	6.621	2.573
	4	1	01.12.2015	30.11.2017	01.12.2017	31.12.2017	32.049	5.661
2d	4	1	01.03.2015	28.02.2017	01.03.2017	31.03.2017	6.862	2.620
	4	1	01.06.2015	31.05.2017	01.06.2017	30.06.2017	5.213	2.283
	4	1	01.09.2015	31.08.2017	01.09.2017	30.09.2017	6.675	2.584
	4	1	01.12.2015	30.11.2017	01.12.2017	31.12.2017	34.136	5.843
3	4	1	01.03.2015	28.02.2017	01.03.2017	31.03.2017	6.962	2.639
	4	1	01.06.2015	31.05.2017	01.06.2017	30.06.2017	5.030	2.243
	4	1	01.09.2015	31.08.2017	01.09.2017	30.09.2017	6.406	2.531
	4	1	01.12.2015	30.11.2017	01.12.2017	31.12.2017	32.564	5.707
4	4	1	01.03.2015	28.02.2017	01.03.2017	31.03.2017	7.244	2.692
	4	1	01.06.2015	31.05.2017	01.06.2017	30.06.2017	5.995	2.449
	4	1	01.09.2015	31.08.2017	01.09.2017	30.09.2017	7.374	2.715
	4	1	01.12.2015	30.11.2017	01.12.2017	31.12.2017	32.423	5.694

Table 4. Comparison between ARIMAX-coefficients and SLP-weights

	ARIMAX	Single Layer Perceptron		
		1a	1b	1c
	Coeff.	*Weights*	*Weights*	*Weigths*
ar1	1,661		0,354	0,180
ar2	-1,085		0,049	0,010
ar3	0,348		-0,155	-0,107
hour_2	-1,847	-0,001		-0,002
hour_3	-3,093	0,004		-0,001
hour_4	-3,723	0,008		0,004
hour_5	-3,276	0,015		0,011
hour_6	-1,503	0,025		0,019
hour_7	4,132	0,051		0,043
hour_8	10,890	0,059		0,048
hour_9	12,958	0,027		0,016
hour_10	11,614	0,003		0,002
hour_11	9,716	-0,001		0,001
hour_12	9,041	0,008		0,008
hour_13	6,561	-0,005		-0,008
hour_14	5,007	0,001		-0,001
hour_15	4,246	0,007		0,005
hour_16	5,300	0,020		0,015
hour_17	7,037	0,024		0,018
hour_18	12,280	0,049		0,042
hour_19	15,243	0,033		0,022
hour_20	15,815	0,016		0,009
hour_21	12,112	-0,013		-0,012
hour_22	8,308	-0,014		-0,010
hour_23	6,666	0,001		0,002
hour_24	1,898	-0,021		-0,024
month_2	0,676	-0,000		-0,000
month_3	0,279	-0,000		-0,000
month_4	9,132	-0,000		-0,001
month_5	4,276	-0,000		0,000
month_6	3,214	-0,000		-0,000
month_7	3,276	-0,000		-0,000
month_8	1,086	-0,000		-0,000
month_9	2,855	-0,000		-0,000
month_10	5,030	-0,000		-0,000
month_11	0,825	-0,001		-0,000
month_12	3,959	-0,000		-0,000
weekday_0	-2,117	0,001		0,001
weekday_1	-1,126	0,001		0,001
weekday_2	0,343	0,001		0,001
weekday_3	0,870	0,001		0,001
weekday_4	0,643	0,001		0,001
weekday_5	0,244	0,001		0,001
ma1	-1,503			
ma2	0,820			
ma3	-0,305			
MSE	121,503	7,823	12,597	7,632
RMSE	11,023	2,797	3,549	2,763

In the MLP-models, the importance of the hidden units can be assessed by their weights towards the output unit. These weights are shown in Table 4. Accordingly, in model 2a, where all three hidden units are specialized on hour, weekday, and month, the above results of a rather weak influence of the monthly pattern is confirmed by a comparatively low weight of the unit that is specialized on the month. The same is true for model 2b, that allows for certain interactions between the binary coded variables. Especially the month-weekday interaction does not seem to play a role in this model, which is fully in line with expectations, as the hourly patterns should change during the year while weekday effects are not supposed to show a similar behavior (Table 5).

Table 5. Comparison of the weights in MLP-networks

	Multi Layer Perceptron							
	2a		2b		2c		2d	
	Weights		*Weights*		*Weights*		*Weigths*	
Hidden Unit 1					*Lag* -	0,716	*Lag*	1,401
Hidden Unit 2	*Hour* -	0,528	*Hour & Weekday* -	0,497	*Hour & Weekday*	0,532	*Lag/Stunde/Tag* -	0,676
Hidden Unit 3	*Weekday*	1,022	*Hour & Month*	0,752	*Hour & Month*	0,167	*Lag/Stunde/Monat*	0,369
Hidden Unit 4	*Month*	0,016	*Month & Weekday*	0,091	*Month & Weekday*	1,046	*Lag/Monat/Tag*	0,854

By introducing the lags in model 2c and 2d, surprisingly the month-weekday interaction gains some relevance, which can be attributed to the fact that technically a counterweight for the negatively weighted lags is needed. Overall, however, it can be stated that the addition of lags makes the models extremely difficult to interpret because no interactions between lag variables and binary-coded data are to be expected.

5 Conclusion

The electricity price at the electricity exchange EEX shows daily, weekly, and annual seasonality patterns. Due to the cyclicality of the considered seasonal components there are non-linear correlative relationships between them. Thus, the present study deals methodologically with non-linear correlative and autocorrelative time series properties of the electricity spot price. We propose a systematic ANN-based approach to address this problem. The usage of different architectures sheds light on the strength of these relationships and their influence on electricity price prediction.

A single layer perceptron shows more or less the same forecasting accuracy than a standard ARIMAX model with binary coded seasonalities used as a benchmark. Possible reasons for the poor predictive quality can be specified: The non-linear activation function of the SLP and, above all, the missing MA term, which smooths the results in the ARIMAX model.

A non-fully connectionist multi-layer perceptron network (MLP) with seasonally specified aggregated units in the hidden layer is not able to improve the forecasting accuracy, as correlative relationships of the seasonality are still prevented. The forecasting accuracy improves slightly with allowing for interactions between the variables. This is an indicator for the fact, that the daily pattern changes with the weekday or the month. Weekdays and months in turn, cannot show autocorrelative effects.

The non-fully connectionist MLP shows only low correlations and a specialization of one unit considering all information. Accordingly, the forecasting accuracy cannot be better than in the single layer perceptron by large extent. This gap is closed by the fully-connectionist MLP, where all interactive relationships between these components find their way into the forecasting model. Last but not least, the long short-term memory (LSTM) model provides the most accurate forecast, which, in addition to the correlative relationships already mentioned, also included autocorrelative relationships on the endogenous side over several periods into the forecast.

The LSTM-network does not perform better than the fully connectionist MLP-network, which can be explained with the fact that both models are already provided with lagged data, so that the LSTM-network cannot fully exploit its advantages to memorize sequential data. The widespread belief that LSTM networks are consistently better suited for the analysis and forecasting of time series cannot be supported, at least in our study.

References

Adebiyi, A.A., Adewumi, A.O., Ayo, C.K.: Comparison of ARIMA and artificial neural networks models for stock price prediction. J. Appl. Math. **2014**, 1–7 (2014)

Barrow, D., Kourentzes, N.: The impact of special days in call arrivals forecasting: a neural network approach to modelling special days. Eur. J. Oper. Res. **264**(3), 967–977 (2018)

Bierbrauer, M., Menn, C., Rachev, S.T., Trück, S.: Spot and derivative pricing in the EEX power market. J. Bank. Finance **31**(11), 3462–3485 (2007)

Box, G.E.P., Jenkins, G.M.: Time Series Analysis: Forecasting and Control. Holden-Day, San Francisco (1971). (2. printing)

Conejo, A.J., Plazas, M.A., Espinola, R., Molina, A.B.: Day-ahead electricity price forecasting using the wavelet transform and ARIMA models. IEEE Trans. Power Syst. **20**(2), 1035–1042 (2005a)

Conejo, A.J., Contreras, J., Espínola, R., Plazas, M.A.: Forecasting electricity prices for a day-ahead pool-based electric energy market. Int. J. Forecast. **21**(3), 435–462 (2005b)

Contreras, J., Espinola, R., Nogales, F.J., Conejo, A.J.: ARIMA models to predict next-day electricity prices. IEEE Trans. Power Syst. **18**(3), 1014–1020 (2003)

Dudek, G.: Multilayer perceptron for GEFCom2014 probabilistic electricity price forecasting. Int. J. Forecast. **32**(3), 1057–1060 (2016)

Filho, J.C.R., de Affonso, C.M., de Oliviera, R.C.L.: Energy price prediction multi-step ahead using hybrid model in the Brazilian market. Electric Power Syst. Res. **117**, 115–122 (2014)

Gajowniczek, K., Ząbkowski, T.: Short term electricity forecasting using individual smart meter data. Procedia Comput. Sci. **35**, 589–597 (2014)

Garcia, R.C., Contreras, J., van Akkeren, M., Garcia, J.B.C.: A GARCH forecasting model to predict day-ahead electricity prices. IEEE Trans. Power Syst. **20**(2), 867–874 (2005)

Ghiassi, M., Saidane, H., Zimbra, D.K.: A dynamic artificial neural network model for forecasting time series events. Int. J. Forecast. **21**(2), 341–362 (2005)

Graves, A., Schmidhuber, J.: Framewise phoneme classification with bidirectional LSTM and other neural network architectures. Neural Netw. **18**(5), 602–610 (2005)

Greff, K., Srivastava, R.K., Koutník, J., Steunebrink, B.R., Schmidhuber, J.: LSTM: a search space odyssey. IEEE Trans. Neural Netw. Learn. Syst. **28**(10), 2222–2232 (2017)

Haykin, S.S.: Neural Networks and Learning Machines, 3rd edn. Pearson Education, Upper Saddle River (2009)
Hochreiter, S., Schmidhuber, J.: Long short-term memory. Neural Comput. **9**(8), 1735–1780 (1997)
Hopfield, J.J.: Neural networks and physical systems with emergent collective computational abilities. Proc. Natl. Acad. Sci. **79**(8), 2554–2558 (1982)
Hu, Z., Yang, L., Wang, Z., Gan, D., Sun, W., Wang, K.: A game-theoretic model for electricity markets with tight capacity constraints. Int. J. Electric. Power Energy Syst. **30**(3), 207–215 (2008)
Hyndman, R.J., Khandakar, Y.: Automatic time series forecasting: the forecast package for R. J. Stat. Softw. **27**(3) (2008)
Isa, A.M., Niimura, T., Sakamoto, N., Ozawa, K., Yokoyama, R.: Electricity market forecasting using artificial neural network models optimized by grid computing. IFAC Proc. **42**(9), 273–277 (2009)
Knittel, C.R., Roberts, M.R.: An empirical examination of restructured electricity prices. Energy Econ. **27**(5), 791–817 (2005)
Kong, W., Dong, Z.Y., Jia, Y., Hill, D.J., Xu, Y., Zhang, Y.: Short-term residential load forecasting based on LSTM recurrent neural network. IEEE Trans. Smart Grid **10**(1), 841–851 (2017)
Koopman, S.J., Ooms, M., Carnero, M.A.: Periodic seasonal reg-ARFIMA–GARCH models for daily electricity spot prices. J. Am. Stat. Assoc. **102**(477), 16–27 (2007)
Koutroumandis, T., Ioannou, K., Arabatzis, G.: Predicting fuelwood prices in Greece with the use of ARIMA models, artificial neural networks and a hybrid ARIMA–ANN model. Energy Policy **37**(9), 3627–3634 (2009)
Krzemien, A., Riesgo Fernández, P., Suárez Sánchez, A., Sánchez Lasheras, F.: Forecasting European thermal coal spot prices. J. Sustain. Min. **14**(4), 203–210 (2015)
Lago, J., de Ridder, F., Vrancx, P., de Schutter, B.: Forecasting day-ahead electricity prices in Europe: the importance of considering market integration. Appl. Energy **211**, 890–903 (2018)
Maciejowska, K., Nowotarski, J., Weron, R.: Probabilistic forecasting of electricity spot prices using Factor Quantile Regression Averaging. Int. J. Forecast. **32**(3), 957–965 (2016)
Malhotra, P., Vig, L., Shroff, G., Agarwal, P.: Long short term memory networks for anomaly detection in time series (2015)
Mandal, P., Senjyu, T., Funabashi, T.: Neural networks approach to forecast several hour ahead electricity prices and loads in deregulated market. Energy Convers. Manag. **47**(15), 2128–2142 (2006)
Maniatis, P.: A taxonomy of electricity demand forecasting techniques and a selection strategy. Int. J. Manag. Excel. **8**(2), 881 (2017)
Marcjasz, G., Uniejewski, B., Weron, R.: On the importance of the long-term seasonal component in day-ahead electricity price forecasting with NARX neural networks. Int. J. Forecast. **34**(1) (2018)
Marín, J.B., Orozco, E.T., Velilla, E.: Forecasting electricity price in colombia: a comparison between neural network, ARMA process and hybrid models. Int. J. Energy Econ. Policy **8**(3), 10 (2018)
Mirakyan, A., Meyer-Renschhausen, M., Koch, A.: Composite forecasting approach, application for next-day electricity price forecasting. Energy Econ. **66**, 228–237 (2017)
Misiorek, A., Trueck, S., Weron, R.: Point and interval forecasting of spot electricity prices: linear vs. non-linear time series models. Stud. Nonlinear Dyn. Econom. **10**(3) (2006)
Panapakidis, I.P., Dagoumas, A.S.: Day-ahead electricity price forecasting via the application of artificial neural network based models. Appl. Energy **172**, 132–151 (2016)

Psaradellis, I., Sermpinis, G.: Modelling and trading the US implied volatility indices. Evidence from the VIX, VXN and VXD indices. Int. J. Forecast. **32**(4), 1268–1283 (2016)

Weron, R.: Electricity price forecasting: a review of the state-of-the-art with a look into the future. Int. J. Forecast. **30**(4), 1030–1081 (2014)

Yamashita, D., Isa, A.M., Yokoyama, R., Niimura, T.: Forecasting of electricity price and demand using autoregressive neural networks. IFAC Proc. **41**(2), 14934–14938 (2008)

Yamin, H., Shahidehpour, S., Li, Z.: Adaptive short-term electricity price forecasting using artificial neural networks in the restructured power markets. Int. J. Electr. Power Energy Syst. **26**(8), 571–581 (2004)

Zareipour, H., Canizares, C.A., Bhattacharya, K., Thomson, J.: Application of public-domain market information to forecast Ontario's wholesale electricity prices. IEEE Trans. Power Syst. **21**(4), 1707–1717 (2006)

Zhang, G.P.: Time series forecasting using a hybrid ARIMA and neural network model. Neurocomputing **50**, 159–175 (2003)

Zhang, G., Patuwo, E.B., Hu, M.Y.: Forecasting with artificial neural networks: the state of the art. Int. J. Forecast. **14**(1), 35–62 (1998)

Zhou, M., Yan, Z., Ni, Y.X., Li, G., Nie, Y.: Electricity price forecasting with confidence-interval estimation through an extended ARIMA approach. IEE Proc. - Gener. Transm. Distrib. **153** (2), 187 (2006)

Complex Systems Theory and Crashes of Cryptocurrency Market

Vladimir N. Soloviev and Andriy Belinskiy(✉)

Kryvyi Rih State Pedagogical University,
54 Gagarina Ave, Kryvyi Rih 50086, Ukraine
vnsoloviev2016@gmail.com, krivogame@gmail.com

Abstract. This article demonstrates the possibility of constructing indicators of critical and crash phenomena in the volatile market of cryptocurrency. For this purpose, the methods of the theory of complex systems have been used. The possibility of constructing dynamic measures of complexity as recurrent, entropy, network, quantum behaving in a proper way during actual pre-crash periods has been shown. This fact is used to build predictors of crashes and critical events phenomena on the examples of all the patterns recorded in the time series of the key cryptocurrency Bitcoin, the effectiveness of the proposed indicators-precursors of these falls has been identified. From positions, attained by modern theoretical physics the concept of economic Planck's constant has been proposed. The theory on the economic dynamic time series related to the cryptocurrencies market has been approved. Then, combining the empirical cross-correlation matrix with the random matrix theory, we mainly examine the statistical properties of cross-correlation coefficient, the evolution of the distribution of eigenvalues and corresponding eigenvectors of the global cryptocurrency market using the daily returns of 24 cryptocurrencies price time series all over the world from 2013 to 2018. The result has indicated that the largest eigenvalue reflects a collective effect of the whole market, and is very sensitive to the crash phenomena. It has been shown that both the introduced economic mass and the largest eigenvalue of the matrix of correlations can act like quantum indicator-predictors of falls in the market of cryptocurrencies.

Keywords: Cryptocurrency · Bitcoin · Complex system · Measures of complexity · Crash · Critical events · Recurrence plot · Recurrence quantification analysis · Permutation entropy · Complex networks · Quantum econophysics · Heisenberg uncertainty principle · Random matrix theory · Indicator-precursor

1 Introduction

The instability of global financial systems with regard to normal and natural disturbances of the modern market and the presence of poorly foreseeable financial crashes indicate, first of all, the crisis of the methodology of modeling, forecasting and interpretation of modern socio-economic realities. The doctrine of the unity of the scientific method states that for the study of events in socio-economic systems, the same methods and criteria as those used in the study of natural phenomena are applicable. Significant

V. Ermolayev et al. (Eds.): ICTERI 2018, CCIS 1007, pp. 276–297, 2019.
https://doi.org/10.1007/978-3-030-13929-2_14

success has been achieved within the framework of interdisciplinary approaches and the theory of self-organization - synergetics. The modern paradigm of synergetics is a complex paradigm associated with the possibility of direct numerical simulation of the processes of complex systems evolution, most of which have a network structure, or one way or another can be reduced to the network. The theory of complex networks studies the characteristics of networks, taking into account not only their topology, but also statistical properties, the distribution of weights of individual nodes and edges, the effects of dissemination of information, robustness, etc. [1–4].

Complex systems are systems consisting of a plurality of interacting agents possessing the ability to generate new qualities at the level of macroscopic collective behavior, the manifestation of which is the spontaneous formation of noticeable temporal, spatial, or functional structures. As simulation processes, the application of quantitative methods involves measurement procedures, where importance is given to complexity measures. I. Prigogine notes that the concepts of simplicity and complexity are relativized in the pluralism of the descriptions of languages, which also determines the plurality of approaches to the quantitative description of the complexity phenomenon [5]. Therefore, we will continue to study Prigogine's manifestations of the system complexity, using the current methods of quantitative analysis to determine the appropriate measures of complexity.

The key idea here is the hypothesis that the complexity of the system before the crashes and the actual periods of crashes must change. This should signal the corresponding degree of complexity if they are able to quantify certain patterns of a complex system. Significant advantage of the introduced measures is their dynamism, that is, the ability to monitor the change in time of the chosen measure and compare it with the corresponding dynamics of the output time series. This allowed us to compare the critical changes in the dynamics of the system, which is described by the time series, with the characteristic changes of concrete measures of complexity. It turned out that quantitative measures of complexity respond to critical changes in the dynamics of a complex system, which allows them to be used in the diagnostic process and prediction of future changes.

Cryptocurrency market is a complex, self-organized system, which in most cases can be considered either as a complex network of market agents, or as an integrated output signal of such a network - a time series, for example, prices of individual cryptocurrency. The research on cryptocurrency price fluctuations being carried out internationally is made more complex by the interplay due to many factors - including market supply and demand, the US dollar exchange rate, stock market state, the influence of crime and the shadow market, and fiat money regulator pressure - that introduces a high level of noise into the cryptocurrency data. Moreover, in the cryptocurrency market, to some extent, the blockchain technology is tested in general. Thus the cryptocurrency prices exhibit such complex volatility characteristics as nonlinearity and uncertainty, which are difficult to forecast and any results obtained are uncertain. Therefore, cryptocurrency price prediction remains a huge challenge.

Among these prediction models, one of the most important models is econometric model such as for example autoregressive integrated moving average (ARIMA) that exploit time series stationarity. Because of the presence of local explosive trends, depicted as bubbles, the Bitcoin exchange rate cannot be modelled by any traditional

ARIMA models (see e.g. [6]). Dassios and Li [7] introduce a new diffusion process to describe Bitcoin prices within an economic bubble cycle. In spite of rather a complicated model, forecast bubble results for December, 2017 are disappointing. Tarnopolski [8] has completed the modelling of Bitcoin price using Monte Carlo method based on model of geometric fractional Brownian motion. The Bitcoin price predicted for the beginning of 2018 turned out to be far from reality.

In addition to the classic econometric approaches, artificial intelligence methods (also known as machine and/or deep learning methods) have been used to uncover the inner complexity of cryptocurrency prices. Separate attempts of using both simple artificial neural networks Elmann [9] and method of Bayesian regression [10] are known, as well as more complex methods based on XGboost [11] or on the long short – term memory (LSTM) algorithm for recurrent neural networks [12, 13]. Nowadays combined classical econometric methods as well as methods of machine learning [14, 15] and those which take into consideration the spirit of social networks regarding the state and tendency of cryptocurrency dynamics [16] are becoming more popular.

Thus, lack of reliable models of prediction of time series for the time being will update the construction of at least indicators which warn against possible critical phenomena or trade changes etc. This work is dedicated to the construction of such indicators – precursors based on the theory of complexity.

In this paper, we consider some of the informative measures of complexity and adapt them in order to study the critical and crash phenomena of cryptomarket.

The paper is structured as follows. Section 2 describes previous studies in these fields. Section 3 presents classification of crashes and critical events on the Bitcoin market during the entire period (16.07.2010 – 08.12.2018). Section 4 describes the technique of quantitative recurrent analysis and recurrent measures of complexity as indicators of crashes. The indicator-precursor of crashes based on the calculation of Permutation Entropy is described in Sect. 5. Network measures of complexity and their effectiveness as indicators of cryptomarket crashes are presented in Sect. 6. In Sect. 7, new quantum indicators of critical and crash phenomena are introduced using the Heisenberg uncertainty principle and the Random Matrix Theory. And finally, we discuss our results in Sect. 8.

2 Analysis of Previous Studies

Throughout the existence of Bitcoin, its complexity became much larger. Crashes and critical events that took place on this market as well as the reasons that led to them, did not go unheeded. We determined that there are a lot of articles and papers on that topic which we will demonstrate.

Donier and Bouchaud [17] found that the market microstructure on Bitcoin exchanges can be used to anticipate illiquidity issues in the market, which lead to abrupt crashes. They investigate Bitcoin liquidity based on order book data and, out of this, accurately predict the size of price crashes.

Bariviera [18] demonstrates the dynamics of the intraday price of 12 cryptocurrencies. By using the complexity-entropy causality plane, authors discriminate three different dynamics in the data set. Another paper [19] compares the time-varying weak-form efficiency of Bitcoin prices in terms of US dollars (BTC/USD) and euro (BTC/EUR) at a high-frequency level by using Permutation Entropy. Their research shows that BTC/USD and BTC/EUR markets have been demonstrating more information at the intraday level since the beginning of 2016, and BTC/USD market has been slightly more efficient than BTC/EUR during the same period. And moreover, their research shows that with the higher frequency we have less price efficiency.

Some papers like this one [20] demonstrate how recurrence plots and measures of recurrence quantification analysis can be used to study significant changes in complex dynamical systems due to a change in control parameters, chaos-order as well as chaos-chaos transitions. Santos et al. [21] discuss how to model activity in online collaboration websites, such as Stock Exchange Question and Answering portals because the success of these websites critically depends on the content contributed by its users. In this paper, they represent user activity as time series and perform an initial analysis of these time series to obtain a better understanding of the underlying mechanisms that govern their creation. For this purpose nonlinear modeling via recurrence plots was used, which gives more granular study and deeper understanding of nonlinear dynamics of governing activity of time series and explaining the activity in online collaboration websites.

Taking to the account studies on network analysis we can notice different papers on this topic [22–24]. Di Francesco Maesa et al. [22] have performed on the users' graph inferred from the Bitcoin blockchain, dumped in December 2015, so after the occurrence of the exponential explosion in the number of transactions. Researchers first present the analysis assessing classical graph properties like densification, distance analysis, degree distribution, clustering coefficient, and several centrality measures. Then, they analyze properties strictly tied to the nature of Bitcoin, like rich-get-richer property, which measures the concentration of richness in the network. Bovet et al. [23] analyzed the evolution of the network of Bitcoin transactions among users and built network-based indicators of Bitcoin bubbles.

In this article [24], authors consider the history of Bitcoin and transactions in it. Using this dataset, they reconstruct the transaction network among users and analyze changes in the structure of the subgraph induced by the most active users. Their approach is based on the unsupervised identification of important features of the time variation of the network. Applying the widely used method of principal component analysis to the matrix constructed from snapshots of the network at different times, they show how changes in the network accompany significant changes in the price of Bitcoin.

Separately, it is necessary to highlight the work of Sornette [25, 26], who built a precursor of crashes based on the generation of so-called log-periodic oscillations by the pre-crashing market. However, the actual collapse point is still badly predicted.

Thus, construction of indicators – precursors of critical and crash phenomena in the cryptocurrency market remains relevant.

3 Data

Bitcoin, despite its uncertain future, continues to attract investors, crypto-enthusiasts, and researchers. Being historically proven, popular and widely used cryptocurrency for the whole existence of cryptocurrencies in general, Bitcoin began to produce a lot of news and speculation, which began to determine its future life. Similar discussions began to lead to different kinds of crashes, critical events, and bubbles, which professional investors and inexperienced users began to fear. Thus, we advanced into action and set the tasks:

(1) Classification of such bubbles, critical events and crashes.
(2) Construction of such indicators that will predict crashes, critical events in order to give investors and ordinary users the opportunity to work in this market.

At the moment, there are various research works on what crises and crashes are and how to classify such interruptions in the market of cryptocurrencies. Taking into account the experience of previous researchers [26–30], we have created our classification of such leaps and falls, relying on Bitcoin time series during the entire period (16.07.2010 – 08.12.2018) of verifiable fixed daily values of the Bitcoin price (BTC) (https://finance.yahoo.com/cryptocurrencies).

For our classification, crashes are short, time-localized drops, with strong losing of price per each day, which are formed as a result of the bubble. Critical events are those falls that could go on for a long period of time, and at the same time, they were not caused by a bubble. The bubble is an increasing in the price of the cryptocurrencies that could be caused by certain speculative moments. Therefore, according to our classification of the event with number (1, 3–6, 9–11, 14, 15) are the crashes that are preceded by the bubbles, all the rest - critical events. More detailed information about crises, crashes and their classification in accordance with these definitions is given in the Table 1.

Accordingly, during this period in the Bitcoin market, many crashes and critical events shook it. Thus, considering them, we emphasize 15 periods on Bitcoin time series, whose falling we predict by our indicators, relying on normalized returns and volatility, where normalized returns are calculated as

$$g(t) = \ln X(t + \Delta t) - \ln X(t) \cong [X(t + \Delta t) - X(t)]/X(t), \qquad (1)$$

and volatility as

$$V_T(t) = \frac{1}{n} \sum_{t'=t}^{t+n-1} |g(t')|$$

Besides, considering that $g(t)$ should be more than the $\pm 3\sigma$, where sigma is a mean square deviation.

Calculations were carried out within the framework of the algorithm of a moving window. For this purpose, the part of the time series (window), for which there were calculated measures of complexity, was selected, then the window was displaced along the time series in a one-day increment and the procedure repeated until all the studied series had exhausted. Further, comparing the dynamics of the actual time series and the corresponding measures of complexity, we can judge the characteristic changes in the dynamics of the behavior of complexity with changes in the cryptocurrency. If this or that measure of complexity behaves in a definite way for all periods of crashes, for example, decreases or increases during the pre-crashes period, then it can serve as an indicator or precursor of such a crashes phenomenon.

Table 1. BTC Historical Corrections. List of Bitcoin major corrections $\geq$ 20% since June 2011

№	Name	Days in correction	Bitcoin high price, $	Bitcoin low price, $	Decline, %	Decline, $
1	07.06.2011–10.06.2011	4	29.60	14.65	50	15.05
2	15.01.2012–16.02.2012	33	7.00	4.27	39	2.73
3	15.08.2012–18.08.2012	4	13.50	8.00	40	5.50
4	08.04.2013–15.04.2013	8	230.00	68.36	70	161.64
5	04.12.2013–18.12.2013	15	1237.66	540.97	56	696.69
6	05.02.2014–25.02.2014	21	904.52	135.77	85	768.75
7	12.11.2014–14.01.2015	64	432.02	164.91	62	267.11
8	11.07.2015–23.08.2015	44	310.44	211.42	32	99.02
9	09.11.2015–11.11.2015	3	380.22	304.70	20	75.52
10	18.06.2016–21.06.2016	4	761.03	590.55	22	170.48
11	04.01.2017–11.01.2017	8	1135.41	785.42	30	349.99
12	03.03.2017–24.03.2017	22	1283.30	939.70	27	343.60
13	10.06.2017–15.07.2017	36	2973.44	1914.08	36	1059.36
14	16.12.2017–22.12.2017	7	19345.49	13664.96	29	5680.53
15	13.11.2018–26.11.2018	14	6339.17	3784.59	40	2554.58

Calculations of measures of complexity were carried out both for the entire time series, and for a fragment of the time series localizing the crash. In the latter case, fragments of time series of the same length with fixed points of the onset of crashes or critical events were selected and the results of calculations of complexity measures were compared to verify the universality of the indicators.

In the Fig. 1 output Bitcoin time series, normalized returns $g(t)$, and volatility $V_T(t)$ calculated for the window size 100 are presented.

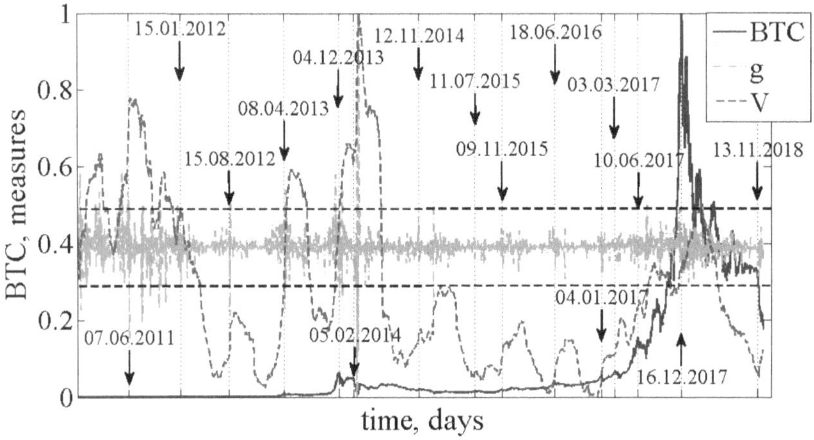

Fig. 1. The standardized dynamics, returns $g(t)$, and volatility $V_T(t)$ of BTC/USD daily values. Horizontal dotted lines indicate the $\pm 3\sigma$ borders. The arrows indicate the beginning of one of the crashes or the critical events.

From Fig. 1 we can see that during periods of crashes and critical events normalized profitability g increases considerably in some cases beyond the limits $\pm 3\sigma$. This indicates about deviation from the normal law of distribution, the presence of the "heavy tails" in the distribution g, characteristic of abnormal phenomena in the market. At the same time volatility also grows. These characteristics serve as indicators of critical and collapse phenomena as they react only at the moment of the above mentioned phenomena and don't give an opportunity to identify the corresponding abnormal phenomena in advance. In contrast, the indicators described below respond to critical and collapse phenomena in advance. It enables them to be used as indicators – precursors of such phenomena and in order to prevent them.

4 Recurrence Quantification Analysis

Recurrence plots (RPs) have been introduced to study dynamics and recurrence states of complex systems. A phase space trajectory can be transformed from a time series $U_i = \{u_1, \ldots, u_n\}$ ($t = i\Delta t$, where Δt is the sampling time) into time-delay structures

$$X_i = (U_i, U_{i+1}, \ldots, U_{i+(m-1)\tau}),$$

where m stands for the embedding dimension and τ for the entire time delay. Both of them can be calculated from the original data using false nearest neighbors and mutual information [31].

A RP is a plot representation of those states which are recurrent. The recurrence matrix and the states are considered to be recurrent if the distance between them within the ε- radius. In this case, the recurrence plot is defined as:

$$R_{ij} = \Theta(\varepsilon - \|x_i - x_j\|), i,j = 1,\ldots,N,$$

and $\|\|$ is a norm (representing the spatial distance between the states at times i and j), ε is a predefined recurrence threshold, and Θ is the Heaviside function (ensuring a binary **R**).

Usually, recurrent plot has a square form and $R \equiv 1$ is included to the representation, but for calculations, it might be useful to remove it [31]. For qualitative description of the system, the graphic representation of the system suits perfectly. For the quantitative description of the system, the small-scale clusters such as diagonal and vertical lines can be used. The histograms of the lengths of these lines are the base of the recurrence quantification analysis developed by Webber and Zbilut and later by Marwan et al. [32–34].

Recurrence rate (RR) is the part of recurrence points in the plot that can be interpreted as the probability that any state of the system will recur. It is the simplest measure, which computes by taking the number of the nearest points forming short, spanning row and columns of the recurrent plot, summarize them and divide by the number of possible points N^2:

$$RR = \frac{1}{N^2}\sum_{1,j=1}^{N} R_{i,j}.$$

The set of recurrence points on the recurrence plots that form line segments of minimal length μ parallel to the matrix diagonal is the measure of determinism (DET):

$$DET^{(u)} = \frac{\sum_{l=\mu}^{N} l \cdot D(l)}{\sum_{i,j}^{N} R_{i,j}} = \frac{\sum_{l=\mu}^{N} l \cdot D(l)}{\sum_{l=1}^{N} l \cdot D(l)},$$

where

$$D(l) = \sum_{i,j}^{N}\left\{(1 - R_{i-1,j-1}) \cdot (1 - R_{i+1,j+1}) \cdot \prod_{k=0}^{l-1} R_{i+k,j+k}\right\}$$

is the histogram of the lengths of the diagonal lines. The understanding of 'determinism' in this sense is of heuristic nature.

The results of calculations of window dynamics of the considered recurrent measures are presented in Fig. 2. Measures RR and DET are calculated for local time series of length in 250 days, with a window of 50 days and a step of 1 day. In this case, the beginning of a crash or critical event is at point 100.

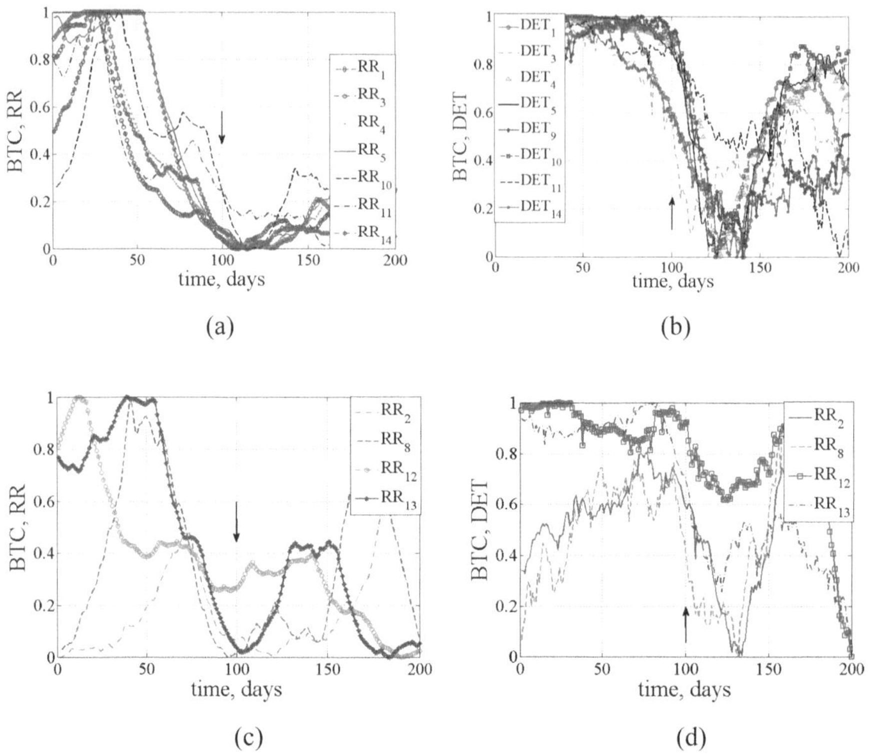

Fig. 2. Dynamics of RR and DET for crashes (a), (b) and for crisis events (c), (d).

It is evident that the two recurrent measures during abnormal periods decrease long before the actual anomaly. The complex system becomes less recurrent and less deterministic which is logical in the periods approaching critical phenomena. And, consequently, RR and DET can be used as precursors of critical and crash phenomena.

5 Permutation Entropy

The Permutation Entropy (PEn) is conceptually simple, computationally a very fast approach which gives an opportunity to quantify complexity in measured time series. Exactly, the measure of entropy is the measure of "randomness". It quantifies the degree of chaos or uncertainty in a system. The uncertainty is associated with a physical process described by the probability distribution

$$P = \{p_i, i = 1, \ldots, M\}$$

is related to the Shannon entropy,

$$S[P] = -\sum_{i=1}^{M} p_i \ln p_i.$$

PEn is based on usual entropy but it is used for the time series analysis of permutation patterns. Bant and Pompe proposed to construct probability distributions using ordinal patterns from recorded time series [35]. These ordinal patterns are constructed based on the relative amplitude of time series values. In this way, if compared with other measures of complexity, this symbolic approach has many advantages over the others as robustness to noise and invariance to nonlinear monotonous transformations [25]. Similar advantages make it particularly attractive for use on experimental data.

If we want to get the ordinal pattern P on which entropy is related, at first we need to define the order of permutations D and ordinal pattern time delay τ. There are $D!$ possible permutations for a vector of length D, so in order to obtain reliable statistics, the length of the time series N should be much larger than $D!$ [35].

The ordinal time delay τ that is responsible for the time scale over which the complexity is quantified can be set by changing. If we change it, we will determine the time separation between values used to construct the vector from which the ordinal pattern is determined. Its value corresponds to a multiple of the signal sampling period. For a given time series $\{u_t, t = 1, \ldots, N\}$, ordinal pattern length D, and ordinal pattern time delay τ, we consider the vector:

$$X_s \rightarrow \left(u_{s-(D-1)\tau}, u_{s-(D-2)\tau}, \ldots, u_{s-\tau}, u_s\right).$$

Relating to the time S equal numbers take their unique symbol according to their position in the time series:

$$\pi = (r_0, r_1, \ldots, r_{D-1}) \text{ defined by } u_{s-r_0\tau} \geq u_{s-r_1\pi} \geq \ldots \geq u_{s-r_{D-2}\tau} \geq u_{s-r_{D-1}\tau}.$$

Then, with all $D!$ possible permutations π_i, the ordinal pattern probability distribution $P = \{p(\pi), i = 1, \ldots, D!\}$ required for entropy calculations is constructed. To take more convenient values, we normalize Permutation Entropy S associating it with the probability distribution P:

$$H_s[P] = \frac{S[P]}{S_{\max}} = \frac{-\sum_{i=1}^{D!} p(\pi_i) \ln p(\pi_i)}{\ln D!}.$$

The values of this normalized permutation have the range $0 \leq H_S \leq 1$ where predictable time series shows a value of zero and absolutely randomize process with a uniform probability distribution presented by a value equal to one. It is important to realize that the Permutation Entropy is a statistical measure and is not able to distinguish whether the observed complexity (irregularity) arises from stochastic or deterministic chaotic processes. It is also important that the PEn provides ways to characterize complexity on different time scales, given by the time delay.

Therefore H_S compared measure of complexity with actual time series under study gives values whose meaning leads to understanding whether we have regular time series or not. Besides, it is understandable that parameter D will not work if we have small values, such as 1 or 2, it is clear that if D is too small, such as 1 or 2, the procedure will not work, because there are only very few distinct states. Enough large parameter D is fine as long as the length of time series can be made proportional to $D!$. The authors of this method recommend using $D = 3, \ldots, 7$. We discovered that $D = 5, 6$, or 7 indicate better result.

Figure 3 shows the PEn calculation results for the entire Bitcoin time series (a) (the window is 100 days, the window offset is 1 day) and also for the local time series of crashes (b) and critical events (c) (the length of the time series is 250 days, the window is 50 days, window offset is 1 day).

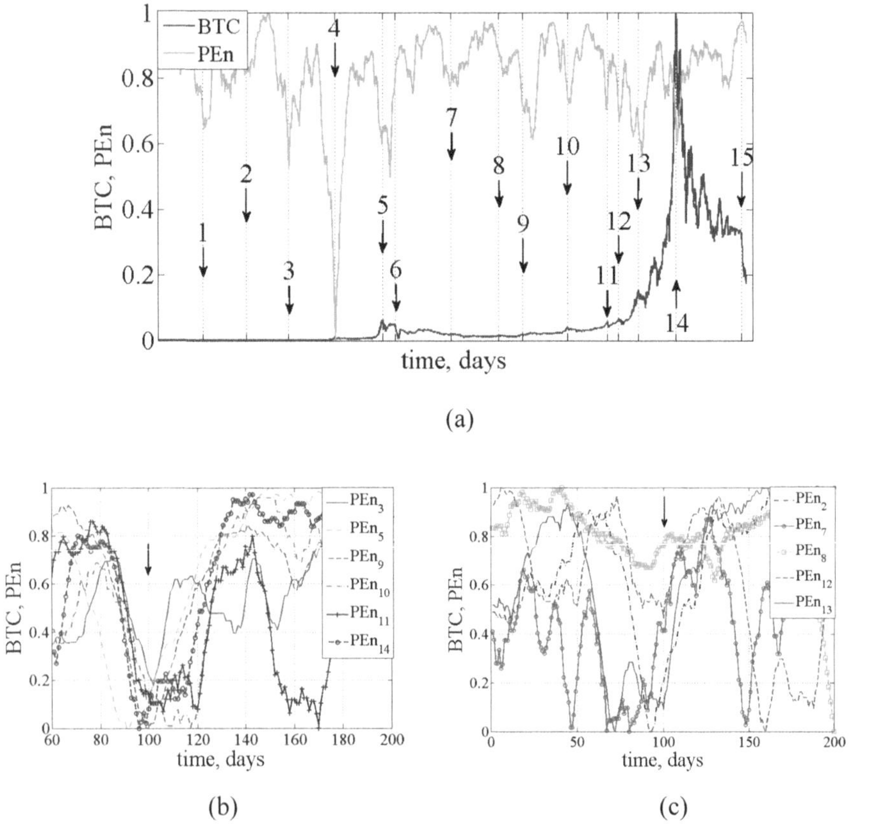

Fig. 3. The dynamics of Permutation Entropy for the entire time series of Bitcoin (a) and for local crashes (b) and events (c). Figure (a) shows the numbers of crashes and critical events in accordance with the Table.

Figure 3 shows that Permutation Entropy decreases for both the entire time series (3a) and for selected crash (3b) or critical (3c) fragments, signaling the approaching of a special state. Comparison of Fig. 3b and c shows that for crash states this behavior is more universal than for critical ones. As in the case of recurrent measures, PEn is an indicator of the precursor of critical and crash phenomena.

6 Complex Network Indicators

The most commonly used methods for converting time sequences to the corresponding networks are recurrent [36], visibility graph [37] and correlation [38]. In the first case, the recurrence diagram is transformed into an adjacency matrix, on which the spectral and topological characteristics of the graph are calculated. The algorithm of the visibility graph is realized as follows. Take a time series $Y(t) = [y_1, y_{2,}, \ldots, y_n]$ of length N. Each point in the time series data can be considered as a vertex in an associated network, and the edge connects two vertices if two corresponding data points can "see" each other from the corresponding point in the time series. Formally, two values of the series y_a (at the time of time t_a) and y_b (at the time of time t_b) are connected, if for any other value (y_c, t_c), which is placed between them (i.e., $t_a < t_c < t_b$), the condition is satisfied:

$$y_c < y_a + (y_b - y_a)\frac{t_c - t_a}{t_b - t_a}$$

To construct and analyze the properties of a correlation graph, we must form a correlation matrix from the set of cryptocurrencies (as is done in Sect. 7), and from it we must pass to the matrix of adjacency. To do this, you must enter a value which, for the correlation field, will be the distance between the correlated assets. Such a distance may be dependent on the correlation coefficients c_{ij} of the value $x(i,j) = \sqrt{2(1 - c_{ij})}$. So, if the correlation coefficient between the two assets is significant, the distance between them is small, and, starting from some critical value, assets can be considered bound on the graph.

For constructed graph methods described above, one can calculate spectral and topological properties. We will show that some of them serve as a measure of the complexity of the system, and the dynamics of their changes allows us to build predictors of crashes or critical events in the financial markets.

Spectral theory of graphs is based on algebraic invariants of a graph - its spectra. The spectrum of graph G is the set of eigenvalues $S_p(G)$ of a matrix corresponding to a given graph. For adjacency matrix A of a graph, there exists an characteristic polynomial $|\lambda I - A|$, which is called the characteristic polynomial of a graph $P_G(\lambda)$. The eigenvalues of the matrix A (the zeros of the polynomial $|\lambda I - A|$) and the spectrum of the matrix A (the set of eigenvalues) are called respectively their eigenvalues λ and the spectrum $S_p(G)$ of graph G. The eigenvalues of the matrix A satisfy the equality $A\bar{x} = \lambda\bar{x}$ ($\bar{x}$ - non-zero vector). Vectors $\bar{x}$ satisfying this equality are called eigenvectors of the matrix A (or the graph G) corresponding to their eigenvalues.

From a multiplicity of spectral and topological measures we will choose only two - the maximum eigenvalue $\lambda_{\max}$ of the adjacency matrix and Average path Length (ApLen). For a connected network of N nodes, the ApLen is equal

$$< l > = \frac{2}{n(N-1)} \sum_{i>j} l_{ij}, \tag{2}$$

where l_{ij} - the length of the shortest path between the nodes i and j.

Figure 4 demonstrates the asymmetric response of the spectral and topological measures of network complexity. For the complete series, the calculation parameters are as follows: window width 100, step is 1 day. For local measures, the length of the fragment is 150, the width of the window is 50 and the step is 1 day.

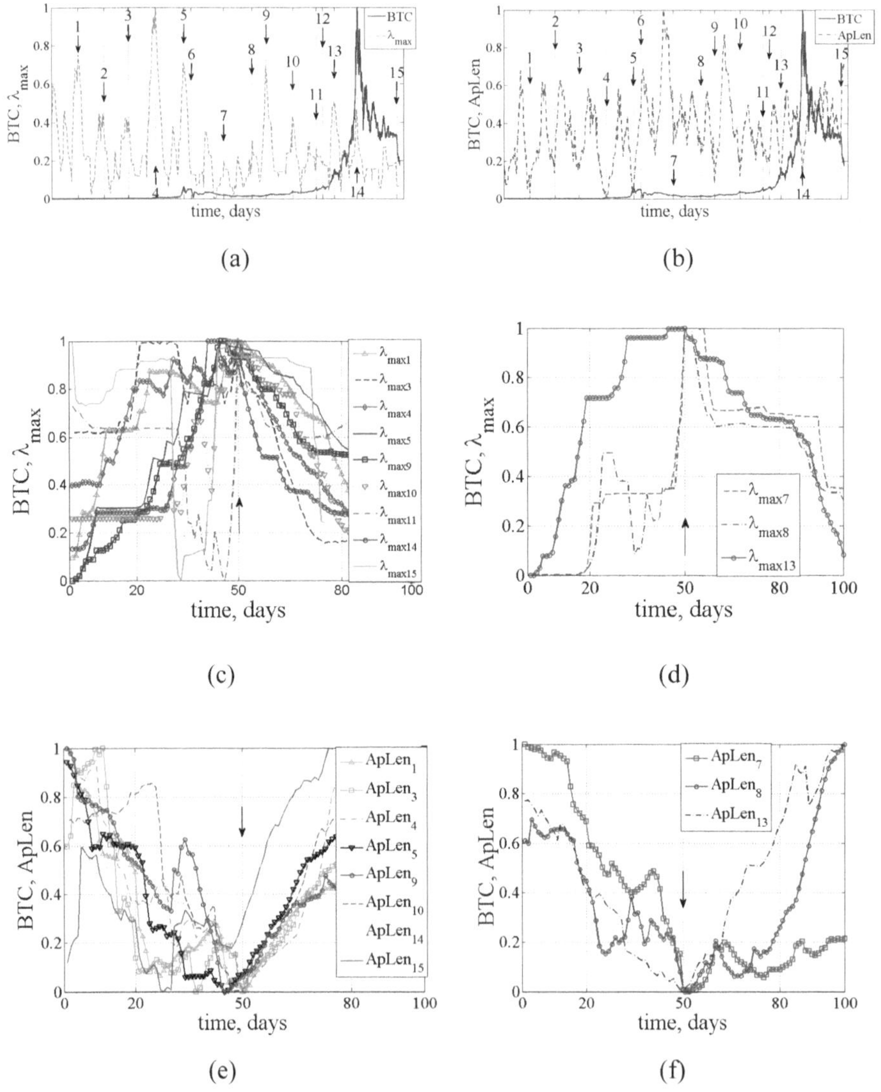

Fig. 4. Visibility graph dynamics of network measures $\lambda_{\max}$ (a), Average path length (b) for all Bitcoin time series. Dynamics of network measures for local crashes (c, e) and crisis events (d, f).

Figure 4 shows the possibility of using both spectral and topological measures of complexity as indicators-precursors of special states in the market of cryptocurrencies. Indeed, the maximum actual value of the adjacency matrix of the visibility graph both for Bitcoin as a whole and for isolated segments of time series containing a crash and critical phenomenon, takes maximum value. It corresponds to the maximum complexity of the system. An especial state of the system leads to a decrease in complexity, and, accordingly, to a decrease in value λ_{max}. Average length of the path on the graph (ApLen) is, on the contrary, minimal for complex systems and increases with the randomization of the system. Such increase during pre-crash and pre-critical states as well as reduce λ_{max} are indicators-precursors of the above mentioned states. You can choose other spectral and topological measures from the calculated ones, e.g. the maximum degree of the vertex and the diameter of the graph, algebraic connectivity and centrality, etc. Network measures of complexity, thus, are the most universal and informative and have obvious advantages in the selection of indicators of special states.

7 Quantum Econophysics Indicators

The attempts to create an adequate model of socio-economic critical events, which, as it has been historically proven, are almost permanent, were, are and will always be made. Actually, it is a super task impossible to solve. However, the potentially useful solutions, local in time or other socio-economic logistic coordinates, are possible. In fact, they have to be the object of interest for a real and effective economic science.

Econophysics is a young interdisciplinary scientific field, which developed and acquired its name at the end of the last century [39]. Quantum econophysics, a direction distinguished by the use of mathematical apparatus of quantum mechanics as well as its fundamental conceptual ideas and relativistic aspects, developed within its boundaries just a couple of years later, in the first decade of the 21st century [40–43].

According to classical physics, immediate values of physical quantities, which describe the system status, not only exist, but can also be exactly measured. Although non-relativistic quantum mechanics doesn't reject the existence of immediate values of classic physical quantities, it postulates that not all of them can be measured simultaneously (Heisenberg uncertainty ratio). Relativistic quantum mechanics denies the existence of immediate values for all kinds of physical quantities, and, therefore, the notion of system status seizes to be algoristic.

In this section, we will demonstrate the possibilities of quantum econophysics on the example of the application of the Heisenberg uncertainty principle and the Random Matrices Theory to the actual and debatable now market of cryptocurrencies.

7.1 Heisenberg Uncertainty Principle and Economic Analogues of Basic Physical Quantities

In our paper [43] we have suggested a new paradigm of complex systems modeling based on the ideas of quantum as well as relativistic mechanics. It has been revealed that the use of quantum-mechanical analogies (such as the uncertainty principle, notion of the operator, and quantum measurement interpretation) can be applied to describing socio-economic processes. Methodological and philosophical analysis of fundamental physical notions and constants, such as time, space and spatial coordinates, mass, Planck's constant, light velocity from the point of view of modern theoretical physics provides an opportunity to search of adequate and useful analogues in socio-economic phenomena and processes.

The Heisenberg uncertainty principle is one of the cornerstones of quantum mechanics. The modern version of the uncertainty principle, deals not with the precision of a measurement and the disturbance it introduces, but with the intrinsic uncertainty any quantum state must possess, regardless of what measurement is performed [44, 45]. Recently, the study of uncertainty relations in general has been a topic of growing interest, specifically in the setting of quantum information and quantum cryptography, where it is fundamental to the security of certain protocols [46, 47].

To demonstrate it, let us use the known Heisenberg's uncertainty ratio which is the fundamental consequence of non-relativistic quantum mechanics axioms and appears to be (e.g. [48]):

$$\Delta x \cdot \Delta v \geq \frac{\hbar}{2m_0}, \tag{3}$$

where Δx and Δv are mean square deviations of x coordinate and velocity v corresponding to the particle with (rest) mass m_0, $\hbar$ - Planck's constant. Considering values Δx и Δv to be measurable when their product reaches its minimum, we derive (from (1)):

$$m_0 = \frac{\hbar}{2 \cdot \Delta x \cdot \Delta v}, \tag{4}$$

i.e. mass of the particle is conveyed via uncertainties of its coordinate and velocity – time derivative of the same coordinate.

Economic measurements are fundamentally relative, are local in time, space and other socio-economic coordinates, and can be carried out via consequent and/or parallel comparisons "here and now", "here and there", "yesterday and today", "a year ago and now" etc.

Due to these reasons constant monitoring, analysis, and time series prediction (time series imply data derived from the dynamics of stock indices, exchange rates, cryptocurrencies prices, spot prices and other socio-economic indicators) becomes relevant for evaluation of the state, tendencies, and perspectives of global, regional, and national economies.

Suppose there is a set of K time series, each of N samples, that correspond to the single distance T, with an equal minimal time step $\Delta t_{\min}$:

$$X_i(t_n),\ t_n = \Delta t_{\min} n; n = 0, 1, 2, \ldots N - 1; i = 1, 2, \ldots K. \tag{5}$$

To bring all series to the unified and non-dimensional representation, accurate to the additive constant, we normalize them, having taken a natural logarithm of each term of the series. Then consider that every new series $x_i(t_n)$ is a one-dimensional trajectory of a certain fictitious or abstract particle numbered i, while its coordinate is registered after every time span $\Delta t_{\min}$, and evaluate mean square deviations of its coordinate and speed in some time window $\Delta T = \Delta N \cdot \Delta t_{\min} = \Delta N, 1 << \Delta N << N$. The «immediate» speed of i particle at the moment t_n is defined by the ratio:

$$v_i(t_n) = \frac{x_i(t_{n+1}) - x_i(t_n)}{\Delta t_{\min}} = \frac{1}{\Delta t_{\min}} ln \frac{X_i(t_{n+1})}{X_i(t_n)} \tag{6}$$

with variance D_{v_i} and mean square deviation Δv_i.

Keeping an analogy with (1) after some transformations we can write an uncertainty ratio for this trajectory [49]:

$$\frac{1}{\Delta t_{\min}} \left(< \ln^2 \frac{X_i(t_{n+1})}{X_i(t_n)} >_{n,\Delta N} - \left(< \ln \frac{X_i(t_{n+1})}{X_i(t_n)} >_{n,\Delta N} \right)^2 \right) \sim \frac{h}{m_i}, \tag{7}$$

where m_i - economic "mass" of an i series, h - value which comes as an economic Planck's constant.

Since the analogy with physical particle trajectory is merely formal, h value, unlike the physical Planck's constant $\hbar$, can, generally speaking, depend on the historical period of time, for which the series are taken, and the length of the averaging interval (e.g. economical processes are different in the time of crisis and recession), on the series number i etc. Whether this analogy is correct or not depends on particular series' properties.

In recent work [50], we tested the economic mass as an indicator of crisis phenomena on stock index data. In this work we will test the model for the cryptocurrency market on the example of the Bitcoin [51].

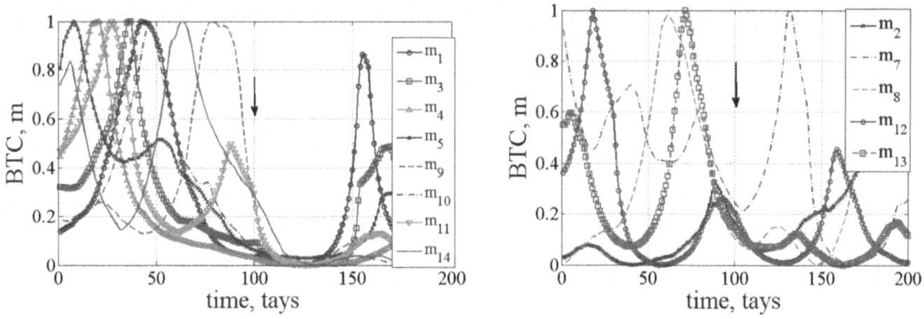

(a) (b)

Fig. 5. Dynamics of measure m for local crashes (a) and critical events (b).

Obviously, there is a dynamic characteristic values m depending on the internal dynamics of the market. In times of crashes known marked by arrows in the Figs. 5(a) and 5(b) mass m is significantly reduced in the pre-crash and pre-critical periods.

Obviously, that the value of m remains a good indicator-precursor even in this case. Value m is considerably reduced before a special market condition. The market becomes more volatile and prone to changes.

The following method of quantum econophysics is borrowed from nuclear physicists and is called Random Matrix Theory.

7.2 Random Matrix Theory and Quantum Indicators-Predictors

Random Matrix Theory (RMT) developed in this context the energy levels of complex nuclei, which the existing models failed to explain (Wigner, Dyson, Mehta, and others [52–54]). Deviations from the universal predictions of RMT identify system specific, nonrandom properties of the system under consideration, providing clues about the underlying interactions.

Unlike most physical systems, where one relates correlations between subunits to basic interactions, the underlying "interactions" for the stock market problem are not known. Here, we analyze cross correlations between stocks by applying concepts and methods of random matrix theory, developed in the context of complex quantum systems where the precise nature of the interactions between subunits are not known.

RMT has been applied extensively in studying multiple financial time series [55–59].

In order to quantify correlations, we first calculate the logarithmic return (1) of the i cryptocurrencies price series over a time scale $\Delta t = 1$ day. It was selected 24 established during the last 5 years the most capitalized cryptocurrencies for the period from 04.08.2013 to 08.12.2018 (https://coinmarketcap.com/all/views/all/). We calculate the pairwise cross-correlation coefficients between any two cryptocurrencies returns time series. For the correlation matrix C we can calculate its eigenvalues, $C = U\Lambda U^T$, where U denotes the eigenvectors, Λ is the eigenvalues of the correlation matrix, whose density $f_c(\lambda)$ is defined as follows, $f_c(\lambda) = (1/N)dn(\lambda)/d\lambda$, where $n(\lambda)$ is the number of eigenvalues of C that are less than λ. In the limit $N \to \infty, T \to \infty$ and $Q = T/N \geq 1$ fixed, the probability density function $f_c(\lambda)$ of eigenvalues λ of the random correlation matrix M has a close form:

$$f_c(\lambda) = \frac{Q}{2\pi\sigma^2}\frac{\sqrt{(\lambda_{\max} - \lambda)(\lambda - \lambda_{\min})}}{\lambda} \tag{8}$$

with $\lambda \in [\lambda_{\min}, \lambda_{\max}]$, where $\lambda_{\min}^{\max}$ is given by $\lambda_{\min}^{\max} = \sigma^2(1 + 1/Q \pm 2\sqrt{1/Q})$ and σ^2 is equal to the variance of the elements of matrix M.

We compute the eigenvalues of the correlation matrix C, $\lambda_{\max} = \lambda_1 > \lambda_2 > \cdots > \lambda_{15} = \lambda_{\min}$. We find that the largest eigenvalue $\lambda_{\max} = 5.48$ and the smallest eigenvalue $\lambda_{\min} = 0.81$. If C is a random matrix, the largest eigenvalue $\lambda_{\max}^{RMT} = 1.45$ and the smallest eigenvalue $\lambda_{\min}^{RMT} = 0.63$, according to Eq. (8). In our case, only one-third of its own values refer to the RMT region.

Eigenvectors correspond to the participation ratio PR and its inverse participation ratio IPR $I^k = \sum_{l=1}^{N} \left[u_l^k\right]^4$, where u_l^k, $l = 1, \ldots, N$ are the components of the eigenvector u^k. Figure 6 shows the comparative characteristics of the eigenvalue distributions for the random matrix (shuffled) and real (a) and the corresponding values of IPR (b). The difference in dynamics is due to the peculiarities of non-random correlations between the time series of individual assets. Under the framework of random matrix theory, if the eigenvalues of the real time series differ from the prediction of random matrix theory, there must exists hidden economic information in those deviating eigenvalues. For cryptocurrencies markets, there are several deviating eigenvalues in which the largest eigenvalue $\lambda_{\max}$ reflects a collective effect of the whole market. As for PR the differences from RMT appear at large and small lambda values and are similar to the Anderson quantum effect of localization [60]. Under crashes conditions, the states at the edges of the distributions of eigenvalues are delocalized, thus identifying the beginning of the crash. This is evidenced by the results presented in Fig. 6 (c).

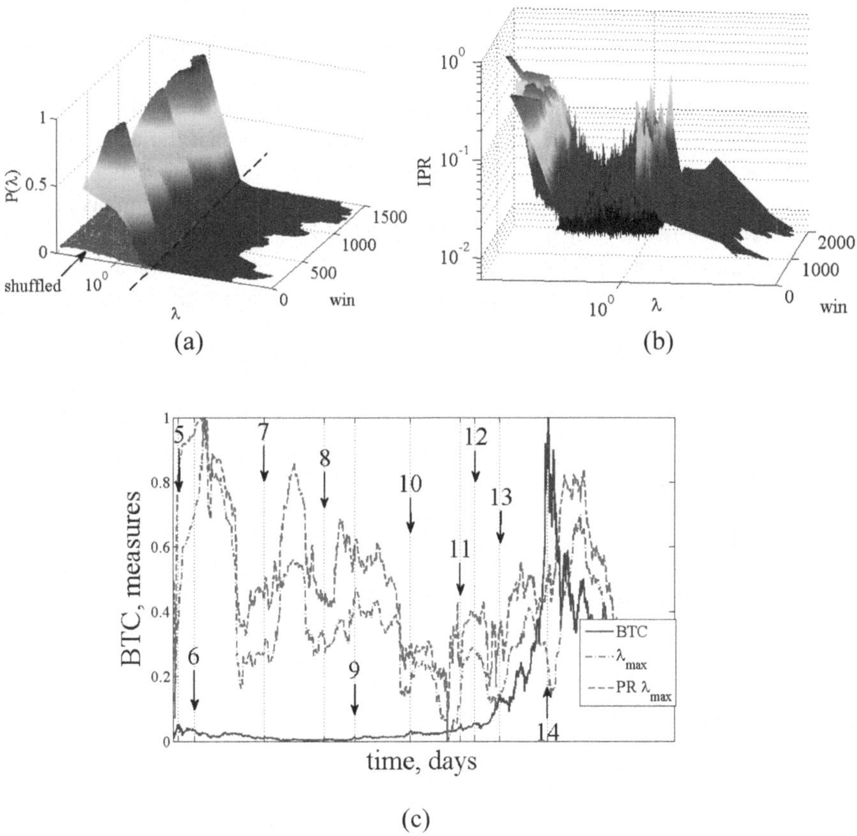

Fig. 6. Window dynamics of the distribution of eigenvalues (a), inverse participation ratio (b) for the initial and mixed (or random) matrices and quantum measures of complexity $\lambda_{\max}$ and its participation ratio

We find that both $\lambda_{\max}$ and *PR* of $\lambda_{\max}$ have large values for periods containing the market crashes and critical events. At the same time, their growth begins in the pre-crashes periods. Means, as well as the economic mass, they are quantum precursors of crashes and critical events phenomena.

8 Conclusions

Consequently, in this paper, we have shown that monitoring and prediction of possible critical changes on cryptocurrency is of paramount importance. As it has been shown by us, the theory of complex systems has a powerful toolkit of methods and models for creating effective indicators - precursors of crashes and critical phenomena. In this paper, we have explored the possibility of using the recurrent, entropy, network and quantum measures of complexity to detect dynamical changes in a complex time series. We have shown that the measures that have been used can indeed be effectively used to detect abnormal phenomena for the time series of Bitcoin.

We have shown that monitoring and prediction of possible critical changes on cryptocurrency is of paramount importance. As it has been shown by us, the quantum econophysics has a powerful toolkit of methods and models for creating effective indicators-precursors of crisis phenomena. In this paper, we have explored the possibility of using the Heisenberg uncertainty principle and random matrix theory to detect dynamical changes in a complex time series. We have shown that the economic mass m, and the largest eigenvalue $\lambda_{\max}$ may be effectively used to detect crisis phenomena for the cryptocurrencies time series. We have concluded though by emphasizing that the most attractive features of the m, $\lambda_{\max}$ and PR of $\lambda_{\max}$ namely its conceptual simplicity and computational efficiency make it an excellent candidate for a fast, robust, and useful screener and detector of unusual patterns in complex time series.

References

1. Halvin, S., Cohen, R.: Complex Networks: Structure, Robustness and Function. Cambridge University Press, New York (2010)
2. Albert, R., Barabási, A.-L.: Statistical mechanics of complex networks. Rev. Mod. Phys. **74**, 47–97 (2002)
3. Newman, M., Watts, D., Barabási, A.-L.: The Structure and Dynamics of Networks. Princeton University Press, Princeton and Oxford (2006)
4. Newman, M.E.J.: The structure and function of complex networks. SIAM Rev. **45**(2), 167–256 (2003)
5. Nikolis, G., Prigogine, I.: Exploring Complexity: An Introduction. W. H. Freeman and Company, New York (1989)
6. Andrews, B., Calder, M., Davis, R.: Maximumlikelihood estimation for α-stable autoregressive processes. Ann. Stat. **37**, 1946–1982 (2009)
7. Dassios, A., Li, L.: An economic bubble model and its first passage time. arXiv:1803.08160v1 [q-fin.MF]. Accessed 15 Sept 2018
8. Tarnopolski, M.: Modeling the price of Bitcoin with geometric fractional Brownian motion: a Monte Carlo approach. arXiv:1707.03746v3 [q-fin.CP]. Accessed 15 Sept 2018
9. Kodama, O., Pichl, L., Kaizoji, T.: Regime change and trend prediction for Bitcoin time series data. In: CBU International Conference on Innovations in Science and Education, Prague, pp. 384–388 (2017). www.cbuni.cz, www.journals.cz, https://doi.org/10.12955/cbup.v5.954
10. Shah, D., Zhang, K.: Bayesian: regression and Bitcoin. arXiv:1410.1231v1 [cs.AI]. Accessed 15 Oct 2018

11. Chen, T., Guestrin, C.: XGBoost: a scalable tree boosting system. In: Proceedings of the 22nd International Conference on Knowledge Discovery and Data Mining, pp. 785–794. ACM, San Francisco (2016)
12. Alessandretti, L., ElBahrawy, A., Aiello, L.M., Baronchelli, A.: Machine learning the cryptocurrency market. arXiv:1805.08550v1 [physics.soc-ph]. Accessed 15 Sept 2018
13. Guo, T., Antulov-Fantulin, N.: An experimental study of Bitcoin fluctuation using machine learning methods. arXiv:1802.04065v2 [stat.ML]. Accessed 15 Sept 2018
14. Albuquerque, P., de Sá, J., Padula, A., Montenegro, M.: The best of two worlds: forecasting high frequency volatility for cryptocurrencies and traditional currencies with support vector regression. Expert Syst. Appl. **97**, 177–192 (2018). https://doi.org/10.1016/j.eswa.2017.12.004
15. Wang, M., et al.: A novel hybrid method of forecasting crude oil prices using complex network science and artificial intelligence algorithms. Appl. Energy **220**, 480–495 (2018). https://doi.org/10.1016/j.apenergy.2018.03.148
16. Kennis, M.: A Multi-channel online discourse as an indicator for Bitcoin price and volume. arXiv:1811.03146v1 [q-fin.ST]. Accessed 6 Nov 2018
17. Donier, J., Bouchaud, J.P.: Why do markets crash? Bitcoin data offers unprecedented insights. PLoS One **10**(10), 1–11 (2015). https://doi.org/10.1371/journal.pone.0139356
18. Bariviera, F.A., Zunino, L., Rosso, A.O.: An analysis of high-frequency cryptocurrencies price dynamics using permutation-information-theory quantifiers. Chaos **28**(7), 07551 (2018). https://doi.org/10.1063/1.5027153
19. Senroy, A.: The inefficiency of Bitcoin revisited: a high-frequency analysis with alternative currencies. Financ. Res. Lett. (2018). https://doi.org/10.1016/j.frl.2018.04.002
20. Marwan, N., Schinkel, S., Kurths, J.: Recurrence plots 25 years later - gaining confidence in dynamical transitions. Europhys. Lett. **101**(2), 20007 (2013). https://doi.org/10.1209/0295-5075/101/20007
21. Santos, T., Walk, S., Helic, D.: Nonlinear characterization of activity dynamics in online collaboration websites. In: Proceedings of the 26th International Conference on World Wide Web Companion, WWW 2017 Companion, Australia, pp. 1567–1572 (2017). https://doi.org/10.1145/3041021.3051117
22. Di Francesco Maesa, D., Marino, A., Ricci, L.: Data-driven analysis of Bitcoin properties: exploiting the users graph. Int. J. Data Sci. Anal. **6**(1), 63–80 (2018). https://doi.org/10.1007/s41060-017-0074-x
23. Bovet, A., Campajola, C., Lazo, J.F., et al.: Network-based indicators of Bitcoin bubbles. arXiv:1805.04460v1 [physics.soc-ph]. Accessed 11 Sept 2018
24. Kondor, D., Csabai, I., Szüle, J., Pósfai, M., Vattay, G.: Infferring the interplay of network structure and market effects in Bitcoin. New J. Phys. **16**, 125003 (2014). https://doi.org/10.1088/1367-2630/16/12/125003
25. Wheatley, S., Sornette, D., Huber, T., et al.: Are Bitcoin bubbles predictable? Combining a generalized Metcalfe's law and the LPPLS model. arXiv:1803.05663v1 [econ.EM]. Accessed 15 Sept 2018
26. Gerlach, J-C., Demos, G., Sornette, D.: Dissection of Bitcoin's multiscale bubble history from January 2012 to February 2018. arXiv:1804.06261v2 [econ.EM]. Accessed 15 Sept 2018
27. Soloviev, V., Belinskiy, A.: Methods of nonlinear dynamics and the construction of cryptocurrency crisis phenomena precursors. arXiv:1807.05837v1 [q-fin.ST]. Accessed 30 Sept 2018
28. Casey, M.B.: Speculative Bitcoin adoption/price theory. https://medium.com/@mcasey0827/speculative-bitcoin-adoption-price-theory-2eed48ecf7da. Accessed 25 Sept 2018

29. McComb, K.: Bitcoin crash: analysis of 8 historical crashes and what's next. https://blog.purse.io/bitcoin-crash-e112ee42c0b5. Accessed 25 Sept 2018
30. Amadeo, K.: Stock market corrections versus crashes and how to protect yourself: how you can tell if it's a correction or a crash. https://www.thebalance.com/stock-market-correction-3305863. Accessed 25 Sep 2018
31. Webber, C.L., Marwan, N. (eds.): Recurrence Plots and Their Quantifications: Expanding Horizons. Proceedings of the 6th International Symposium on Recurrence Plots, Grenoble, France, 17–19 June 2015, vol. 180, pp. 1–387. Springer, Heidelberg (2016). https://doi.org/10.1007/978-3-319-29922-8
32. Marwan, N., Wessel, N., Meyerfeldt, U., Schirdewan, A., Kurths, J.: Recurrence plot based measures of complexity and its application to heart rate variability data. Phys. Rev. E **66**(2), 026702 (2002)
33. Zbilut, J.P., Webber Jr., C.L.: Embeddings and delays as derived from quantification of recurrence plots. Phys. Lett. A **171**(3–4), 199–203 (1992)
34. Webber Jr., C.L., Zbilut, J.P.: Dynamical assessment of physiological systems and states using recurrence plot strategies. J. Appl. Physiol. **76**(2), 965–973 (1994)
35. Bandt, C., Pompe, B.: Permutation entropy: a natural complexity measure for time series. Phys. Rev. Lett. **88**(17), 2–4 (2002)
36. Donner, R.V., Small, M., Donges, J.F., Marwan, N., et al.: Recurrence-based time series analysis by means of complex network methods. arXiv:1010.6032v1 [nlin.CD]. Accessed 25 Oct 2018
37. Lacasa, L., Luque, B., Ballesteros, F., et al.: From time series to complex networks: the visibility graph. PNAS **105**(13), 4972–4975 (2008)
38. Burnie, A.: Exploring the interconnectedness of cryptocurrencies using correlation networks. In: The Cryptocurrency Research Conference, pp. 1–29. Anglia Ruskin University, Cambridge (2018)
39. Mantegna, R.N., Stanley, H.E.: An Introduction to Econophysics: Correlations and Complexity in Finance. Cambridge University Press, Cambridge (2000)
40. Maslov, V.P.: Econophysics and quantum statistics. Math. Notes **72**, 811–818 (2002)
41. Hidalgo, E.G.: Quantum Econophysics. arXiv:physics/0609245v1 [physics.soc-ph]. Accessed 15 Sept 2018
42. Saptsin, V., Soloviev, V.: Relativistic quantum econophysics - new paradigms in complex systems modelling. arXiv:0907.1142v1 [physics.soc-ph]. Accessed 25 Sept 2018
43. Colangelo, G., Clurana, F.M., Blanchet, L.C., Sewell, R.J., Mitchell, M.W.: Simultaneous tracking of spin angle and amplitude beyond classical limits. Nature **543**, 525–528 (2017)
44. Rodriguez, E.B., Aguilar, L.M.A.: Disturbance-disturbance uncertainty relation: the statistical distinguishability of quantum states determines disturbance. Sci. Rep. **8**, 1–10 (2018)
45. Rozema, L.A., Darabi, A., Mahler, D.H., Hayat, A., Soudagar, Y., Steinberg, A.M.: Violation of Heisenberg's measurement-disturbance relationship by weak measurements. Phys. Rev. Lett. **109**, 100404 (2012)
46. Prevedel, R., Hamel, D.R., Colbeck, R., Fisher, K., Resch, K.J.: Experimental investigation of the uncertainty principle in the presence of quantum memory. Nat. Phys. **7**(29), 757–761 (2011)
47. Berta, M., Christandl, M., Colbeck, R., Renes, J., Renner, R.: The uncertainty principle in the presence of quantum memory. Nat. Phys. **6**(9), 659–662 (2010)
48. Landau, L.D., Lifshitis, E.M.: The Classical Theory of Fields. Course of Theoretical Physics. Butterworth-Heinemann, Oxford (1975)
49. Soloviev, V., Saptsin, V.: Heisenberg uncertainty principle and economic analogues of basic physical quantities. arXiv:1111.5289v1 [physics.gen-ph]. Accessed 15 Sept 2018

50. Soloviev, V.N., Romanenko, Y.V.: Economic analog of Heisenberg uncertainly principle and financial crisis. In: 20-th International Conference SAIT 2017, pp. 32–33. ESC "IASA" NTUU "Igor Sikorsky Kyiv Polytechnic Institute", Ukraine (2017)
51. Soloviev, V.N., Romanenko, Y.V.: Economic analog of Heisenberg uncertainly principle and financial crisis. In: 20-th International Conference SAIT 2018, pp. 33–34. ESC "IASA" NTUU "Igor Sikorsky Kyiv Polytechnic Institute", Ukraine (2018)
52. Wigner, E.P.: On a class of analytic functions from the quantum theory of collisions. Ann. Math. **53**, 36–47 (1951)
53. Dyson, F.J.: Statistical theory of the energy levels of complex systems. J. Math. Phys. **3**, 140–156 (1962)
54. Mehta, L.M.: Random Matrices. Academic Press, San Diego (1991)
55. Laloux, L., Cizeau, P., Bouchaud, J.-P., Potters, M.: Noise dressing of financial correlation matrices. Phys. Rev. Lett. **83**, 1467–1470 (1999)
56. Plerou, V., Gopikrishnan, P., Rosenow, B., Amaral, L.A.N., Guhr, T., Stanley, H.E.: Random matrix approach to cross correlations in financial data. Phys. Rev. E **65**, 066126 (2002)
57. Shen, J., Zheng, B.: Cross-correlation in financial dynamics. EPL (Europhys. Lett.) **86**, 48005 (2009)
58. Jiang, S., Guo, J., Yang, C., Tian, L.: Random matrix analysis of cross-correlation in energy market of Shanxi, random matrix analysis of cross-correlation in energy market of Shanxi, China. Int. J. Nonlinear Sci. **23**(2), 96–101 (2017)
59. Urama, T.C., Ezepue, P.O., Nnanwa, C.P.: Analysis of cross-correlations in emerging markets using random matrix theory. J. Math. Financ. **7**, 291–307 (2017)
60. Anderson, P.W.: Absence of diffusion in certain random lattices. Phys. Rev. **109**, 1492 (1958)

Implementation of Robo-Advisor Services for Different Risk Attitude Investment Decisions Using Machine Learning Techniques

Oleksandr Snihovyi, Vitaliy Kobets(✉), and Oleksii Ivanov

Kherson State University, 27, Universitetska st., Kherson 73000, Ukraine
snegovoy@hotmail.com, vkobets@kse.org.ua,
sink2385@gmail.com

Abstract. In this paper we have researched how to use machine learning in the financial industry on the example of robo-advisor; defined the basic functionality of robo-advisor, an implementation of robo-advisor based on analysis of the most popular financial services, such as Betterment, FutureAdvisor, Motif Investing, Schwab Intelligent and Wealthfront. We have also compared their functionality, formulated a list of critical features and described our own high-level architecture design of a general robo-advisor tool for private investors. Our goal is to build three application modules for a single robo-advisor which combines its architecture and modern financial instruments – cryptocurrencies for the first time. The first module is a Long short-term memory (LSTM) neural network, which forecasts cryptocurrencies prices daily. As a result of simulation experiment through the application using real data from open sources, we have found that the combination of criterion can explain 61% of cryptocurrencies prices variation. The second module uses robo-advising approach to build an investment plan for novice cryptocurrencies investors with different risk attitude investment decisions. The third module is ETL (Extract-Transform-Load) for a statistics dataset and neural networks models. Results of the investigation show that investing in cryptocurrencies can give 23.7% per year for risk-averse, 31.8% per year for risk-seeking investors and 16.5% annually for investors of hybrid type.

Keywords: Robo-advisor · Markowitz model · Financial instruments · Neural networks · Machine learning

1 Introduction

Intelligent data analysis is one of the areas of artificial intelligence which solves the problem of learning automatic systems without their explicit programming, focuses on developing algorithms that are self-learning based on the proposed data [1].

Financial corporations that need to adapt quickly to the environment have realized that it is more efficient to develop self-learning systems that manually improve existing systems as needed. It saves the resources of the company and optimizes the process of developing a financial software product. However, according to a Bloomberg survey in

V. Ermolayev et al. (Eds.): ICTERI 2018, CCIS 1007, pp. 298–321, 2019.
https://doi.org/10.1007/978-3-030-13929-2_15

2017 in New York, only 16% of firms have introduced Machine Learning into their investment strategies and software [2].

Robo-advisor is a favorite tool for novice investors, which builds an investment plan and calibrates it from time to time based on a current market state. Robo-advisor is digital platforms consisting of interactive and intelligent user assistance components [3]. Investment goals of a client, risk preferences are quantified by algorithms using automated processes. Robo-advisor differs from existing online investment platforms with two issues: customer assessment (online questionnaires) and customer portfolio management (which includes several financial instruments that require no active decision concerning portfolio management like Exchange Traded Funds) [4]. This combination of financial instruments and algorithms can considerably reduce management cost through full automation. At the same time, architecture of existing robo-advisors is hidden as commercial classified information.

Generally, it allows investing in stocks, bonds, ETFs and mutual funds. However, now there is one additional way to invest money – cryptocurrencies. They had become famous from the beginning of 2017, when most of the cryptocurrencies were created and their prices started increasing. Investments in cryptocurrencies are considered too risky because of hype. For example, Bitcoin (the most popular cryptocurrency) has raised more than $19,000 only to fall sharply within minutes at the end of 2017. Currently, a price of Bitcoin for USD is much less than it was (according to Coindesk.com the price of Bicoin on September 6, 2018 is $6,402.69). Most novice cryptocurrencies investors do not know which coin they should invest to and how much. It leads to losing money in a long-run period.

The **purpose** of the paper is to develop the architecture of a robo-advisor which can work with cryptocurrencies to help novice investors to increase their capital without fear of losing money. The **task** is to use robo-advising approach through a minimum viable product (MPV) for cryptocurrencies trading composed of three modules: the first will be responsible for prices forecasting via LSTM neural network; the second will build an investment plan based on forecasted data from the first module; the third will parse and manage models for the first module.

The paper is organized as follows: part two describes related works; part three describes application modules for our robo-advisor; part four includes criteria of influence on cryptocurrencies prices using machine learning techniques; part 5 introduces modules of a robo-advisor; part six describes the results of experiment for robo-advisor service; the last part concludes.

2 Related Works

2.1 Robo-Advisor

The first cryptocurrency was born in 2008. In the same year, the first robo-advisor were launched. Both of them were invented to develop and improve the financial industry. Cryptocurrencies can be used to prevent fraud [5], de-corrupt charities [6], and many other things. Robo-advisor helps to automate investment portfolio rebalancing, wealth management and get rid of brokers and human financial advisors. Also, they are great for risk-averse investors, who do not want to invest much money in the beginning by offering a minimum investment sum. Some of them do not have the minimum investment sum at all.

Kobets, Yatsenko, Mazur, and Zubrii have analyzed robo-advisor to find its strengths, weaknesses, opportunities, and threats [7]. As the result of their research, robo-advisor has more strengths and opportunities than weaknesses and threats. The primary challenge for robo-advisor now is a necessity of face-to-face interaction with an advisor for users, especially for novice investors and robo-advisor inability to follow up with questions and make recommendations based on the answers. Cocca in his research [8] says that currently robo-advisor can manage only low-complexity financial decisions.

From our side in the previous research, we defined the basic functionality for a robo-advisor and described its general high-level architecture [9]. We defined a couple of modules there, but the most important are: Investment plan module, Calculations module, and Parser module. These three modules are major. The most crucial job belongs to them: Parser module gets and updates data regularly, which is vital for keeping users investment plans up to date; Calculations module predicts prices based on up to date data; Investment plan module uses data from Calculations module to rebuild users investment plans daily.

Fein studied robo-advisor in 2015 [10] and described that it offers an advice based on responses to specific questions, portfolio rebalancing or reallocating of investments that primarily represents the main functionality of such tools.

Investors who plan to use robo-advisor have to take into account both aspects – advantages and risks. In fact, the European Banking Authority (EBA) states that robo-advisor poses potentially a great risk to both wealth management firms and customers [11]. According to the survey of [12], there is a drop in the percentage of end-user who are highly willing to try Robo-Advisors. Furthermore, there are indications that financial advisors do not find that robo-advisors have a strong effect on their businesses. Echoing this view, Nanalyze, an investor information service group, expressed disappointment with Robo-Advisors, stating that the embedded technology does not live up to the hype [13].

Based on studied works, we have found out that separating on modules, as we did in our previous research, will help robo-advisor to do its job better in term of performance. So, we have used our previous approach here in part three.

2.2 Cryptocurrencies

We studied criteria which affect prices of cryptocurrencies [14] and found out that combination of supply, mining difficulty, trading volume, and news reaction for each date can predict more than 70% of the price (we used Bitcoin for research).

Philips and Gorse studied how to predict cryptocurrency prices bubbles using epidemic modeling and human reaction on social media [15].

Also, Colianni, Rosales, and Signorotti investigated cryptocurrencies algorithmic trading techniques based on Twitter sentiments analysis [16]. Lamon, Nielsen, and Redondo studied cryptocurrency price changes based on news and Reddit sentiments

[17]. Kim et al. in 2016 did significant research about how users activities in communities affect prices of cryptocurrencies [18].

All researches we have mentioned above show that users activities affect prices. However, we applied a different approach in this research. Our idea was to predict cryptocurrencies prices based on their daily trading volume. The results are in part four.

2.3 Machine Learning Role in the Financial Industry

The ability of computer programs to learn and improve themselves has become a conventional technology continuously growing in all industries. Large companies like Google, Facebook, Amazon use Machine Learning (ML) to improve performance, user experience, and data security. In the financial industry, the following areas were affected by ML [19]:

- Fraud Prevention;
- Risk Management;
- Customer Service;
- Virtual Assistant;
- Network Security;
- Algorithmic Trading;
- Investment Portfolio Management.

All of these areas combine such a process as forecasting. Also, they all carry a vast array of data that can be combined to create a detailed view [19]. It is the primary component of ML. Having an extensive multi-layered data where each layer affects others the goal is to find a pattern and to forecast the next values, or based on found values provide the most profitable solution. Also, it is not the only one advantage because the ML's knowledge base always increases, the later forecasts will be much accurate than they were in the beginning. Let's consider some examples of situations and possible scenarios of using ML in FinTech. For example, Virtual Assistant (VA) is an integral part of any high-quality product, especially financial software like online banking. VA can save bank's money and minimize the cost of real assistance. In case of regular money receipt to a customer's account and positive account balance after all withdrawals, well-trained VA can propose a profitable type of deposit. Also, if the bank has an assignment in partnership with MasterCard (or any other company), VA may offer to all owners of MasterCard cards some unique bonus. However, if the customer regularly rejects the same bonuses in the past VA can mark such customer as not a part of the target audience (but, of course, he can find all information about bonuses by himself). Another excellent example of using intelligent data analysis is algorithmic trading – a method of executing a large order using a programmed algorithm based on trading instructions. Usually, to succeed such software should have a big dataset with all values even those which affect the main one (for example, goods prices, the costs of raw materials, the costs for creating and sale) for an extended period. Having so much information to learn, the ML algorithm can forecast numbers, and traders will know either they need to buy, to sell, or to wait.

However, constant living income (CLI) can be the most common usage of ML in the financial industry. It is a type of income that does not depend on daily activities.

(e.g. investment, ownership or deposits). CLI combined all ML areas used in financial industry. ML can automate the process of getting CLI (through offering new types of income, different forecasts, and metrics) and this process will be improved continuously.

3 Robo-Advisor as Financial Software

A good example of financial software for making passive income and managing a financial investment portfolio is robo-advisor (RA). Now, this software is common, but until 2008, this term did not even exist.

RA is a set of algorithms, which calibrates investment portfolio based on customer's goals and risks. The customer enters his goal, age, current income and financial assets. For example, 30 years old man with a salary of $120 000 per year has saved $100 000, and he plans to retire at the age of 50 with $10 000 000 savings. The system begins to offer the expansion of investment between classes of assets and financial instruments to achieve customer's goals. Also, it calibrates the expansion based on changes to the customer's goals and market changes in real time. So, RA always tries to find what is most closely related to the clinet's goals [20, 21]. Unfortunately, RAs algorithms are unknown to the public because they are a commercial secret. We try to overcome this gap using our own approach with open data about course of financial instruments.

There are many of RAs, but only 5 of them were chosen as the most popular, to review and to define the main functionality. The comparison of some features is shown in the following Table 1.

Table 1. RAs features comparison

Feature	Betterment	FutureAdvisor	Motif investing	Schwab intelligent	Wealthfront
The user can create their own account	+	+	+	+	+
Two-factor authentication	+ (sms only)	–	–	–	–
Portfolio rebalancing	+	+	–	+	+
Advice	+ (Human)	+ (Automated)	+ (Automated)	+ (Huma)	+ (Automated)
Customer Service	+	+	+	+	+
Mutual funds	+	+	–	–	+
Fees	Digital -0.25%/ year; Premium – 0.40%/year	0.50%/year	$9.95/trade	0.28%; $900 quarterly cap	0.25%, but first $10,000 is free

(continued)

Table 1. *(continued)*

Feature	Betterment	FutureAdvisor	Motif investing	Schwab intelligent	Wealthfront
Retirement planning	+	+	+	+	+
Automated investments	+	+	+	+	+

Betterment is the one of the oldest RA (dashboard screen is shown in the Fig. 1). The company has developed reliable software to help novice investors. The user should set how much they plan to invest into ETFs (Exchange Trade Fund, the investment fund), and how much into ETFs bonds. There is no minimum deposit to open an account. One commission is charged in the range of 0.15% to 0.35% based on the balance of the account. RA has easy-to-use tools, which help investors decide on the distribution of stocks, bonds, and other financial instruments as cryptocurrencies [22, 23].

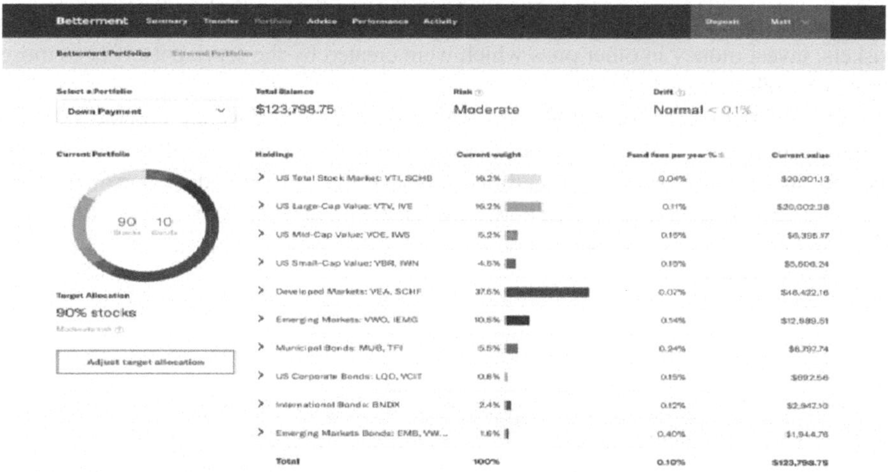

Fig. 1. Betterment dashboard screen [24]

FutureAdvisor is the RA that works with Fidelity, an American holding company, one of the largest asset management companies in the world, and TD Ameritrade, an American company that set up an electronic trading platform. This RA offers a reliable investment evaluation tool. Users can associate existing investment account in the system for free. It assesses the investments feasibility based on productivity, diversification, commissions, and taxes. Also, this product may provide guidance on changing the investor's assets distribution [25, 26].

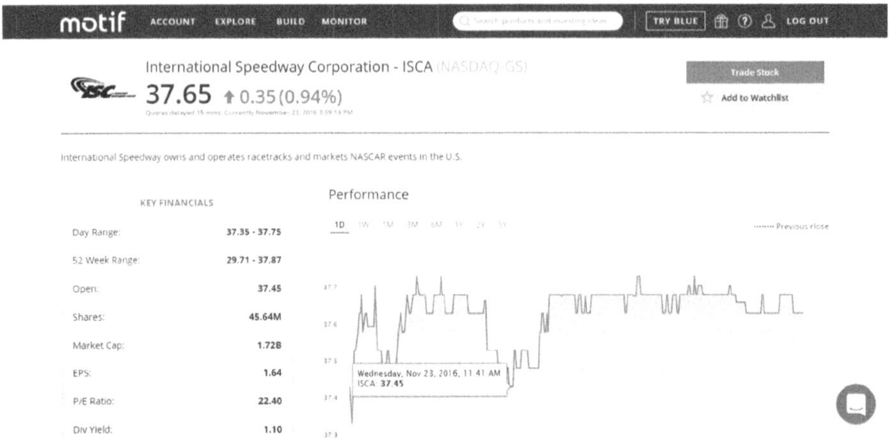

Fig. 2. Motif investing dashboard screen [29]

Motif Investing is a product for active traders, which allows users to create stock baskets and ETFs (dashboard screen is shown in the following Fig. 2). After the creating, the user can buy up to 30 stocks of ETFs for $9.95. Investors can create their baskets, invest money in other ones which were created by the service itself or in those created by other users [27, 28].

Schwab Intelligent is one of the best RA according to NerdWallet's review [30, 31] at the beginning of 2018. It offers advisory service with automated portfolio management and unlimited access to certified financial planners including personal financial guidance. However, generated portfolios have high allocation in cash (means a part of a customer's money remains not invested permanently) and investors must be comfortable with it.

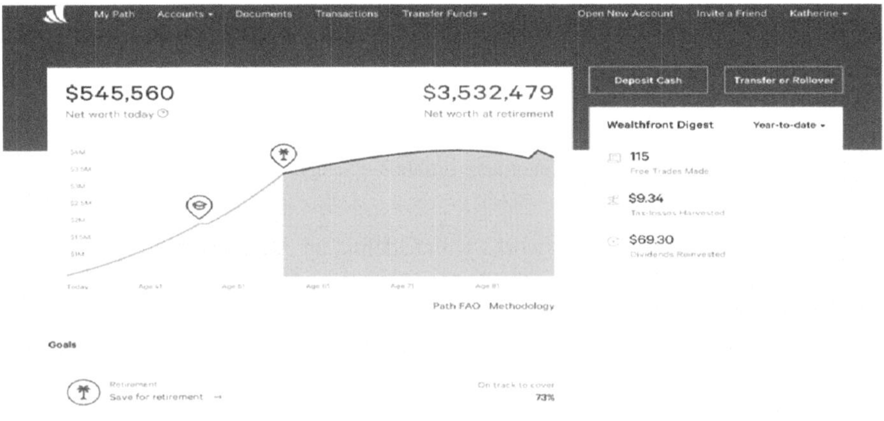

Fig. 3. Wealthfront dashboard [33]

Wealthfront (dashboard is shown in the following Fig. 3) is a crucial force in the online advisor industry that offers one of the most robust tax-optimization services available with no human advice offering at all (strict robo-advisor as the opposite to the Betterment) [32].

Based on the selected RAs the following functionality can be defined as the basic:

- Account creation and goals setup;
- Personal data analysis;
- Recommendations for investing and distributing assets;
- Communications between users for mutual investments;
- Active trading and investing in ETFs, stocks, bonds;
- User's data protection;
- Portfolio rebalancing;
- Retirement planning.

As described above, the basic functionality is quite suitable for ML due to their perfect matching to the areas of usage. RAs even should include all possible ML use cases for Financial Industry. This fact makes RA one of the most difficult systems from development perspective.

4 Cryptocurrencies Prices Forecasting Using Machine Learning Techniques

4.1 Criteria of Influence on Cryptocurrencies Prices

The creation of cryptocurrencies has changed FinTech industry and it continues to change it today. Whereas people think that during 9 years nobody has found the real use cases for blockchain technology [34]. Now people still depend on banks, because most countries did not define cryptocurrencies as national currencies, but in the future the decentralized systems such as Bitcoin, can substitute traditional currencies. Also, due to continuously increasing digital society, financial services providers are looking to offer their customers the same services to which they are accustomed but in a more efficient, secure and cost-effective way.

In addition to mining (the process of extraction of the cryptocurrency), trading with cryptocurrencies is popular nowadays. It is risky but on the other hand, it is a fast way to get a great sum of money [35]. For example, at the beginning of 2017, Bitcoin cost lower than $1000 but in December 2017 it cost almost $20000. The research question is: Which criteria affect the price of the cryptocurrency and how? We have found the next points:

1. Total number of mined coins;
2. Mining difficulty level;
3. Cryptocurrency's trading volume;
4. Acceptance of the cryptocurrency's value by the society;
5. Price of Bitcoin;

4.2 Total Number of Mined Coins

Fundamental economic factors, such as demand and supply, affect the price of many things, cryptocurrencies are not exception. The supply is also created in at a constant rate and is unchangeable due to the conscious rules (Fig. 4). It creates an amount that is limited, and thus, people will pay more to get coins they think have value.

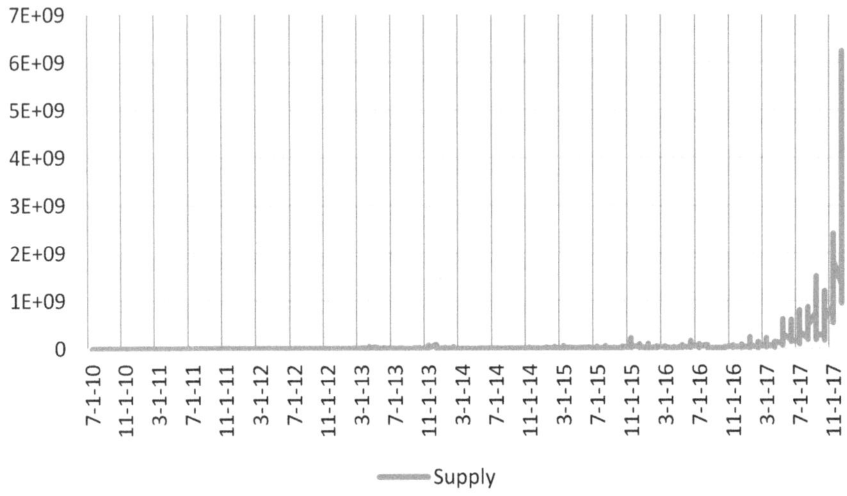

Fig. 4. Bitcoin supply curve from 2010 to the end of 2017 [36]

4.3 Mining Difficulty Level

Difficulty concerning cryptocurrencies is a measure of how hard to find the hash below a given difficulty value. For example, Bitcoin has fixed global difficulty per block of hashes. The formula of difficulty is:

$$D = \frac{t^1}{c} \tag{1}$$

where D – difficulty per block, t^1 – a hash the leading 32 bits are 0 and the rest are 1 (it is also known as pool difficulty), c – a 256-bit number that all cryptocurrency's network client share (the lower the target the more difficult it is to generate a block).

For example, Bitcoin's difficulty changes every two weeks (every 2016 blocks) (Fig. 5).

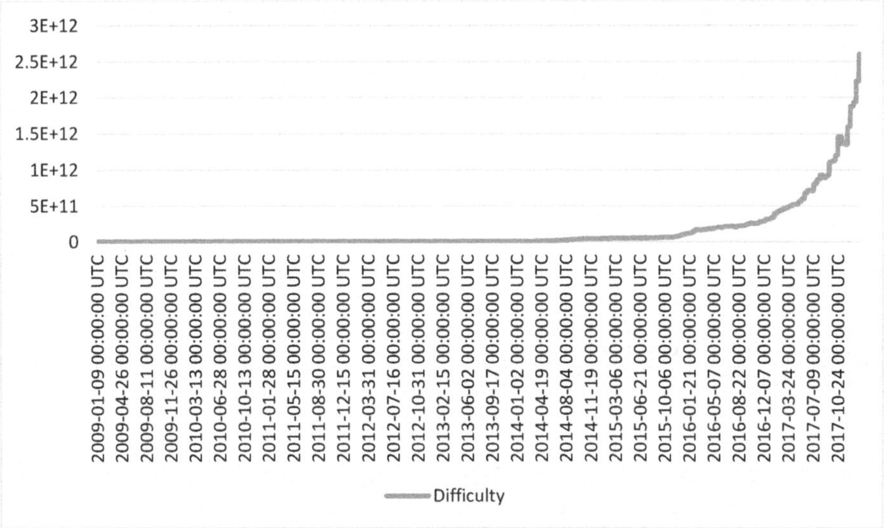

Fig. 5. Bitcoin's difficulty chart overall [37]

4.4 Cryptocurrency's Trading Volume

The coin's volume depends on trader's trading style [38]. However, each change of volume immediately affects the price (the higher the volume, the easier it is to get in and get out).

4.5 Society Perceptions

Different statements regarding cryptocurrencies from different famous people in the financial industry or criminal scandals with the cryptocurrency can affect the price in both good and bad ways.

Also, legal and government issues can affect the price. If a government is oppressive with tax or assets laws it can be trivial to hide assets in a cryptocurrency. Government regulation of cryptocurrencies can substantially cause changes in their prices. Cryptocurrency's rate changes due to its status: if the cryptocurrency is official it can have a positive effect whereas banning can do the opposite.

4.6 Price of Bitcoin

Bitcoin is the most expensive cryptocurrency now, so each change of its price affects prices of other cryptocurrencies as every currency is priced [39] with the base of US Dollar, any change in Dollar would inevitably affect the price of currency [40].

Similarly, all cryptocurrencies are initial prices on the bases of Bitcoins and any change in the value of Bitcoin will automatically change the value of other cryptocurrencies.

Based on the graphic above and coefficient of determination, we proved that these criteria could not be used (in such combination) to predict the price of the cryptocurrency and there should be more of them to make the forecasting more accurate.

For our investigation, we used ML LR algorithm which allows calculating. The algorithm of LR in Python using Pandas and Scikit-learn libraries looks like:

```
import pandas as pd
from sklearn import linear_model
from sklearn.metrics import mean_squared_error, r2_score
# load dataset
dataset = pd.read_csv('./dataset_prices.csv', sep=',')
# split on 2 parts
y = dataset['price'].values
x = dataset[['supply', 'difficulty', 'trading_volume',
'reaction']].values
# split the data into training and testing sets
# 32 means the number of rows what should be excluded
from the dataset
# to create training and testing sets
x_train = x[:-32]
x_test = x[-32:]
# split the target into training and testing sets
y_train = y[:-32]
y_test = y[-32:]
# create linear regression object and set that criteria
will be normalized
regression = linear_model.LinearRegression(normal-
ize=True)
# train the model
regression.fit(x_train, y_train)
# make prediction based on test set
y_predict = regression.predict(x_test)
# diplay stats
print('Coefficients: \n', regression.coef_)
print('Intercepts: \n', regression.intercept_)
print("Mean squared error: %.2f" % mean_squared_er-
ror(y_test, y_predict))
print('R^2 score: %.2f' % r2_score(y_test, y_predict))
```

After running this code, we received the next output:

$$y = 6917.6 - 5.66 \cdot 10^{-4} \cdot x_1 + 6.06 \cdot 10^{-9} \cdot x_2 - 1.68 \cdot 10^{-6} \cdot x_3 + 6.58 \cdot 10^{2} \cdot x_4 \quad (2)$$

Parameter $b_1 = -5.66 \cdot 10^{-4}$ indicates how much the price of bitcoin will decrease with the growth of one more mined coin. Parameter $b_2 = 6.06 \cdot 10^{-9}$ demonstrates how much the price of bitcoin will rise due to the unit growth of difficulty per block. Parameter $b_3 = -1.68 \cdot 10^{-6}$ presents how much the price of bitcoin will decrease due to the growth of bitcoin supply by one more coin. Parameter $b_4 = 6.58 \cdot 10^2$ means that social perceptions have a positive impact on the price of bitcoin. Four factors x_1, x_2, x_3 can explain about 61% of bitcoin price variation. It means that our model can predict price of bitcoin in 61 cases out of 100.

In the chart below the results of prediction are displayed (the last 32 dates in the period from 18th of June 2017 to 22 of January 2018) (Fig. 6).

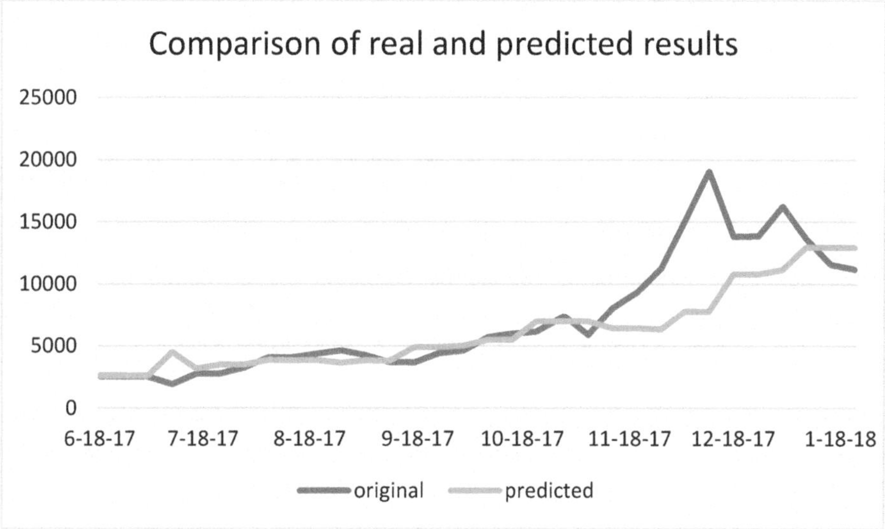

Fig. 6. Comparison of real Bitcoin price and predicted one

5 Modules of a Robo-Advisor

We used our previous architecture approach with small modifications:

- To make things look clear, we have merged Parser module and Clean Up module in ETL (Extract-Transform-Load) because this is what it does, Parsed Data on Dataset & Models and User Data to Investment Plans;
- We have connected Investment Plan modules with Investment Plans storage.

The reason is that we want to keep the high-level design simple. We also understand that each particular implementation will separate these modules on more, because of a specific set of features, specific infrastructure, and other things. Also, the current architecture does not have a security module, and the reason for this is that we decided to move it from the application level to the infrastructure level. It provides simplified architecture and connection between services in general.

In Fig. 7 high-level architecture is described.

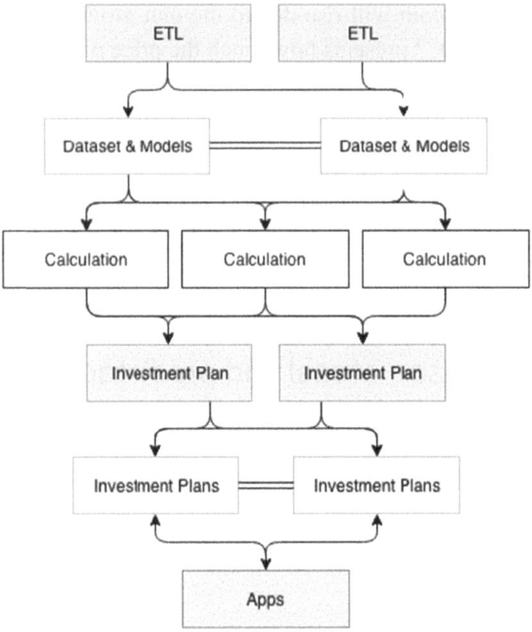

Fig. 7. High-level MVP architecture. (Color figure online)

Some of the modules are replicated databases modules and CRUD (Create-Read-Update-Delete) APIs (yellow color blocks), some are running by schedule (green color blocks), some do computations (while color blocks), and some are apps (blue color block).

ETL here is a parser which grabs cryptocurrencies statistics from open sources and saves it to the database called Dataset & Modules in Fig. 7. Also, it analyzes which cryptocurrencies are "dead" using services such as DeadCoins and CryptoCompare and removes them from the database. We have found that CryptoCompare provides API for all the statistics we need (has data from more than 90 exchanges and does updates regularly). Our goal was to get data for each day from the beginning of trading for each live cryptocurrency. We managed to get data for more than 500 live cryptocurrencies at all using the next algorithm for ETL services:

1. Get the list of coins;
2. Walk through all coins:
   a. If the coin appears as "dead", remove it from the database; if it exists, go the next coin;
   b. If a coin is not in the database, get statistics for the longest period, otherwise for the previous day;
3. Save all the values to the database;
4. If there is no LSTM model in the database for this coin then create one for trading volume and one for a price depending on trading volume, otherwise get these models and continue training with the newest data;

5. Save these models to the database;
6. Shutdown till the next run.

Our high-level architecture demonstrates logic of applying of robo-adviser's modules. The implemented ETL module is responsible for collection of real-time cryptocurrency data for our robo-advisor. Using CryptoCompare website (Fig. 8) we can get relevant data, such as a list of cryptocurrencies, historical data about rate and volume of cryptocurrencies. CryptoCompare service provides an open REST API, so we can use the Python 3.6 requests library to get data from this resource.

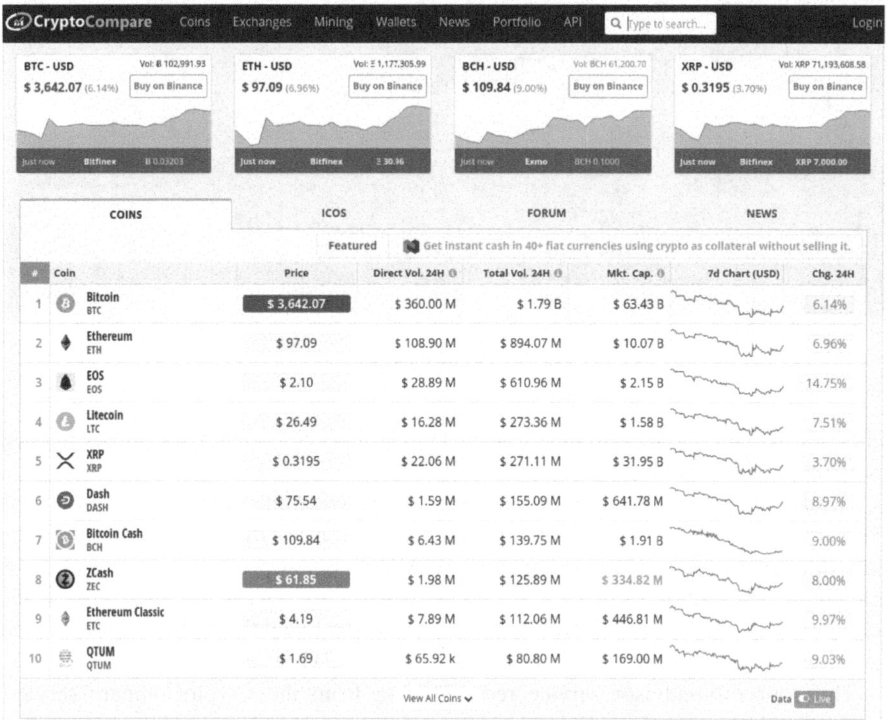

Fig. 8. CryptoCompare service

At the same time second task of ETL module is data updating. CryptoCompare service does not provide data about reliability and relevance of cryptocurrencies which are presented on this web site. Therefore we need another service DeadCoins to achieve second task.

DeadCoins service (Fig. 9) allows us to check if this cryptocurrency was noticed in any financial fraud and if it is reliable then we can also check whether it is still being traded. The complete test is quite simple, because this service contains information about unreliable cryptocurrencies, so we can check if cryptocurrency is blacklisted. If it

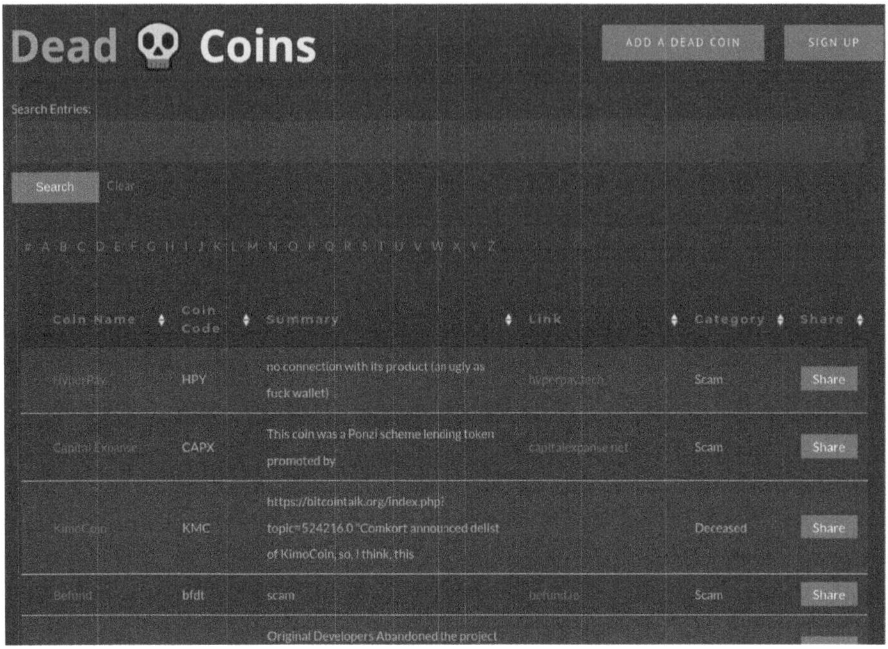

Fig. 9. DeadCoins service

is true our RA service will exclude it from the list of cryptocurrency for processing and preparing of investment plan.

DeadCoins service has website, but does not provide any API. Thus we need use indirect way to get data - web scrapping. The site itself is not difficult to scrap, since the main difficulty as it turned out is parsing HTML code, but for such purposes Python contains a BeatifulSoup library that allows to parse the DOM tree and retrieve the necessary data from the provided HTML code. In order to get the HTML code, we used the same requests library which allows us to process HTTP requests.

Thus our robo-advisor service receives data from the CryptoCompare service, checks this data on the DeadCoins service, and then collects historical data of verified list of cryptocurrencies. After that we can use Investment Plan module for investors with different risk attitude.

There are many coins, so by default it is good to provide the most common for portfolio, such as Bitcoin (BTC), Litecoin (LTC), Etherium (ETH), BitcoinCash (BCH) and others. The mentioned cryptocurrencies have more than a billion USD market capitalization because they are trusted.

Calculation modules forecast price of selected cryptocurrency for few days related to the US Dollar based on forecasted volume. It is required for choosing the most diverse and sustainable options for the investment portfolio rebalancing. Generally, these modules use LSTM neural network (because it is well-known, suitable for financial predictions neural network) developed with Keras (high-level neural networks API developed on top of TensorFlow). The algorithm has several steps:

1. Trading Volume forecasting:
   a. Get a trading volume LSTM model from Dataset & Models for the selected cryptocurrency;
   b. Predict for some days;
   c. Go to the last step if fails;
   d. Save the result for the price prediction temporary;
2. Price forecasting:
   a. Get a price LSTM model from Dataset & Models for the selected cryptocurrency;
   b. Predict for some days using previously predicted trading volume;
   c. Go to the last step if fails;
   d. Return predicted price;
3. Finish.

For the testing purpose we got BTC dataset [41, 42] and predicted values for both trading volume and price from June 24th, 2018 to June 30th, 2018. Results of forecasting are in Figs. 10 and 11 below:

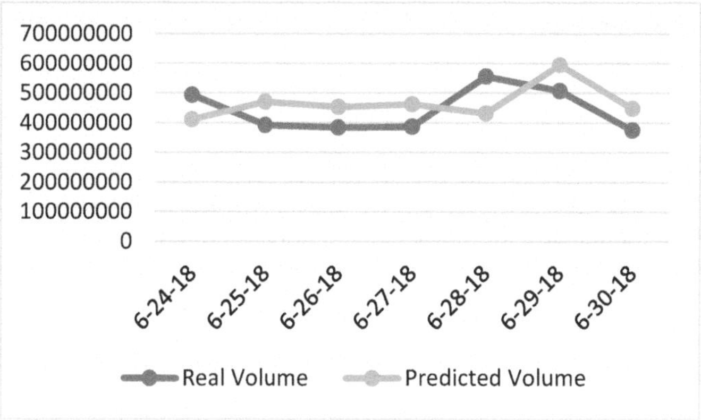

Fig. 10. BTC trading volume chart [43]

Fig. 11. BTC prices chart [44]

Investment Plan services use Markowitz model described by below Eq. (3), to build the investment portfolio for risk-averse, risk-seeking and risk-neutral investors based on forecasted prices via Calculation services:

$$E(R_p) = \sum_i w_i E(R_i) \tag{3}$$

where R_p is a return of the portfolio, R_i is a return on asset and w_i is the weighting of a component asset i. The algorithm and experiment for the Investment Plan services are described in the following part.

6 Experiment

The purpose of an investor is to increase its capital through the formation of a set of financial instruments (investment portfolio) [45, 46]. Portfolio value is formed as the total value of all components of financial instruments. If the value of the portfolio is P, then through the time interval t the profitability of the portfolio will be $\frac{P' - P}{P}$. Let x_i to be a share of capital spent on the purchase of a financial instrument i; d_i is a return of the financial instrument per 1 invested dollar. Then the return on investment portfolio will be the following:

$$d_p = \sum_{i=1}^{n} d_i \tag{4}$$

The profitability and risk of the investment portfolio is measured by the mathematical expectation m_p and the variance $\sigma = r_p$ respectively, where:

$$m_p = x_1 \cdot E(d_1) + \ldots + x_n \cdot E(d_n) = \sum_{i=1}^{n} x_i \cdot m_i \tag{5}$$

$$r_p = \sum_{i=1}^{n} \sum_{j=1}^{n} x_i \cdot x_j \cdot v_{ij} \tag{6}$$

where v_{ij} is the covariance of financial instruments. Since the returns of financial instruments are random, then the return of the portfolio is also a random variable.

Consider the initial data of quotations of the two most popular cryptocurrencies (https://finance.yahoo.com/cryptocurrencies).

Adverse risk investor can achieve a certain level of return under minimal risk:

$$\begin{cases} r_p = \sum_{i=1}^{n} \sum_{j=1}^{n} x_i \cdot x_j \cdot v_{ij} \rightarrow min, \\ \sum_{i=1}^{n} x_i \cdot d_i = m_p, \\ \sum_{i=1}^{n} x_i = 1, x_i \geq 0. \end{cases} \tag{7}$$

The initial distribution of the financial instruments will be set at the level $x_1 = x_2 = 0.5$. The objective function in Markowitz model (7) is the quadratic form $r_p = X^T VX$, where X^T is the transposed matrix, V is the covariance matrix.

If investors are risk seeking and strive to maximize their returns under acceptable risk level (8), then their goal and restrictions are described as follows:

$$\begin{cases} \sum_{i=1}^{n} x_i \cdot d_i = m_p \rightarrow max, \\ r_p = \sum_{i=1}^{n}\sum_{j=1}^{n} x_i \cdot x_j \cdot v_{ij}, \\ \sum_{i=1}^{n} x_i = 1, x_i \geq 0. \end{cases} \tag{8}$$

Over the time of investment portfolio formation using RA, these findings for each investor will significantly depend on the availability of alternative financial instruments and the volatility of their rates.

For experiment [47, 48] we have selected five popular cryptocurrencies for investors with different risk attitude, such as risk-averse, risk-seeking and risk-neutral ones. The primary goal of RA is to support investors by converting their specific requirements into an adequate portfolio of financial instruments without human intervention. We used active and passive robo-advisor because our investment strategy and portfolio construction are not fixed (dynamic approach), and only RA decides about the actual execution of investment (rebalancing) process (passive approach).

We can describe, on the following Fig. 12, robo-advisory process as a sequence of following steps: configuration (proposals for different ration average income-risk), matching (correspondence between type of client and type of proposal) and maintenance (algorithm of execution for chosen investment proposal).

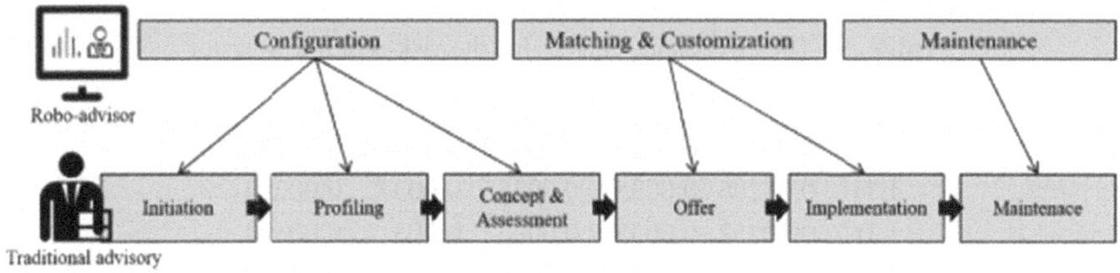

Fig. 12. Robo-advisory process

We have chosen the following cryptocurrencies:

- BTC;
- ETH;
- LTC;
- NEO;
- BCH.

Also, we have chosen days for which data is available for all of the cryptocurrencies. This is a period from August 2017 to June 2018 including the last week of June 2018 which were forecasted using Calculation module in the previous section.

The steps were the following:

- Calculate a percentage change of an exchange rate for each day;
- For each chosen cryptocurrency calculate the average rate per day for chosen time period August, 05, 2017 – June, 23, 2018 (average d_i).

After performing the steps above, we got the following values: for BTC it was 0.36%; for ETH it was 0.38%; for LTC it was 0.48%, for NEO it was 0.75%, and for BCH it was 0.89%.

Then we calculated a covariance matrix 5 × 5 (diagonal elements are variations of corresponding cryptocurrencies rates, nondiagonal elements are covariance of corresponding cryptocurrencies) for the selected cryptocurrencies for the risk-averse investor in the Table 2.

Table 2. The covariance matrix for the risk-averse investor

	BTC	ETH	LTC	NEO	BCH
BTC	0.00288	0.00206	0.00252	0.00238	0.00188
ETH	0.00206	0.00359	0.00355	0.00342	0.00294
LTC	0.00252	0.00355	0.00623	0.00314	0.00304
NEO	0.00238	0.00342	0.00314	0.01024	0.00350
BCH	0.00188	0.00294	0.00304	0.00350	0.01062
X'	29%	0%	7%	25%	39%
$X'V$	0.00234	0.00285	0.00313	0.00483	0.00579

We have also made calculations for the risk-seeking and risk-neutral (hybrid) investors and the following results described in the Tables 3 and 4 below.

Table 3. The covariance matrix for the risk-seeking investor

	BTC	ETH	LTC	NEO	BCH
BTC	0.00288	0.00206	0.00252	0.00238	0.00188
ETH	0.00206	0.00359	0.00355	0.00342	0.00294
LTC	0.00252	0.00355	0.00623	0.00314	0.00304
NEO	0.00238	0.00342	0.00314	0.01024	0.00350
BCH	0.00188	0.00294	0.00304	0.00350	0.01062
X'	0%	0%	0%	0%	100%
$X'V$	0.00188	0.00294	0.00304	0.00350	0.01062

In the end we calculated daily yield for each cryptocurrency using (9):

$$Y = d_i * X'V \tag{9}$$

We got the following results (earnings $x_i \cdot d_i$ from each cryptocurrencies per day) for the risk-averse investor: for BTC it was 0.00103 (or $0.103 per day for each

Table 4. The covariance matrix for hybrid type of investor

	BTC	ETH	LTC	NEO	BCH
BTC	0.00288	0.00206	0.00252	0.00238	0.00188
ETH	0.00206	0.00359	0.00355	0.00342	0.00294
LTC	0.00252	0.00355	0.00623	0.00314	0.00304
NEO	0.00238	0.00342	0.00314	0.01024	0.00350
BCH	0.00188	0.00294	0.00304	0.00350	0.01062
X'	47%	24%	9%	9%	12%
$X'V$	0.00249	0.00278	0.00323	0.00351	0.00341

invested $100), for ETH it was 0, for LTC it was 0.00033, for NEO it was 0.00187, and for BCH it was: 0.00348.

For the risk-seeking investor the results were following: for BTC, EHT, LTC, and NEO it was 0, but for BCH it was 0.00865.

For the hybrid type of investor the results were following: for BTC, EHT, LTC, and NEO were correspondingly 0.00158, 0.000859, 0.00043, 0.00062, 0.00102. For hybrid type of investor both risk level and profitability is lower than for 2 previous types of investors.

Thus, profitability of the risk-averse investor is 0.67% daily or 24.5% annually with 0.44% daily risk.

For the risk-seeking investor the yield will be higher – 0.87% daily or 31.6% annually with 1.06% daily risk.

For hybrid type of investor the yield will be 0.45% daily or 16.5% annually with 0.28% daily risk.

If we divide data on a monthly basis, we get the following schedule of optimal investments in financial instruments on the Fig. 13 below.

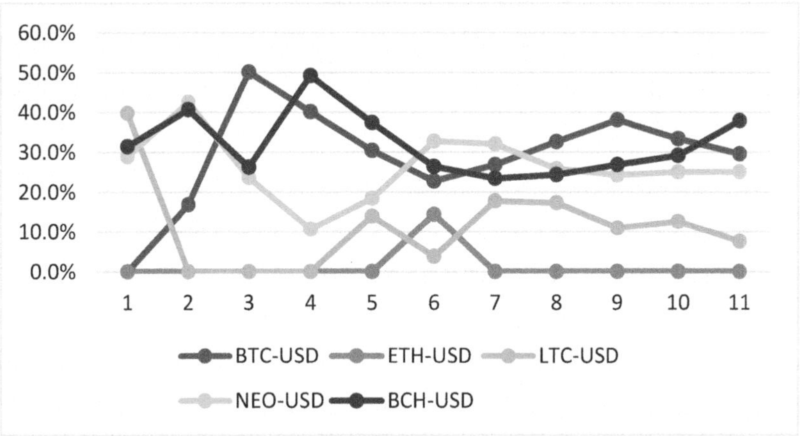

Fig. 13. Dynamic of optimal investments in financial instruments

In the ETH it is better not to invest at all, it is better to invest in BTC on the level of 30%. NEO and BCH are promising financial instruments for investing at a level of 25–30% each.

7 Conclusion

Machine Learning shows new ways of developing various areas of the financial industry. It also can give a new life to old tools which help companies and individuals to invest, to trade and use robo-advisor more. Even if it is not a new idea, it is still developing the financial industry. For example, it can use personal data to prevent fraud (such as accounts duplicates, or a pre-arrangement for investing); it can do investment and asset allocation guidelines, which is the ideal task for ML because of a significant amount of data to process and analyze. Despite the fact that RAs are often criticized, they make investments easier and provide new tools that can radically change an investment landscape. We implemented a proof of concept of the robo-advisor application for investment portfolio formation with financial instruments under adverse risk attitude and risk-seeking behavior of investors.

We have analyzed selected criteria combination based and tested how they affect the price of Bitcoin. For our investigation, we used ML LR algorithm implemented in scikit Python library in Anaconda Data Science tool.

So, we have developed MVP of a cryptocurrencies robo-advisor with Investment plan, Calculations and ETL modules. ETL module used data from CryptoCompare and DeadCoins to continue LSTM models training and clean database from scam, Calculation module did forecasts for trading volume and for price for each cryptocurrency. Investment plan module built investment plans using Markowitz model for risk-averse, risk-seeking and risk-neutral investors.

As a result, we have got 5 cryptocurrencies and found out that if the user is risk-averse, they should invest into 4 of them to gain 23.7% annually with 0.44% risk. If the user is risk-seeking, they should invest into 1 of them to gain 31.7% annually with 1.06% risk. Finally, if the user is risk-neutral, they should invest into all of the five to gain 16.5% annually with 0.28% risk.

On this base, we plan to improve general robo-advisor architecture to the real-market product and to study cryptocurrencies and robo-advisor more and how they can be combined together.

References

1. Kohavi, R., Provost, F.: Glossary of terms. Mach. Learn. **30**, 271–274 (1998)
2. The implications of machine learning in finance. Mode of access. https://www.bloomberg.com/professional/blog/implications-machine-learning-finance/. Accessed 24 Feb 2018
3. Maedche, A., Morana, S., Schacht, S., Werth, D., Krumeich, J.: Advanced user assistance systems. Bus. Inf. Syst. Eng. **58**(5), 367–370 (2016)

4. Jung, D., Dorrner, V., Glaser, F., Morana, S.: Robo-advisory – digitalization and automation of financial advisory. Bus. Inf. Syst. Eng. **60**, 81–86 (2018). https://doi.org/10.1007/s12599-018-0521-9
5. Firth, N.: Want to make your vote really count? Stick a blockchain on it. 6 September 2017. Mode of access. https://www.newscientist.com/article/mg23531424-500-bitcoin-tech-to-put-political-power-in-the-hands-of-voters/. Accessed 30 June 2018
6. Galeon, D., Reedy, C.: Blockchain Is Helping Us Feed the World's Hungriest Families, 21 March 2017, Mode of access. https://futurism.com/blockchain-is-helping-us-feed-the-worlds-hungriest-families/. Accessed 30 June 2018
7. Kobets, V., Yatsenko, V., Mazur, A., Zubrii, M.: Data analysis of private investment decision making using tools of Robo-advisers in long-run period. In: Proceedings of the 14th International Conference on ICT in Education, Research and Industrial Applications. Integration, Harmonization and Knowledge Transfer, Kyiv, Ukraine, 14–17 May 2018, vol. 2104, pp. 144–159 (2018)
8. Cocca, T.: Potential and limitations of virtual advice in wealth management. J. Financ. Transf. **44**, 45–57 (2016)
9. Ivanon, O., Snihovyi, O., Kobets, V.: Implementation of robo-advisors tools for different risk attitude invesment decisions. In: Proceedings of the 14th International Conference on ICT in Education, Research and Industrial Applications. Integration, Harmonization and Knowledge Transfer, Kyiv, Ukraine, 14–17 May 2018, vol. 2104, pp. 195–206 (2018)
10. Fein, M.L.: Robo-Advisors: A Closer Look, 30 June 2015. http://dx.doi.org/10.2139/ssrn.2658701. Accessed 30 June 2018
11. Robo-advice poses a substantial risk to wealth management firms and customers alike – EBA warns. 15 November 2018, Mode of access. https://thewealthnet.com/page_fullstory.php?articleid=58571&categoryid=2&interestid=15. Accessed 1 Dec 2018
12. The Real Truth Behind The Rise of Robo-Advisors. 15 November 2018, Mode of access. https://www.nanalyze.com/2016/01/the-real-truth-behind-the-rise-of-robo-advisors/. Accessed 1 Dec 2018
13. Reuba, K.: Robo-Advisors: Early Disruptors in Private Wealth Management. 5 December 2017, Mode of access. https://www.allianzgi.com/en/insights/investment-themes-and-strategy/robo-advisors-early-disruptors-in-private-wealth-management. Accessed 1 Dec 2018
14. Ivanov, O., Snihovyi, O., Kobets, V.: Cryptocurrencies prices forecasting with anaconda tool using machine learning techniques. In: Proceedings of the 14th International Conference on ICT in Education, Research and Industrial Applications. Integration, Harmonization and Knowledge Transfer, Kyiv, Ukraine, 14–17 May 2018, vol. 2105, pp. 453–456 (2018)
15. Phillips, R.C., Gorse, D.: Predicting cryptocurrency price bubbles using social media data and epidemic modelling. In: 2017 IEEE Symposium Series on Computational Intelligence (SSCI), Honolulu, HI, pp. 1–7 (2017)
16. Colianni, S., Rosales, S.M., Signorotti, M.: Algorithmic Trading of Cryptocurrency Based on Twitter Sentiment Analysis (2015). Mode of access. http://cs229.stanford.edu/proj2015/029_report.pdf
17. Lamon, C., Nielsen, E., Redondo, E.: Cryptocurrency price prediction using news and social media sentiment. SMU Data Sci. Rev. **1**(3), 1–22 (2017)
18. Kim, Y.B., Kim, J.G., Kim, W., Im, J.H., Kim, T.H., Kang, S.J., et al.: Predicting fluctuations in cryptocurrency transactions based on user comments and replies. PLoS One **11**(8), e0161197 (2016). https://doi.org/10.1371/journal.pone.0161197
19. Faggella, D.: Machine Learning in Finance – Present and Future Applications. Mode of access. https://www.techemergence.com/machine-learning-in-finance/. Accessed 24 Feb 2018

20. Kashner, E.: Ghosts In The Robo Advisor Machine. Mode of access. http://www.etf.com/sections/blog/22973-ghosts-in-the-robo-advisor-machine.html. Accessed 24 Feb 2018
21. Markowitz, H.M.: Portfolio selection. J. Finance **7**(1), 77–91 (1952). https://doi.org/10.2307/2975974
22. Black, F., Litterman, R.: Global portfolio optimization. Financ. Anal. J. **51**, 133–138 (1995)
23. Betterment Review 2018. Mode of access. https://www.nerdwallet.com/blog/investing/betterment-review/. Accessed 24 Feb 2018
24. Betterment Adds Institutional Platform, Inks Deal With Fidelity. Mode of access. https://bankinnovation.net/2014/10/betterment-adds-institutional-platform-inks-deal-with-fidelity/. Accessed 3 Mar 2018
25. Future Advisor Review 2017. Mode of access. https://www.nerdwallet.com/blog/investing/futureadvisor-review/. Accessed 24 Feb 2018
26. YC Alum FutureAdvisor Is Now Managing $600 Million In Assets. Mode of access. https://techcrunch.com/2015/06/24/yc-alum-futureadvisor-is-now-managing-600-million/. Accessed 3 Mar 2018
27. Thangavelu, P.: Motif Investing Broker Review: Easy Thematic Investing. Mode of access. https://www.investopedia.com/articles/active-trading/030415/motif-investing-broker-review-easy-thematic-investing.asp. Accessed 24 Feb 2018
28. Motif Investment Review 2018. Mode of access. https://www.nerdwallet.com/blog/investing/motif-investing-review-1/. Accessed 24 Feb 2018
29. Motif Investing Review. Mode of access. http://www.investmentzen.com/motif-investing-review. Accessed 3 Mar 2018
30. Schwab Intelligent Portfolios Review 2018. Mode of access. https://www.nerdwallet.com/blog/investing/charles-schwab-intelligent-portfolios-review/. Accessed 3 Mar 2018
31. Schwab Intelligent Portfolios Review 2018 – A Free Robo Advisor? Mode of access. https://investorjunkie.com/39634/schwab-intelligent-portfolios-review/. Accessed 3 Mar 2018
32. Wealthfront Review 2018. Mode of access. https://www.nerdwallet.com/blog/investing/wealthfront-review/?trk_content=brokerage_compare_module. Accessed 3 Mar 2018
33. Wealthfront Review. Mode of access. https://www.stockbrokers.com/review/wealthfront. Accessed 3 Mar 2018
34. Ten years in, nobody has come up with a use for blockchain. https://hackernoon.com/ten-years-in-nobody-has-come-up-with-a-use-case-for-blockchain-ee98c180100. Accessed 28 Jan 2018
35. Want to Be a Millionaire? Two Main Rules of Bitcoin Investing. https://cointelegraph.com/news/want-to-be-a-millionaire-two-main-rules-of-bitcoin-investing. Accessed 28 Jan 2018
36. Data set for supply. https://drive.google.com/open?id=1FZowiJQUZokrf98FfJiirrvNnBqVAUri
37. Data set for mining difficulty. https://drive.google.com/open?id=1fqG37woV4Zqja2W_NK6iVkI1je1UK3hM
38. Is Trading Volume important? https://steemit.com/cryptocurrency/@cryptopy/is-trading-volume-important. Accessed 28 Jan 2018
39. Why Do Currencies Fluctuate? http://www.xe.com/moneytransfertips/why-do-currencies-fluctuate.php. Accessed 28 Jan 2018
40. Why the Dollar Is the Global Currency. https://www.thebalance.com/world-currency-3305931. Accessed 28 Jan 2018
41. Bitcoin's dataset used for forecasting. https://www.dropbox.com/s/dswj09nn3wc2crb/bitcoin_dataset.csv?dl=0. Accessed 30 June 2018
42. Bitcoin's dataset used for comaprison real trading volume and forecasted. https://www.dropbox.com/s/e1fgz8oyuo1bfq4/bitcoin_trading_volume_predicted_dataset.csv?dl=0. Accessed 30 June 2016

43. Bitcoin's dataset used for comparison real price and forecasted. https://www.dropbox.com/s/qzikltpfaud63mb/bitcoin_price_predicted_dataset.csv?dl=0. Accessed 30 June 2017
44. Bitcoin's dataset used for comparison real price and forecasted. https://www.dropbox.com/s/qzikltpfaud63mb/bitcoin_price_predicted_dataset.csv?dl=0. Accessed 30 June 2018
45. Kobets, V., Poltoratskiy, M.: Using an evolutionary algorithm to improve investment strategies for industries in an economic system. In: CEUR Workshop Proceedings, vol. 1614, pp. 485–501 (2016). (Indexed by: Sci Verse Scopus, DBLP, Google Scholar). CEUR-WS.org/Vol-1614/ICTERI-2016-CEUR-WS-Volume.pdf
46. Kobets, V., Yatsenko, V.: Adjusting business processes by the means of an autoregressive model using BPMN 2.0. In: CEUR Workshop Proceedings, vol. 1614, pp. 518–533 (2016). (Indexed by: Sci Verse Scopus, DBLP, Google Scholar). CEUR-WS.org/Vol-1614/ICTERI-2016-CEUR-WS-Volume.pdf
47. Markowitz model Python implementation. https://gist.github.com/alekseysink/fca420a5e4e60bb010fbb1d21f628bf. Accessed 1 July 2018
48. Kobets, V., Yatsenko, V., Poltoratskiy, M.: Dynamic Model of Double Electronic Vickrey Auction. In: CEUR Workshop Proceedings, vol. 1356, pp. 236–251 (2015). (Indexed by: Sci Verse Scopus, DBLP, Google Scholar). CEUR-WS.org/Vol-1356

Author Index

The manufacturer's authorised representative in the EU is Springer Nature Customer Service Centre GmbH, Europaplatz 3, 69115 Heidelberg, Germany. If you have any concerns regarding our products, please contact ProductSafety@springernature.com

Printed and bound by CPI Group (UK) Ltd, Croydon, CR0 4YY
05/10/2026
02228140-0015